THE FRONTIERS COLLECTION

T0207406

THE FRONTIERS COLLECTION

Series Editors:
A.C. Elitzur M.P. Silverman J. Tuszynski R. Vaas H.D. Zeh

The books in this collection are devoted to challenging and open problems at the forefront of modern science, including related philosophical debates. In contrast to typical research monographs, however, they strive to present their topics in a manner accessible also to scientifically literate non-specialists wishing to gain insight into the deeper implications and fascinating questions involved. Taken as a whole, the series reflects the need for a fundamental and interdisciplinary approach to modern science. Furthermore, it is intended to encourage active scientists in all areas to ponder over important and perhaps controversial issues beyond their own speciality. Extending from quantum physics and relativity to entropy, consciousness and complex systems – the Frontiers Collection will inspire readers to push back the frontiers of their own knowledge.

Brigitte Falkenburg

PARTICLE METAPHYSICS

A Critical Account
of Subatomic Reality

With 30 Figures

 Springer

Professor Dr.Dr. Brigitte Falkenburg
Universität Dortmund
Institute of Philosophy
Emil-Figge-Straße 50
44227 Dortmund
e-mail: falkenburg@fb14.uni-dortmund.de

Series Editors:

Avshalom C. Elitzur
Bar-Ilan University,
Unit of Interdisciplinary Studies,
52900 Ramat-Gan, Israel
email: avshalom.elitzur@weizmann.ac.il

Mark P. Silverman
Department of Physics, Trinity College,
Hartford, CT 06106, USA
email: mark.silverman@trincoll.edu

Jack Tuszynski
University of Alberta,
Department of Physics, Edmonton, AB,
T6G 2J1, Canada
email: jtus@phys.ualberta.ca

Rüdiger Vaas
University of Gießen,
Center for Philosophy and Foundations of Science
35394 Gießen, Germany
email: Ruediger.Vaas@t-online.de

H. Dieter Zeh
University of Heidelberg,
Institute of Theoretical Physics,
Philosophenweg 19,
69120 Heidelberg, Germany
email: zeh@urz.uni-heidelberg.de

Cover figure: Image courtesy of the Scientific Computing and Imaging Institute,
University of Utah (www.sci.utah.edu).

ISSN 1612-3018

ISBN 978-3-642-07029-7   e-ISBN 978-3-540-33732-4

Springer is a part of Springer Science+Business Media

springer.com

© Springer-Verlag Berlin Heidelberg 2007
Softcover reprint of the hardcover 1st edition 2007

Cover design: KünkelLopka, Werbeagentur GmbH, Heidelberg

In memory of my mother

Preface

Metaphysics, with which, as fate would have it, I have fallen in love but from which I can boast of only a few favours, offers two kinds of advantage. The first is this: it can solve the problems thrown up by the enquiry of mind, when it uses reason to spy after the more hidden properties of things. But hope is here all too often disappointed by the outcome. And, on this occasion, too, satisfaction has escaped our eager grasp. [...] The second advantage of metaphysics is more consonant with the nature of the human understanding. It consists [...] in knowing what relation the question has to empirical concepts, upon which all our judgements must at all times be based. To that extent metaphysics is a science of the *limits of human reason*. [...] Thus, the second advantage of metaphysics is at once the least known and the most important, although it is also an advantage which is only attained at a fairly late stage and after long experience.

Immanuel Kant[1]

The tradition of the particle concept goes back to traditional metaphysics and ancient philosophy. The idea that matter is made up of microscopic constituent parts stems from ancient atomism. At the very beginnings of modern physics, it was taken up by Galileo, Descartes, and Newton. Newton thought that there are atoms of matter and light, but with the methods of Newtonian mechanics and optics they were beyond the reach of experiments. In the 19th century, the wave theory of light succeeded and electrodynamics, thermodynamics, statistical mechanics, and the kinetic theory of heat were developed. Two hundred years after Newton's *Principia*, the experimenters discovered cathode rays, X rays, and radioactivity, and began to investigate the atomic and subatomic structure of matter. At the same time, however, the empiricist Ernst Mach still doubted the existence of the unobservable microscopic constituent parts of matter and asked for more natural ideas about the structure of matter.[2]

---

[1] Kant 1766, Akad. 2.385–2.386.
[2] Mach 1883, 466.

Modern particle physics started with the development of the cloud chamber (1895) and the discovery of the electron (1897).[3] But it turned out that light comes in quanta and that the atom has a very small nucleus. The light quantum hypothesis contradicted the well-confirmed wave theory of light, whereas the atomic nucleus seemed to preclude any stable classical model of the atom. With respect to the classical views about the constitution of matter and light, these discoveries initiated a history of disillusion. The structure of matter and light no longer fitted in with a classical picture of subatomic reality. The more accessible the subatomic realm became, the less it could be explained in terms of familiar physical concepts. The classical particle concept as well as the classical theory of radiation were at odds with the quantum phenomena of subatomic physics. Hence, when it became clear that there is no classical construal of quantum phenomena, the old doubts about the existence of atoms were rekindled. After the rise of quantum mechanics, never-ending philosophical debates about the reality of subatomic processes and particles added to the debate about the existence of unobservable entities.

The present book attempts to unravel these old and new philosophical debates about subatomic reality. It tries to explain in detail how one may still talk about particles today. But what are the particles of today's particle physics, in the aftermath of the quantum revolution and the failure of the classical particle concept? And how legitimate is the claim that modern particle physics refers to subatomic matter constituents and other subatomic particles? In the following, these questions will be answered in the light of the following ideas:

1. The classical particle concept is based on strong metaphysical assumptions about physical objects and their interactions. But quantum physics is its strongest critic, and any view of subatomic reality should take this criticism into account.
2. In order to substantiate this criticism, the phenomena of subatomic physics, the ways in which they are generated in experiments, and the language in which they are expressed have to be investigated in detail.
3. The measurement methods of particle physics are far from straightforward. With the development of particle physics they became increasingly sophisticated and theory-laden. A historical account that investigates the growth of measurement theories is preferable for their analysis.
4. In high energy physics, larger and larger accelerators explore smaller and smaller structures, giving rise to the metaphor of *looking into the atom*. The rationale behind this metaphor deserves attention.
5. Many classical laws enter the measurement methods of particle physics. This results in a disunified physics. Nevertheless, its pieces are connected

---

[3]For the invention of the cloud chamber by C.T.R. Wilson see Millikan 1917, 46, and Pais 1986, 86. For the discovery of the electron see Sect. 3.2.

by several unifying principles, amongst them a generalized version of Bohr's correspondence principle.

6. The 'particles' of today's particle physics are *not classical*. Hence, a shift of meaning must have happened to the particle concept. Indeed, the quantum revolution has given rise to *several* meaning shifts.

7. In physical practice, the quantum concepts of subatomic particles are associated with explicit or implicit reference to wave–particle duality. The rationale behind this talk deserves much more attention than recent philosophy has accorded it.

There are many books on the history of particle physics, on the history and theoretical foundations of quantum physics, on the interpretation of quantum mechanics or quantum field theory, and on the question of whether subatomic entities do really exist. But there is no systematic account of the connections between these topics. The present book attempts to fill this gap by spelling out the above ideas one by one.

Chapters 1 and 2 deal with philosophical questions of principle. Chapters 3, 4, and 5 give a detailed account of the measurement methods of particle physics and the empirical basis of the current particle concept(s). Chapters 6 and 7 detail the meaning shifts that the particle concept has undergone, from the discovery of the electron to the detection of the top quark, from Einstein's light quanta and de Broglie's matter waves to recent ways of preparing quantum waves and detecting quantum particles. At the beginning of each chapter, a survey of the specific contents is given. The main topics are as follows.

Chapter 1 attempts to shed new light on the philosophical debate on *scientific realism*, that is, on questions of whether physical theories aim at truth and whether the forces, fields, and particles they postulate really do make up the world. The chapter ends with a plea for a moderate realism concerning the contingent events measured in the experiments of physics and the actual values of physical quantities attributed to them. This version of realism is associated with the view that physical quantities correspond to the properties of natural kinds. It is weaker than classical realism (i.e., the assumption of full-fledged microscopic objects carrying all physical properties that may be measured), but stronger than empiricism or instrumentalism (i.e., the view that there is nothing but empirical phenomena, experimental devices, and the use of physical theories for technological goals).

Chapter 2 investigates the constructive features of modern physics. The experimental method aims at extending our access to physical reality beyond the limitations of sensory experience. It is explained how the scales of physical quantities are constructed and what kinds of idealizations are involved in the experiments of modern physics. In order to explain why physicists have good reasons for adhering to quite a generous view of empirical reality, various constructivist arguments against *scientific realism* and a generalized concept of observation are discussed. In particular, Eddington's provocative question of whether the entities of physics are manufactured or discovered is taken up and a distinction between the observation of entities and the measurement of physical properties is proposed.

Chapter 3 introduces an intuitive mereological and causal particle concept: Particles are the constituent parts of matter or light and the causes of local effects in particle detectors. The detection of a particle is a position measurement. It requires certain amounts of the dynamic properties of matter or light to be localized in some experimental device. According to this criterion, the discoveries of the electron and the photon needed much more time than the usual histories of particle physics claim. After demonstrating this, it is shown how position measurements make up particle tracks, particle tracks make up scattering events, and scattering events make up resonances, introducing more and more theory into the empirical basis of particle physics. Here, the distinction between safe background knowledge and theories under test is crucial. It shifts with time, as the discovery of the positron and the problems of particle identification in the 1940s show. Nevertheless, according to the mereological and causal particle concept, the particles discussed so far were indeed discovered rather than manufactured.

Chapter 4 investigates the models behind high energy scattering experiments that probe the structure of the atom. There is a chain of models of subatomic scattering processes, moving from classical Rutherford scattering to the quark momentum distributions measured in deep-inelastic nucleon-lepton scattering. Even though the quantum field theory behind the quark model does *not* refer to subatomic matter constituents in any obvious sense, the above-mentioned chain of models connects Rutherford's discovery of an atomic nucleus to the quark–parton model of the nucleon. The connection is given in terms of pointlike structures observed at a given energy scale. It establishes *some* correspondence between quantum mechanical models of scattering and the classical model of scattering at a well-defined scattering center. Only this correspondence (which may be understood in the sense of Bohr's correspondence principle) establishes reference of the quark model to subatomic structures. Nevertheless, the metaphor of looking into the atom by means of a particle detector has its limitations.

Chapter 5 investigates the heterogeneous measurement theory of particle physics. Even though it is based on non-relativistic and relativistic, classical and quantum laws (which were considered to be based on incommensurable concepts by T.S. Kuhn and his followers[4]), it expresses a hidden unity of physics. The scales of physical quantities (i.e., the length, time, and mass scale) are constructed from the quarks to the whole universe. The measurement methods of particle physics contain redundancies, making it possible to cross-check the coherence of this construal. Finally, there are several unifying principles that connect the classical and the quantum domain, the most important being a generalized correspondence principle in Bohr's sense; the

---

[4]Kuhn 1962, 1970.

symmetries, invariances, and conservation laws of physics; and the assumption that in a macroscopic environment quantum states decohere. All these principles support the data analysis of particle tracks on the grounds of classical measurement laws *plus* quantum corrections and the construction of scales.

Chapter 6 investigates the features of classical particles and their fate in quantum theory. There are as many particle concepts as physical theories, or even more. The meaning shifts begin with quantum mechanics and the light quantum hypothesis. They result in a modest operational particle concept, according to which particles are collections of dynamic properties that may be localized independently in a particle detector. Field quanta, Wigner's group theoretical particle definition, the virtual particles of quantum field theory, the quasi-particles of solid state physics, and the quark–parton constituent model of matter give rise to further meaning shifts and to a further weakening of the original particle concept. What remains is the operational particle concept, on the one hand, and some further particle features such as symmetries, conservation laws, and sum rules for the dynamic properties of the constituent parts of matter, on the other hand. Taken together, these features more or less agree with the intuitive particle features discussed in Sect. 3.1, but with one crucial qualification: the causal particle concept is *not really* supported by current quantum theories. Particles in the operational sense are local effects rather than causes, whereas the causes of these local effects are non-local.

Chapter 7 discusses the predominant common feature of all quantum particles, namely wave–particle duality. In order to explain this concept, a historical approach is once again chosen. Wave–particle duality can be traced back to de Broglie's matter waves and the Einstein–de Broglie relation, Born's probability waves, Bohr's Como lecture, and Heisenberg's 1930 book on quantum theory. Unfortunately, these are four *different* accounts of wave–particle duality. Today, a pragmatic view of preparing waves and detecting particles, a confusing discussion about Bohr's complementarity principle and Heisenberg's uncertainty relations, and a generalized account of unsharp physical properties and corresponding generalized Heisenberg–Robertson uncertainty relations coexist. In order to clarify these concepts, several experiments of recent quantum optics and their interpretation are discussed. In accordance with the results of Chap. 7, these experiments show that a unique causal particle concept is no longer tenable, since too many different concepts of causality are involved in their interpretation.

Chapter 8 summarizes my philosophical conclusions about subatomic reality. In contradistinction to constructivist and empiricist positions and the related instrumentalist views, I maintain that subatomic particles and quantum processes belong to empirical reality. Matter is made up of subatomic constituent parts; the mereological particle concept is empirically supported by the sum rules for mass, charge, and other dynamic properties of subatomic particles. This amounts to an entity realism. However, the corresponding entities are neither objects in a classical sense nor substances in a metaphysical sense. In particular, the causal particle concept is at odds with quantum theory. Particles in the operational sense of single-particle detections are local effects of non-local quantum processes. In contradistinction to classical metaphysical realism, I finally defend the view that the reality of subatomic particles and quantum processes is not a reality in its own right. Rather, it is relational. It only exists relative to a macroscopic environment and to our experimental devices. The quantum entities are processes, dynamic structures, conserved physical properties, and event probabilities in the macroscopic world.

This book is a completely revised and considerably extended version of my German *Teilchenmetaphysik*, first published in 1994. What began as a translation finally became a new book. All chapters have been rewritten and a lot of new material has been added. The detailed discussion of the quantum particle concepts in Chap. 6 and the final chapter on wave–particle duality are completely new. Some of my philosophical views have changed, too. In particular, in this book I emphasize the constructive features of physics more than I did in the German edition. This Kantian theme is also indicated by the subtitle of the book. It will be brought into play occasionally in the subsequent chapters and taken up in the general conclusions of the book. Concerning the interpretation of quantum theory, I remain as abstinent as possible again, except in one regard. I now see the function of a generalized correspondence principle for the semantic unity of classical and quantum physics much more clearly than when I was writing the German version of the book. In my present understanding, quantum mechanics and quantum field theory only refer to individual systems due to the ways in which the quantum models of matter and subatomic interactions are linked by semi-classical models to the classical models of subatomic structure and scattering processes. All these links are based on tacit use of a generalized correspondence principle in Bohr's sense (*plus* other unifying principles of physics). Today, I would no longer insist on a quantum theory of individual systems. A quantum theory on its own is a probabilistic theory. Apart from conservation laws and superselection rules, it only applies at the ensemble level. The question of how this view fits in with the belief of practising physicists that they prepare waves and measure particles at the level of individual systems will be discussed in my critical conclusions about subatomic reality (Chap. 8).

The present book has grown over such a long time that it is impossible to thank all colleagues and friends who have made critical remarks, hints to the literature, or explanations of crucial concepts on the occasion of talks, workshops, and conferences. Here I can only mention some of them. In particular, I am grateful for a decade of fruitful and inspiring discussions with Peter Mittelstaedt. I would like to thank Claus Beisbart, Patrick Grete, Reiner Hedrich, Hernán Pringe, and Erich Runge for reading large parts of the book and making many helpful comments. Obviously, their criticism has helped me to improve the book, but they are not responsible for the remaining errors. Florian Braun helped me to make the reference list and to keep it largely consistent with the footnotes. Angela Lahee from Springer supported this edition with a lot of patience. Stephen Lyle improved my English and did the editing. Finally, without being accompanied over so many years by the love and comprehension of my husband Friedrich Fulda, our son Johannes Falkenburg, and my mother Edel Falkenburg the present edition would not have been possible.

Dortmund,   *Brigitte Falkenburg*
January 2007

Contents

Appendices

1 Scientific Realism

Most physicists have a realistic view of physical entities. They assume that they investigate real forces, electromagnetic fields and subatomic particles such as electrons, protons or neutrons. They presuppose that these entities exist in nature, outside as well as inside a physics laboratory. They consider them to be the true causes and constituents of experimental and natural phenomena, even if any physical theory describes them in terms of mathematical idealizations. They subscribe to an unquestioned realism regarding subatomic particles when they detect and analyze particle tracks in the scattering experiments of high energy physics; and when they assume that the detection is due to electromagnetic effects in the electronic devices of their experiments.

Particle physics no longer investigates the question of *whether* subatomic particles exist, but *which kinds* of particles exist and what the fundamental constituents of matter and forces are. Between 1932 and 1956, for example, it was unclear whether Pauli's neutrino hypothesis was assumed ad hoc to save the principle of energy conservation in β decays, or whether there really are neutrinos. In 1956, the neutrino hypothesis was confirmed by the inverse β-decay.[1] Similarly, between 1964 and 1974 most physicists did not believe in quarks. They did not think that there were constituent parts of the proton or neutron with fractional charges. In 1974, with the detection of the J/Ψ resonance, the scientific community of particle physics finally became convinced of the existence of quarks.[2] Nowadays, particle physicists are investigating the question of whether there are supersymmetric partners of the particles of the current standard model of particle physics and they are searching for effects of weakly interacting massive particles (WIMPs) in high energy cosmic rays.

Many philosophers of science do not subscribe to this kind of realism. They call it *scientific realism* in contradistinction to the old philosophical debate on the reality of the external world that was doubted for example in Descartes' *Meditations* and Berkeley's idealism. Empiricist or instrumentalist philosophers of science do not doubt the reality of the external world. They

---

[1] Pais 1986, 138 and 569.

[2] Pais 1986, 605–606. For the history of this so-called November revolution, see Pickering 1984 and Riordan 1987.

only doubt the existence of physical objects that cannot be immediately perceived. In particular, they doubt that subatomic particles exist. The recent philosophical debate on scientific realism is about two kinds of beliefs: the belief in the existence of non-observable physical entities (*entity realism*) and the belief in the existence of structures in the world to which the laws of mathematical physics may approximately correspond (*structural realism*). Realist combatants defend the existence of forces, fields and subatomic particles and of true laws of nature, while anti-realists argue against it.[3] The anti-realist party obviously does not criticize the results of theory formation in physics as such. Rather, they argue against naively interpreting these results in everyday language, as if forces and subatomic particles were as evident and substantial as tables or chairs. Indeed, with regard to the subject matter of quantum theory a naive everyday realism produces crude misunderstandings.

Nevertheless, in the recent debate on scientific realism old philosophical problems come back disguised in new clothes. In 20th century physics new questions arose which became interwoven with unresolved traditional philosophical topics. The three most important roots of the debate are as follows:

(1) The debate on *realist or instrumentalist* views of physical theories is very old. Anti-realism traces back to the very beginnings of modern science, namely to Osiander's unauthorized preface to Copernicus' book *De revolutionibus orbium coelestis*. He recommended understanding the Copernican theory not literally but only as a useful tool for saving the phenomena by means of a mathematical description that does not claim truth. Galileo had to defend his new physics and the Copernican view of the world against this instrumentalism and 17th century Aristotelism. Three centuries later, Mach's empiricism and Duhem's instrumentalism stood against Boltzmann's atomism and Planck's realism. Recently, the debate was again nurtured by *Kuhn's incommensurability thesis*, giving rise to *social constructivism* and related post-modern views of science. Obviously, these old and new debates are about the question of what mathematics and physics can tell us about nature.

(2) Logical empiricism initiated the search for criteria of a *demarcation between empirical knowledge and metaphysics*. The search for such criteria is rooted in the traditional controversy of empiricism and rationalism: the empiricist epistemologies of Locke or Hume take sensory perception as a touchstone of our theories, whereas according to the rationalist views of Descartes or Leibniz, sensory perception may be illusory and only mathematics is reliable. The traditional controversy dealt with the epistemological question of how we may acquire reliable knowledge of the external world, or of objective reality. Regarding *physical* reality, however, both

---

[3]See Leplin 1984, van Fraassen 1980, Churchland and Hooker 1985, Psillos 1999. Structural realism (Worrall 1989) is a new and more modest variant of the old belief in the existence of true laws of nature; see Psillos 1999, 146–161 and the literature quoted there.

traditional epistemologies are at odds with the experimental method of modern physics and with the very structure of physical theories.

(3) The debate on the *interpretations of quantum theory* deals with the fact that the laws of subatomic physics collide with the classical view of a completely determined structure of reality. Thus the question arose: Does quantum theory really deal with an objective subatomic reality, or is it grist for the mills of empiricism and instrumentalism? Typical non-local quantum effects are at odds with the classical account of physical reality, which is indeed rooted in traditional metaphysics. The debate on the interpretations of quantum theory is substantially tied to the traditional controversies on empiricism or rationalism (respectively instrumentalism or realism), even though their topics are currently most often split off from each other in highly specialized philosophical discourses.

In current philosophy of science these three kinds of problems are subject to separate discussions, even though they actually belong together. In order to investigate their relations, I will give an outline of the debate on *scientific realism* that deliberately exceeds the scope of the current discussion. I attempt to unravel the subject matter of the debate starting from the claim of a two-fold mismatch between physics and philosophy. Epistemologically, empiricism cannot cope with the methods of 20th century physics (Sects. 1.1–1.3), whereas ontologically, traditional metaphysics cannot cope with the structure of quantum theory (Sects. 1.4–1.6). The founders of logical empiricism neglected the first, traditional problem in their quest for a demarcation between empirical science and metaphysics. Many defenders of *scientific realism*, however, disregard the second, new problem in their defence of subatomic reality.

Carnap and Reichenbach relied on relativity and quantum theory when they criticized all *a priori* concepts as metaphysical. They defended a verification theory of meaning and proposed to reconstruct the empirical contents of physical theories laid bare of any metaphysical assumptions. But in doing so, they neglected the fact that since Galileo's day the experimental methods of physics *have extended* and *drastically changed* our understanding of empirical knowledge. The use of observation instruments and measuring devices favors a liberal or generalized understanding of observation and empirical evidence. In physics, measurement results matter whereas sense data do not. Hence, logical empiricism gave rise to the following questions. What is *empirical knowledge*, in view of the methods of modern physics? And where does empirical science end and metaphysics begin, according to this kind of knowledge? As far as the recent debate on *scientific realism* arises from these epistemological questions, it is an inheritance of 20th century logical positivism (Sect. 1.1). Indeed, the main current positions of the debate are obtained by demarcating the empirical against the metaphysical in various ways. Naive, global realism considers all the mathematical structure of physical theories to have a counterpart in reality. Constructivism believes in nothing that is constructed by means of mathematical tools and technological instruments. In between are several positions of critical realism, moderate empiricism, and strict empiricism (Sect. 1.2). Considered so far, the debate struggles for epistemic criteria that help to decide which concepts of physical theories describe something real, and which do not. Is it possible to find an independent criterion for what we know to be real, a criterion that depends neither on our physical theories nor on our metaphysical beliefs? This question is settled in terms of the physical events that actually happen in the measurement devices. Their epistemic hallmark is contingency (Sect. 1.3).

At this point, the ontological (or metaphysical) problems raised by quantum theory come into play. It is a brute fact that quantum theory is needed in order to explain the contingent events and measurement results of subatomic physics. But quantum theory is *at odds* with the kind of realism that has been defended for centuries, namely *classical realism*. Traditionally, realism comes in terms of physical objects with completely determined properties and deterministic laws of nature. The underlying metaphysics is the idea that physical objects are something like Aristotelian substances, or individuals, carrying causal powers as dynamic properties. However, quantum theory tells us that physical reality is *not* made up of substances like this (Sect. 1.4). The kind of realism challenged in this way is the metaphysics behind classical physics. In order to see what remains of it in view of quantum theory, the metaphysical and non-empirical assumptions of physics have to be examined. It turns out that some of them are indispensable for doing physics, even though quantum theory seems to be at odds with them (Sect. 1.5). Finally, a realism of physical properties based on the scales of physical quantities is suggested. It is strong enough to support the mathematical and experimental methods of physics, but weak enough to be compatible with the structure of quantum theory (Sect. 1.6).

1.1 Empirical Knowledge and Metaphysics

Around 1900, when the attempts to complete classical physics finally resulted in its abandon, the traditional debate about Galilean science began to interweave with the new challenges of scientific realism. Mach had given an excellent analysis of classical mechanics, and Duhem explained the aim and structure of physical theories.[4] They defended empiricist and instrumentalist positions, respectively. According to Mach, neither atoms nor forces exist. In his view, these hypothetical entities are only assumed in order to describe the phenomena in an economical way and they should better be replaced by a more natural description.[5] Based on detailed investigations of the way in which physical quantities relate to experimental phenomena, Duhem gave an

---

[4]Mach 1883, Duhem 1906.
[5]Mach 1883, 466.

influential instrumentalist account of classical physics. Both positions were opposed to Maxwell and Boltzmann's kinetic theory of gases and its atomistic foundations, respectively. Around the time when Boltzmann despaired, however, atoms and their constituents became accessible to indirect experimental observations and measurements in the laboratories of Thomson and Rutherford. In 1903, the light flashes produced by α rays were directly observed for the first time. Considerations about atomism, the kinetic theory of gases, and the statistical foundations of thermodynamics also entered the formulation of the physical problems that finally gave rise to Planck's law of black body radiation and Einstein's hypothesis of light quanta. Thomson's or Rutherford's experiments with massive charged particles were still explained completely in classical terms, but Planck's and Einstein's theoretical descriptions of the interaction of matter and radiation were different. They put an end to classical physics, which neither Planck nor Einstein was willing to accept.

Carnap and Reichenbach became the most influential philosophers of 20th century physics. Impressed by the scientific revolutions which led to relativity and quantum theory, they dispensed with neo-Kantianism and turned to Mach's empiricist view of reality. The Vienna circle of logical empiricism adopted Mach's disdain of metaphysics and began to run after criteria for the demarcation between empirical knowledge and metaphysics. This search dominated the epistemology of modern physics for the following decades. The philosophy of science focused on the reconstruction of the empirical content of physical theories. However, the search for such demarcation criteria failed. Empiricism had to confront the Duhem–Quine thesis, according to which the empirical parts of physical theories cannot be isolated, and the changes of our world view due to the big scientific revolutions indicated by Kuhn.[6] In these views, empiricism had to face the problem that empirical science is not as neatly separable from metaphysics as supposed. The recent debate on scientific realism arose at the time when Kuhn's famous book on the structure of scientific revolutions appeared. As we shall see, it is indeed a debate on the metaphysical aspects of modern science.

For 20th century empiricist philosophy of science, *metaphysics* is a system of *a priori* propositions about the world. From an empiricist point of view, 'metaphysical' is contrary to 'empirical'. This understanding of metaphysics is indebted to Kant's view of metaphysics as an account of the world which is neither mathematical nor empirical.[7] In accordance with Carnap's and Reichenbach's neo-Kantian background, it is rooted in traditional philosophy, particularly in the opposition of empiricism and rationalism. In contradistinction to Kant's views about metaphysics,[8] however, it goes hand in hand with Mach's disdain of metaphysics.

---

[6]Quine 1951; Kuhn 1962, 1970. See also below.

[7]See Kant 1781/87, A 837–850/B 865–878.

[8]See Falkenburg 2004.

Modern objections against the existence of 'theoretical' entities and the assumption of true laws of nature are still indebted to this origin. They are related to the traditional logical and epistemological objections against non-perceivable entities. Amongst such non-perceivable entities include: Gassendi's or Galileo's atoms, which were inherited from ancient natural philosophy; Descartes' mechanical corpuscles, which were similar to atoms; Newton's absolute space, which was defended by Clarke against Leibniz; the causal link between events exhibiting law-like regularities, which was doubted by Hume; and substances in the sense of traditional metaphysics, that is, the carriers of the phenomenal properties of material objects. According to Kant, the traditional debates about these entities aim at metaphysical concepts. In his view, such metaphysical concepts do not have objective reality, or do not apply to empirical objects in space and time.[9]

The theories of current physics deal with atoms and their internal structure, with forces, space and time, and the universe. Nominally, they refer to entities to which traditional natural philosophy, cosmology, and metaphysics had also referred. In contradistinction to traditional metaphysics, however, modern physics has *experimental access* to electromagnetic radiation and field quanta, to subatomic matter constituents and to the forces which hold them together, and to the cosmic background radiation of the universe. The laws of physical theories enable scientists to make precise quantitative predictions about empirical and experimental phenomena. Physical objects, systems and structures which can be measured are taken away from the traditional metaphysical twilight of a realm of unobservables which are postulated but neither proved nor disproved. The atoms as constituent parts of matter stayed within such metaphysical twilight as long as there was no experimental evidence for electrons and the atomic nucleus. As their spatio-temporal background, absolute space and time were postulated as independent entities. These entities remained in the same metaphysical twilight until Einstein developed his operational concept of simultaneity and his principles of equivalence, giving rise to special and general relativity. And the universe as a whole was subject to *a priori* proofs for or against the possibility of a finite world age, whereas modern physical cosmology is based on empirical observations as a touchstone of distinct cosmological models.

But the empirical basis of modern physics is not given in terms of immediate sensory perception, as empiricism would have liked. Instead, it is based on theory-laden experimental observations and measurements. Therefore, in the view of modern physics the traditional distinction of empirical knowledge and metaphysics lost its innocence. In order to establish a new distinction, the following considerations may be helpful. The objects of physical theories

---

[9] ... as far as they are not necessary conditions of the very possibility of experience; see Kant 1781/87. Kant's theory of space and time as forms of intuition *a priori* was rejected by neo-Kantianism and logical empiricism alike; see Falkenburg 2000, Chap. 7.

escaped from the metaphysical twilight of traditional natural philosophy due to two epistemological features of modern physics:

(i) The objects of physical theories were never regarded as unquestionable or ultimate.[10] In contradistinction to the dogmatic theories of traditional metaphysics, empirical theories are open for revision. New experiments may reveal unexpected properties of the phenomena. Only when physics aims at ultimate explanations does it grow suspicious of falling back into metaphysics. The quest for final explanations carries a heavy metaphysical burden. It is suspicious of either giving rise to circularity or to an infinite regress. The quest for a final physical theory embraces the meta-theoretical requirements of having no rivals and eliminating all contingent parameters, including the fundamental constants of nature (the gravitational constant γ, the speed of light c, and Planck's constant $\hbar$).

(ii) The subject matter of physical theories counts as *empirical* as far as its concepts are experience-based. Here, the meaning of 'empirical' and 'experience-based' may be understood in a strict sense or in a more liberal way. *Either* the contents of an acceptable physical theory are considered to be grounded in sensory perception and logic alone, as logical empiricism assumed, *or* alternatively one has to admit that the interplay of theory and experiment gives rise to empirical knowledge in a generalized sense, which is on an equal footing with sensory perception.

All current mature physical theories satisfy these criteria. In particular, the views about the constituent parts of matter are repeatedly revised. The atoms of atomic physics are not atoms in the literal sense of ancient natural philosophy. They have parts and they are not indivisible. Experiments show that atoms are composed of electrons, protons, and neutrons. Protons and neutrons are composed of quarks. The quarks cannot be isolated, but they give rise to fragmentation processes of protons or neutrons. Whether the quarks have constituent parts (and in which sense they may have), remains open. The current standard model of particle physics is not considered to be definitive. Up to now, all physical theories are open for revision. The criterion suggests that the current search for an *ultimate* theory of physics, say, quantum gravitation and quantum cosmology, may be *metaphysical*.[11] Many physicists would agree – but mainly because the current attempts at unification also violate the *second* criterion of being experience-based.

From a strict empiricist point of view, however, the *second* criterion is problematic. Logical empiricism pursued the project of reconstructing the

---

[10]That is, they are not unconditioned or noumenal objects in Kant's sense, which may give rise to cosmological antinomies. See Kant 1781/1787, B 432–596/A 406–567; and Falkenburg 2000, Chap. 5.

[11]In terms of the metaphysical elements of physical theories described in Sect. 1.5, it is in fact metaphysical. It relies on the indispensable meta-theoretical principles of theory formation in physics.

contents of physical theories based on sensory perception. But Carnap's original project had to be weakened step by step. Finally, it failed to specify unambiguous criteria for the empirical significance of theoretical concepts,[12] and in the end, only a *generalized concept of experience* remained for the demarcation between empirical and metaphysical propositions. But this turned out to be fatal for the empiricist project of providing criteria for the demarcation of empirical knowledge and metaphysics. *Any generalized* concept of experience necessarily implies *non*-empirical propositions, which are metaphysical from an empiricist point of view. In this way, a successor of the traditional metaphysical debates about atoms or absolute space and time emerged, namely the question of whether there is a legitimate generalized sense in which we may have empirical knowledge of atoms and subatomic particles. While the questions of whether there are atoms, or what the nature of space and time may be, acquired empirical content within physics, within philosophy they reappeared in new clothes.

Any philosophical attempt to clarify the question of whether or not knowledge of atoms is possible confronts the expert's dilemma. Non-experts have no access to the physicist's empirical methods and concepts. In cases where the layman cannot observe any phenomenon at all, the physicist talks of empirical evidence. Non-experts have no criteria for deciding whether the physicists obtain *their* theory-laden evidence in a completely arbitrary and circular way, or not. The only experts, namely physicists themselves, most often do not care about such apparently *minor* philosophical problems arising from their results. Obviously, it is not their job to be interested in the question of which generalized meaning of 'empirical' agrees with the experimental results that make their theories empirical. For them, the question simply does not arise; it is already answered in terms of experimental results and theoretical predictions. But any non-physicist is hardly able to understand the experiments of atomic and subatomic physics, and to see in which way they depend on safe background knowledge. However, the arguments of the realist as well as the anti-realist party concerning the existence of unobservable entities and the truth of physical laws will necessarily miss the point, if they do not take into account the complex structure of modern physics. Indeed, within the philosophy of science several approaches have been developed to do so. None of them has turned out to be fruitful for improving the dialogue between physics and philosophy. Indeed the best of them are either very complicated or very specific.[13]

---

[12]See Carnap 1956 and Stegmüller 1970, Chap. V.

[13]The so-called 'syntactic' approach in first order logic (Ramsay 1929; Carnap 1956, 1966) turned out to be impracticable. Its successor, the model-theoretic approach, was based on naive set theory and Tarski's semantics (Suppes 1969); it was more successful; van Fraassen (1980, 1987) called it the 'semantic' view of theories. It was worked out in more detail in the structuralist approach (Sneed 1971); for an overview see Balzer et al. 1987; Balzer and Moulines 1996. The structuralist approach has been elaborated in a much more sophisticated way in Ludwig 1990

Without going into any details of these approaches, the problem of whether to count physical objects as empirical or not may be pinned down as follows. Any doubts concerning the empirical contents of physical theories arise from the *constructive* character of theory formation and experiment. The main achievement of Galileo's new science was to replace sensory perception by experiment and measurement. Modern physics has not only put a network of theoretical constructs over the empirical phenomena. Under laboratory conditions, experiments also generate many new phenomena which do not occur in nature. The experimental method of physics has not only *extended* but also *drastically changed* our empirical access to reality. Experiments aim at investigating isolated systems under approximately ideal conditions. In order to do so, the phenomena are analyzed on the lines of Galileo's resolutive–compositive (or analytic–synthetic) method.[14] For example, Galileo analyzed the behavior of a falling body under the effect of gravity and the disturbance of air resistance. In addition, he investigated it under varying experimental conditions. Are we entitled to say that the network of physical theories represents something in nature if it structures and extends in this way our empirical access to reality? If not, most physical objects are not empirically given. If yes, our empirical access to reality is highly theoretical. Then we need new philosophical criteria for the demarcation between the safe (or empirical) and the uncertain (or metaphysical) parts of our theories.

Indeed, a closer look at the debate on scientific realism shows that it does *not* deal with metaphysical pseudo-problems in the sense of logical empiricism.[15] Quite the contrary, in a very specific sense, its participants debate about the question of which law-like statements and theoretical structures can be interpreted literally and which cannot. The realist and the anti-realist parties have the same goal: they want to clarify to what extent physical theories give an adequate description of real objects and processes in nature, or to what extent they permit a realistic interpretation.

At this point, another conceptual clarification is needed. Philosophers use the word 'realistic' in a way that differs substantially from the way in which scientists talk about the realistic description of a phenomenon or a system. Scientists call a theoretical model realistic if it is good, that is, if it is

---

and Scheibe 1997b, 1999. During the last decade it became fashionable to reconstruct not physical theories and their empirical content but only models with a restricted domain and the way in which they relate to the phenomena; see Morgan and Morrison 1999; Bailer-Jones 2004.

[14]See Losee 1993, 57–62, and Sect. 2.2

[15]Sometimes, however, the positions in the debate are characterized in a way that support this suspicion; e.g., Fine 1984, 97: "So, when the realist and the anti-realist agree, say, that there are really electrons and that they really carry a unit negative charge and really do have a small mass (of about 9.1×10^{-28} grams), what the realist wants to add is the emphasis that it is really so. 'There really are electrons, really!'" Fine introduces his *natural ontological attitude* in order to avoid such pseudo-problems.

not based on inadequate ('unrealistic') idealizations. Philosophers interpret a model as realistic if it has reference. The difference between the two uses is as follows. The philosopher of science puts into question *whether* a theory refers to something in nature or not. The scientist, on the other hand, asks *how well* a model describes (or predicts the behavior of) the system to which it refers, without questioning the existence of the latter. From a pragmatic point of view, most scientists are realists in the philosopher's sense. In physical practice, when they develop models and perform experiments, they usually *act* as realists. Usually, that is, in phases of normal science, they do not put their theories and the reference of their theoretical terms in question. For them, reference is not at stake in principle but only with regard to specific physical concepts which have not yet been confirmed by experiment. (Above all, this is true for the quantum mechanical wave functions and wave packets that they employ in the concrete models of quantum physics and prepare in their experiments.[16]) They mainly struggle for adequate theoretical descriptions and explanations of the phenomena. This pragmatic attitude towards reference is shared by all scientists, independently of the specific epistemological views they would assert when asked epistemological questions, say, about quantum objects.

To make a philosophical decision for or against a realistic interpretation of the laws and entities of physics is something different. It depends crucially on philosophical or metaphysical presuppositions, on distinct assumptions about reality. From an empiricist point of view, reality consists only in empirical data, and in their causes as far as they derive from the data.[17] In contradistinction to empiricism, scientific realism is the assumption that the material world is something like a sum total of the objects of our best accepted scientific theories. In the last few decades, the dissent has been disputed on several levels, abstract or concrete, general or specific, for whole theories and in terms of many case studies. However, the dispute was separated from the philosophical debate on the ways in which modern physics questions our traditional view of reality. This view is itself metaphysical, and it has been shaken by quantum theory.[18] Nevertheless, we may take it for granted that the objects of mature physical theories escape the metaphysical

---

[16]See Sect. 7.3.

[17]Empiricism may be more liberal or more strict. Cartwright 1983 admits causes, van Fraassen 1980 does not.

[18]The ongoing debate on the interpretations of quantum theory has almost no overlap with the recent debate on scientific realism, even though it is relevant for the positions that can be defended in accordance with modern physics. Scientific realism in a classical sense of reality is at odds with the fact that subatomic particles exhibit no classical behavior, or no spatio-temporal existence in the usual, classical sense. Strict empiricism gets rid of this problem – but it is at odds with the fact that without belief in the existence of subatomic particles the very project of investigating them in the typical experiments of quantum physics makes no sense at all. This crucial problem of any *particle metaphysics* is taken up in Sect. 1.4.

twilight of traditional natural philosophy according to the two criteria mentioned above: (i) any knowledge of them is open for revision, and (ii) their concepts are experience-based.

In this sense, they belong to *empirical or phenomenal reality* in the widest sense, that is, to those aspects, parts or structures of the material world, or of nature, which are the subject matter of physical theories (or natural science in general). Empirical reality is an embodiment of anything that is empirically given and may be investigated by the methods of modern science. Such a general concept of empirical reality concerns our *epistemic* account of reality. In terms of it, the debate on scientific realism may be simply characterized as follows. It deals with the question of *what on earth we should call 'empirical reality' in view of modern science.* How weakly or how strongly may a conception of empirical reality be infiltrated by theory, if we require it to accord with modern physics? To what extent do the contents of our best theories determine our views about empirical reality, and conversely to what extent do the latter determine the empirical contents of the former? Let us now look at what empiricism can teach us about these epistemological questions.

1.2 More or Less Empiricist Demarcations

Strict empiricism admits only a narrow view of empirical reality, and our *empirical knowledge* of reality. Concerning what we do not know based on sensory experience, it is sceptical or agnostic. The empiricist epistemology of science has a big philosophical tradition that goes back to Aristotle. To what extent is empiricism tenable, if we want to explain empirical phenomena? Traditional empiricism is still inclined to explain the phenomena in terms of unobservable entities: Aristotle's essences, Locke's atoms and primary qualities; Mach's simple elements of sensory perception. The only radical and agnostic sceptic was Hume.

Carnap's original project of reconstructing the empirical content of physical theories was based on the contents of sensory experience and may be understood as an attempt to *reconcile theoretical explanations with strict empiricism.* Later, Carnap developed a weakened version of this project, extending his earlier views about empirical reality. In 1956, he thought that empirical reality embraces all objects to which theoretical terms refer, as far as these theoretical terms belong to accepted theories and have empirical significance, i.e., refer to some empirical data.[19]

Finally, Carnap's project of reconstructing physical theories failed due to the very structure of physical theories. This structure does not allow one to

---

[19]See Carnap 1956. In his writings, Carnap translates Frege's terms 'Sinn' (sense) and 'Bedeutung' (reference) by 'meaning' and 'significance', respectively. Hence, his requirement of empirical significance is the quest for reference to some empirical data.

give unambiguous criteria for empirical significance in Carnap's sense[20] and leads one to suspect that the demarcation line between the empirical and the metaphysical runs through the midst of empirical theories. In a crucial correction of Carnap's project, philosophers of science admitted that this demarcation line does indeed run through science, if they did not reject the very idea that there is such a demarcation line. In the tradition of Duhem, Quine emphasized that empirical theories are *empirically underdetermined*. That is, they are never completely determined by the data that make up their empirical basis. In addition, Quine defended the view that empirical theories are testable only as a whole, not in isolated parts. This is the famous Duhem–Quine thesis.[21] Duhem himself did not defend such a strong thesis. He believed that, when an experiment fails, it is up to the scientist's experience to decide which parts of the underlying theoretical system are questionable or not. Duhem's belief can be explained in terms of the distinction between background knowledge and theoretical assumptions under test. However, accepting the insight that all theories are empirically underdetermined, the empiricist philosophy of science developed the follow-up project of reconstructing empirical theories in such a way that their empirical substructure is isolated from their non-empirical parts.

The empiricist research program may be saved (*against* Carnap) by means of the distinction of the empirical and the non-empirical parts of a theory. According to this distinction, it is admitted that empirical theories have non-testable, metaphysical parts. As van Fraassen puts it, only the empirical substructure of a theory definitely corresponds to something in empirical reality.[22] What belongs to the empirical substructure of a theory, however, depends again on what may count as 'empirical'. Strict empiricism must claim that it is bound to sensory perception. Whether this attempt to save strict empiricism is tenable or not may be doubted.[23] But more liberal views of the empirical part of a theory are also indebted to an empiricist account of reality. The demarcation line between the empirical and the non-empirical (or metaphysical) parts of a theory may be drawn in several ways. These more or less liberal demarcations shed new light on the debate on scientific realism. They unfold a spectrum of epistemic positions which are indeed found in the debate.

The demarcation line depends on the *kind of experience* which is regarded as *relevant* for making the content of a theory empirical. For logical empiricism, it was the pointer position of the measuring device which can be

---

[20] See Stegmüller 1970, Chap. V.

[21] See Quine 1953, Duhem 1906.

[22] See van Fraassen 1980, 1987. Indeed, it is well known that physical theories contain non-empirical, conventional elements such as Einstein's convention of 1905 concerning the two-way speed of light and the definition of simultaneity.

[23] See my discussion of van Fraassen's version of strict empiricism in Sects. 2.6 and 3.6.

recorded in the logbook of an experiment.[24] For philosophers of science who generalize the empiricist criterion of observability it is any object which can be made visible by means of optical or electronic devices such as the telescope, the microscope or the electron microscope. For practising physicists it is any physical quantity that appears in a theory and can be measured. For most physicists and for philosophers of science who have good knowledge of physical experiments, it is in addition any experimental result based on data analysis and safe theoretical background knowledge. These positions differ only in making the cut between the empirical and the non-empirical at different theoretical levels, according to the degree of trust in the theoretical assumptions allowed to go into the data. At the extremes of the spectrum there are two more positions. They correspond to admitting *all*, respectively, *no* parts of an accepted physical theory as having a correlate in reality. In this way, we obtain a crude classification of five distinct epistemological attitudes towards unobservable physical objects and the laws of physics: (1) global realism, (2) critical realism, (3) moderate empiricism, (4) strict empiricism, (5) constructivism. Obviously, the differences between these five epistemic positions are not very sharp. They are a matter of degree rather than of principle. There are also mixed positions, such as van Fraassen's 'constructive empiricism'.

1. **Global realism** is the maximum position. The assumption is that all parts of a mature theory correspond to something in reality. According to this, anything which is subject to current physics exists in nature: the particles of the current standard model of particle physics including the Higgs boson, the real and virtual field quanta of quantum field theories, quantum mechanical wave packets and black holes. Global realism corresponds to what is often called 'naive' realism. Amongst its defenders there are mainly unreflecting scientists and laymen.

2. **Critical realism** is the Kantian belief that all and only the objects of possible physical experience exist. Here, 'critical' is meant in the sense of Kant's criticism of traditional metaphysics, and physical experience has to be understood in a broad sense of experimental results and empirical observations. Hence, what matters is possible empirical evidence in the sense of significant experimental data and measurement results. According to this view, neither Newton's absolute space nor its modern correlate in a substantivalist account of general relativity actually exist,[25] and nor do the virtual particles of quantum field theory. Physical objects, systems, and processes exist if the scientific community of physics either agrees on having experimental evidence for them or if it has a measurement method for them, as in the case of the Higgs boson. In addition, the laws that describe these entities may be considered to be at least approximately true.

---

[24]However, Nagel 1931 has already shown that the relation between a counter position and the value of a physical quantity is very complex.

[25]See, e.g., Earman 1989.

From a critical realist's point of view the following questions may still be debated: Does the perturbation expansion of a quantum field theory at least legitimate the belief in *clouds* of virtual field quanta as approximately separate parts of reality?[26] Does the superstring program belong to physics rather than metaphysics even though it is still void of any specific empirical content?[27] But such unclear cases of presumed entities of mature or unmature theories do not shake the critical realist's belief that the gravitational field, electromagnetic waves, atoms, electrons, protons, photons, and neutrinos exist. If asked long enough, probably most physicists would defend such a position. Philosophers of science who have detailed knowledge of some part of physics also tend to adopt this position, or one of the two sub-positions described below. Critical realism has the advantage that it takes empiricism seriously in its sceptical attitude but also remains open for the growth of theoretical background knowledge. Several philosophers of science uncouple the critical realist's assumptions about physical objects and the laws of physics. Even though they are closely related, they make up the following two sub-positions of critical realism:

- *Structural realism* is a recent name for the belief that the laws of physics surviving scientific revolutions are approximately true, or correspond to invariant structures in nature.[28] Structural realism admits that a scientific revolution may result in the disappearance of postulated entities such as absolute space, the ether, or phlogiston, from scientific discourse. If a scientific revolution leads to the insight that certain entities do not exist, the resulting theory no longer contains terms referring to them and the scientific community becomes convinced that the corresponding terms in the former theory cannot have any reference. But if this happens repeatedly, we cannot be sure whether the terms in our currently accepted theories refer. Structural realism is an answer to this kind of reasoning which was called *pessimistic (meta-)induction*.[29] Paradigmatically, this answer holds for

---

[26]See Sects. 6.4.3 and 6.6.

[27]Hedrich 2006.

[28]See Psillos 1999, 146–161, and the literature quoted there; in particular, Worrall 1989. On the basis of historical case studies on theory change in physics and chemistry, Carrier 1991, 1993 develops a similar position. He argues that it is reasonable to defend a realistic interpretation of the *structures* in nature, which are grasped by the concepts and laws of empirical science. But as the case of phlogiston shows, these concepts and laws may be associated with misleading assumptions about the underlying natural kinds. Therefore, structural realists do not believe in the ontology behind these concepts and laws. Theoretical assumptions about reference may change whereas the empirical properties of natural kinds survive scientific revolutions. Poincaré defended a variant of structural realism; see Worrall 1989; Huber 1999.

[29]See Laudan 1981; Psillos 1999, 32, 81, 101–108.

the current quantum theories. They have symmetries which seem to be invariants of nature. A structural realist should believe in them. But the way in which the quantum mechanical wave function refers beyond Born's probabilistic interpretation is as unclear as it was in the 1930s. Similarly, the Maxwell equations of classical electrodynamics (that are kept in quantum electrodynamics, *pace* quantization) can be taken for true without any belief in the reality of the single components of the Fourier expansion of an electromagnetic field. Structural realism is mainly a position of theoreticians and philosophers of science with a background in theoretical philosophy.

- *Causal realism* is the traditional view that the unobservable causes of observable effects exist. This view gave rise to modern physics. Newton explained it in his *Principia* as the first *rule of reasoning*, according to which physics aims at finding the necessary and sufficient "true causes" of the phenomena.[30] Many physicists tend to subscribe to this kind of causal realism – at least, as long as they do not care about the foundations of quantum theory.[31] Above all, experimenters *must* have *some* pragmatic attitude of causal realism. Today, causal realism is defended by those philosophers of science who have detailed knowledge of physical experiments, observation instruments, and phenomenological laws. As an attempt to overcome the philosophical problems of scientific revolutions, causal realism is complementary to structural realism. A causal realist typically believes that electrons exist because they make observable tracks in a bubble chamber. She may believe in electrons without claiming that quantum electrodynamics is true. Such a belief is supported by the matter of fact that the electrons in the storage ring of a particle accelerator obey the laws of classical electrodynamics rather than quantum theory. As the phenomenon of synchrotron radiation shows, in contradistinction to electrons inside the atom, they radiate. There are several versions of causal realism. Cartwright's and Hacking's positions are most prominent. Cartwright defends a causal realism according to which nature has causal powers or capacities that are subject to causal analysis. Her account of Nature's capacities is coupled with belief in the (approximate) truth of phenomenological laws.[32] Hacking suggests a stronger version of causal realism in order to obtain a sufficient criterion for the existence of an entity. He couples causal realism with the requirement

---

[30]Newton 1687, 794. Here, the relation between cause and effect is assumed to be real and expressed in terms of *some* kind of necessary relation between cause and effect. This presupposition is obviously stronger than an empiricist regularity view of causality. In the following, I tend to a Kantian position, but my argument does not depend on Kant's specific philosophical assumptions about causality.

[31]Quantum theory gives rise to several arguments *against* causal realism; see Sects. 2.5, 5.3, and 7.3–7.6.

[32]Cartwright 1983, 1989.

of successfully using an entity as a technological device. According to
his reality criterion, electrons exist because they can be successfully
used as experimental devices with observable effects in a scattering
experiment. Or, as Hacking put it, "if you can spray them then they
are real".[33]

3. **Moderate empiricism** is the liberal empiricist view of physicists that,
in addition to the immediately observable phenomena, all measured values of physical quantities have an empirical correlate in reality. I do not
know of any physicist who would not subscribe to that. According to this
view, the empirical content of theories lies in predictions of the values
of physical quantities such as position, time, mass, momentum, energy,
charge, temperature, and so on. The measured values of these quantities
are considered to be true (relative to some fixed unit of the respective
quantity), up to measurement errors. They are regarded as the physical
properties of natural kinds, but the theories to which they belong are not
necessarily interpreted as true and the question for the carriers of these
properties may be left open. Once again the best argument for moderate empiricism is to point to quantum theory. If we talk about electrons,
quantum theory tells us that we must not regard them as carriers of
sharp values of position and momentum at the same time. In addition,
we are neither committed to believe in classical electrodynamics nor in
quantum electrodynamics. All we have to believe according to experiment
and measurement is that the electron is a certain quantity of charge, and
that the so-called electron is nothing but a collection of certain values of
mass and charge that go constantly together in the sense of a Lockean
empirical substance.[34] This position also lies at the heart of the empiricist theory of measurement, as it has been developed in terms of model
theory by Suppes and his coworkers.[35]

4. **Strict empiricism** is the heroic view that empirical reality consists only
in phenomena which are observable or perceivable in a strict sense. According to this, the empirical content of theories lies in observables in
a literal sense, that is, in magnitudes which correspond to sensory data.
Physical magnitudes belong to these observables only insofar as they have
a spatio-temporal representation. This Machian position, which has recently been defended by van Fraassen,[36] is heroic. Nevertheless, it is not
metaphysically abstinent at all. The claim that, e.g., the standard model
of current particle physics does indeed have some empirical content actually relies on a very strong trust in the unity of the current heterogeneous
measurement theories of particle physics.[37] Carnap's original distinction

---

[33]Hacking 1983, 23. See Sect. 2.4.

[34]Locke 1689, Book II, beginning of Chap. XXIII.

[35]See Suppes 1980; Krantz 1971. See also Falkenburg 1997.

[36]See van Fraassen 1980; see Sect. 3.6.

[37]See Chap. 5.

of observational concepts and theoretical concepts was not really of any help for understanding these problems.[38]

5. **Constructivism** embraces many positions that range from the neo-Kantian idealism of the Marburg school[39] and the Erlangen school[40] to views based on current cognitive science[41] and cultural studies; the latter gave rise to social constructivism.[42] As far as the philosophy of science is concerned, constructivist positions adhere to an Aristotelian view of nature, according to which neither experimental phenomena nor the theories which describe their causes have a counterpart in nature because they are all generated by human activities. Both constructivism and strict empiricism only count the immediately observed as real, be it the stars in the sky, the flowers in my garden or my bubble chamber photograph of particle tracks. In contradistinction to the strict empiricist, however, the constructivist only takes the stars in the sky and the flowers as true or real (or natural). For him, the bubble chamber photograph is a technological artefact rather than a part of empirical reality or nature. Hence, constructivism subscribes to the ancient conception of φύσις. Constructivism is the minimum position in our spectrum of the debate. It does not admit that physical theories which are grounded in experiment may bring out anything about reality or about the world in which we live. According to constructivism, physical theories are mere instruments and physics only aims at technological devices.

1.3 The Real and the Actual

By considering empirical reality to be an embodiment of anything that is empirically given and interpreting the term 'empirical' in different ways, we obtained the above spectrum of positions. Any of these positions presupposes another account of empirical knowledge that determines what mature physical theories tell us about reality, whereas conversely the belief in what mature physical theories tell us about reality determines the respective account of empirical knowledge. So how can we decide between the options? In order to escape from the circle, we need an *independent* criterion of what belongs to empirical reality in view of modern physics: a criterion of reality

---

[38]See Carnap 1956. Today the distinction is most often understood relative to a given theory, e.g., in terms of Sneed's T-theoretical quantities; see Sneed 1971 and Balzer et al. 1987.

[39]Cohen 1896; Natorp 1910.

[40]Janich 1996; Butts and Brown 1989; Tetens 1987; Lorenzen 1974.

[41]von Foerster 1981.

[42]Latour and Woolgar 1979; Knorr-Cetina 1984; Pickering 1984. The positions of Tetens 1987 and Pickering 1984 are discussed in Sects. 2.1 and 2.2. These versions of constructivism misunderstand the scientist's account of the aims of science; see Scheibe 1997a.

which is independent of the above positions here and the structure of physical knowledge there.

Unfortunately, the current debate on scientific realism does not provide such a reality criterion. The spectrum of all positions sketched above is indebted to one and the same empiricist account of reality. The positions differ mainly in understanding the *empirical* in a more or less liberal way regarding the theory-ladeness of data, by making the distinction between the empirical and the non-empirical differently. In doing so, they seem to depend on the empiricist view of metaphysical concepts. According to 20th century empiricism, the metaphysical is what is *not* empirical, whereas only the empirical counts as scientific. Global realism claims that *no* concepts of our best scientific theories are metaphysical. Constructivism claims that *all* of them are metaphysical. Critical realism, moderate empiricism, and strict empiricism try to find their ways between these extremes, claiming that some concepts of empirical science are metaphysical and others are not. Because the non-empirical is metaphysical and the empirical is controversial, the debate on scientific realism is circling within itself, as far as it deals with the question of which scientific concepts stem from empirical knowledge as opposed to metaphysics.

The circle can only be broken by dispensing with the empiricist view of metaphysics. This view is untenable for the simple reason that it is impossible to neatly separate the empirical and the non-empirical parts of scientific theories. According to the Duhem–Quine thesis and the related insight in the holistic structure of science, there is *no empirical science without metaphysics*.[43] All scientific theories unavoidably contain metaphysical elements. Therefore, the empiricist account of metaphysics as *opposed* to empirical science is at odds with the structure of physical theories. (Long ago, Hegel emphasized that empiricism is the unconscious metaphysics.[44]) The crucial question is then, how much of metaphysics is *indispensable* for modern physics, *dispensable*, or even *too much*? In other words, what are the features which qualify the sense data of strict empiricism as well as the theory-laden experimental data of physics as information about reality?

At this point, reality comes into play in the sense of an independent external world. The debate on scientific realism is struggling over what we know of it due to modern science, *taking it for granted that reality exists in this sense*. Is there any *independent epistemic criterion* for what is real, which does not depend on some arbitrary option concerning the question of what empirical knowledge may be?

Obviously, 'reality' is a metaphysical concept. Any attempt to find out what is real, in contradistinction to ideal, fictitious, or non-existing, faces the following metaphysical dilemma. On the one hand, the real is conceived to

---

[43] Quine 1951.

[44] Hegel 1830, § 38. His general point was that scientific empiricism itself, like any philosophical theory, makes use of non-empirical concepts and logic.

be *independent of human activities*. But an absolutely independent reality is epistemically inaccessible. On the other hand, in whatever terms we conceive of the real, we *cannot conceive it independently* of our language, conceptual schemes, categories, or whatever theoretical ideas. No subject matter of our thinking can be conceived independently of our thoughts. This dilemma has provoked a never-ending debate between realism and idealism in general, which is indeed as old as philosophy itself and cannot be discussed here. Only one point is clear: the above dilemma is an *epistemic* problem. Therefore, the question of what the real *is* is better replaced by the question of what we *know* about the real as something that is independent of all human concepts and activities. What we certainly know about reality is that it is *not arbitrarily at our disposal*. A *sufficient condition for something to be real* is that it is a brute fact that disappoints our prejudices and expectations.

This sufficient reality criterion comes close to Popper's requirement that empirical theories should be falsifiable. It is complementary to Hacking's reality criterion: "If you can spray them, they exist" mentioned above, since the latter requires that something acts in accordance with our ideas rather than foiling them. However, it is more general. It is independent of critical realism and its sub-positions, to which Hacking's and Popper's views about physics belong. By disappointing our expectations, reality proves its existence, *independently* of our account of empirical knowledge. Brute facts are quite often not such as we would like them to be: this happens in everyday life, in science in general, and in physics. Everywhere, we face reality as if it were an independent actor that does not obey our will. But usually this independent actor does not act in a completely chaotic way. In physics the law-like structure of what actually happens is investigated under well-defined experimental conditions.

Several traditional philosophical concepts of reality take this power of acting independently into account. Empirical reality may disappoint our theoretical expectations, and due to this matter of fact it may give rise to objective knowledge. According to Kant, nature is a witness who answers 'yes' or 'no' to the questions put by our experiments.[45] The point was taken up by Peirce, the founder of pragmatism, and by the neo-Kantian Edgar Wind who developed a pragmatistic theory of experiment. According to Peirce and Wind, nature is a metaphysical entity which lets some of our theories and experiments fail, and others succeed.[46] But the tradition of this account of reality is much older. The paradigm of reality, or nature, as an independent actor with her own causal powers is Aristotle's concept of ενεργεια. The modern physical concept of energy still has its etymological roots in this con-

---

[45]See Kant 1787, B XIII.

[46]See Wind 1934 and Falkenburg 2001. Wind was influenced by Peirce's pragmatism and metaphysical realism. By Wind as well as in the following, it is presupposed that empirical science has to adopt correspondence *criteria* of truth, whatever *theory* of truth one may finally defend from a philosophical point of view.

cept. And Planck's quantum of action indicates that in the quantum domain nature does not act as expected in classical physics.

In order to make this point about reality more precise, Leibniz' and Kant's distinction of *reality* and *actuality* is helpful. According to Leibniz and Kant, the real consists in the *qualia*, that is, in the features or properties of things; whereas the actual, as distinct from the possible and the necessary, is a modal category.[47] Kant's philosophy teaches that any experience depends on *a priori* concepts and forms of intuition. According to his epistemology, empirical reality is on the one hand structured by the categories and principles of pure reason, whilst on the other hand it is made up of sense data. For him, the structure of *reality* defines our theoretical universe of discourse, whereas *actuality* is made up of the actual empirical properties of the things we perceive. For him, empirical reality is made up of both ingredients, the real and the actual. Both come together in the *qualia* of sense data. They fall under the categories of the real (in particular, the pure concept of an intensive quantity) and the actual, obeying the *a priori* principle that intensive quantities have a degree. This degree, which Kant calls the "real of the sensation", is the actual that is *a posteriori* given in the sense data.[48]

In distinguishing the categories of reality and actuality, Kant stands in the tradition of Leibniz' logic and epistemology.[49] Leibniz distinguishes logical and real possibility. Here, logical possibility obeys the logical principle of contradiction and concerns consistency, whereas real possibility obeys the metaphysical principle of sufficient reason. For Leibniz, the real consists in all real possibilities, or all possible worlds, whereas the actual is the actually existing world.[50]

In modern philosophy, these traditional distinctions between reality and actuality are stripped of their metaphysical connotations and reproduced as follows. Reality is something like a logical space of properties that makes up our theoretical universe of discourse. Actuality consists of the *contingent* properties of the phenomena, as they are perceived by us in sensory experience or measured by our experimental devices in physics. The term 'contingent' is crucial. The actual is what *could* be otherwise, according to *our* theoretical universe of discourse and *our* knowledge. Here, Leibniz's epistemological distinction of *verités de raison* and *verités de fait* is taken up.

---

[47]See the categories and principles of pure reason in Kant 1781/87. The English term 'actual' preserves the idea that something is acting, as does the German term 'wirklich'.

[48]Kant 1787, B 207. Kant's account of *objective reality* in addition underlies the conditions of his theory of space and time as forms of pure intuition, which are neglected here.

[49]See, e.g., Poser 1981.

[50]Leibniz' "real possibility" concerns the 'compossibility' or conceptual compatibility of all substances in the world, the monads or subjects of monadic predication. The actual world underlies an additional metaphysical principle of Leibniz', the principle that it be the *best* one. See Leibniz 1714; Poser 1981.

The truths of logic and mathematics are due to human reason (to which Leibniz attributed the capacity of getting insight into the objective structure of the world), whereas the truths of empirical science are due to the facts and the way in which we experience them. For us, the actual is contingent, as far as human reason does *not* have insight into the structure of the world and the causal powers of reality. Reality acts independently of our will in the facts that actually happen. As far as the actual *cannot* be predicted by our theoretical knowledge, it is *contingent* for us. As far as it *can* be predicted by us on the basis of logical and mathematical theories, it belongs to our *knowledge* about the world, and we have succeeded in reducing contingency by establishing law-like connections between the facts. As far as these connections are based on logic and mathematics, anybody except the defenders of strict empiricism or constructivism will consider them to be necessary.

Hence, in order to understand what we can know about reality, we should take *two* distinctions into account: (i) the *modal* distinction of the real and the actual, and (ii) the *epistemic* distinction of the necessary and the contingent. The real is our universe of discourse, a logical space of properties and theoretical connections which permits many possible worlds. The actual is the *independent* entity called 'reality', 'external world', or 'Nature' which we attempt to capture in *our* concepts. It gives rise to occurrences that sometimes do *not* fit in with our theoretical expectations. A sufficient (though not necessary) feature of the actual are contingent occurrences such as we *cannot* anticipate or predict. Hence, whatever fact is contingent (or once *was* contingent, when we could not yet predict it), belongs to the actual world; and whatever theoretical concepts, structures, or causal explanations we find in order to predict the facts and to reduce their contingency are good candidates for adequate descriptions of the actual world. At this point, several questions discussed in the debate on scientific realism may come up. They concern the confirmation of our theories and the relation between truth and warranted assertion. They may be left open here because they do *not* concern the epistemic criterion that the contingent, or what once *was* contingent, is a feature of the actual world, or some past state of it, respectively. We may deceive ourselves about what we consider to be necessary or law-like. But we cannot be mistaken about our expectations being refuted or confirmed.

Part of the actual is that the sky is one day blue and another day cloudy, or that Millikan's measurement of oil drops resulted in a ratio of mass and charge which (in electrostatic units) has the value 1.761011 A s/kg. Any empirical or experimental matter of fact that might be otherwise and may surprise us belongs to the contingent features of empirical reality, to the actual world. In everyday experience the knowledge of such matters of fact is commonly shared by all speakers of ordinary language. In physics, the actual is conceived in terms of events, observed qualitative phenomena, pointer positions of measuring devices, recorded data, and numerical measurement results. Thus, the universes of discourse of everyday experience and of physics do not differ in principle, even though the former is expressed in terms of ordinary language and the latter in terms of mathematics and physical quantities. Both universes have in common that they embrace *qualitative* and *modal* features. They are made up of a logical space of properties (which in physics has an algebraic structure) *plus* the contingent values of these properties which can be attributed to things or events or to spatio-temporal locations.

The main epistemic problem concerning physical science is that there are *several* overlapping universes of discourse. There are several classical and quantum, non-relativistic and relativistic theories behind what has been called theoretical background knowledge. It is indeed a manifold of theoretical discourses which make up the accepted background knowledge of science in general and of physics. There is no unified theory. Science is pluralistic, and the background knowledge provided by current physics is very heterogeneous. In the data analysis of any recent experiment of physics, many theories, laws, and models are used. However, every matter of fact which is *not* completely explained from such theoretical background knowledge belongs to the contingent features of empirical reality. Closest to the criterion of happening contingently and not being at our disposal is the notion of an *event*. An event is something that occurs locally in such a way that it marks an observable difference from its environment: a light flash, a 'click' in a Geiger counter, a spot on a photographic plate. The concepts of physical phenomena, data, pointer positions, and measurement results are much more complex: they depend on theoretical background knowledge. Only the experts who know the experiments have epistemic access to the latter as constituent parts of the actual. The former are unstructured, they are simply observed to occur. The way in which such unstructured events give rise to theoretical background knowledge, which in turn gives rise to concepts of physical phenomena, to models, to the construction of measuring devices, to data recording, to reading out pointer positions, and to measurement results, is *very* complicated. The line between the contingent matters of fact and theoretical background knowledge shifts whenever a theory or model has been accepted but it is also open for revisions. For the events, concepts, and measurement theories of particle physics, this will be investigated in the subsequent chapters of the book.

The events, data, phenomena, and measurement results of physics are matters of fact as far as the way they are derives neither from theories nor from conventional elements in theories that may be arbitrarily chosen. Whether such events, etc., occur or not, and what their contingent features may be, can perhaps be *predicted* by theory, but as a brute matter of fact it is *not at our disposal*. The contingent matters of fact are blanks in the theoretical map of our universe of discourse, which have to be filled by observation, measurement and experiment, as van Fraassen put it in view of

Millikan's measurement of e/m.[51] This is a case for empiricism – and against any radical variant of constructivism, according to which the empirical reality of science is *nothing but* a theoretical construct. But strict empiricism gives too weak a determination of the structure of empirical reality. It does not enable us to count contingent phenomena or data as empirical if they depend substantially on the theoretical background knowledge of physics. (To take an example, the mass of an unstable subatomic particle which was measured from the width of a resonance found in a scattering experiment of high energy physics is surely contingent.) Therefore, strict empiricism does not come to grips with many contents of modern physics.

Hence, there is a pragmatic way out of the above dilemma that reality is *either* independent but epistemically inaccessible *or* known by us but dependent on the way we conceive of it. We may escape from it with the help of the epistemic criterion that the actual is in many cases contingent *for us*. What *might* occur in another way than it does has the hallmarks of the empirical reality we are after. Nature may actually put limitations on applying our theories, and these limitations are independent of our cognitive capacities as well as of our theoretical universes of discourse. The contingent is what is not controlled by theory. Empirical reality consists of events, data, phenomena, and so on which are given insofar as they are contingent. They are empirical as far as they *could* be (or could have been) otherwise, relative to the present (or past) state of knowledge.

In this way we get a version of realism which is based on modal distinctions. We might call it *modal realism*, but this should not be confused with modal realism in van Fraassen's sense, the only realistic position admitted by his constructive empiricism. For him, 'modal realism' means the sceptical view that, *beyond* the empirical phenomena, nature *could* be just as the non-empirical parts of an accepted theory propose.[52] He draws the demarcation line between the empirical substructure and the non-empirical parts of a theory in a strict empiricist sense and applies the modal category of the possible to the non-empirical. In contradistinction to his views, I defend a realism which draws the demarcation line between the empirical and the non-empirical parts of a theory in terms of the contingent (i.e., events and measurement results) and the necessary (i.e., laws that put theoretical constraints on the data). Here, contingency is regarded as a sufficient epistemic criterion for actuality, whereas necessity comes along in the clothes of mathematical physics.

---

[51] "[...] the experiment shows that unless a certain number (or a number in a certain interval) is written in the blank, the theory will become empirically inadequate." van Fraassen 1987, 119–120.

[52] See van Fraassen 1991, 4: "When we come to a specific theory, the question: *how could the world possibly be the way this theory says it is?* concerns the content alone. This is the foundational question *par excellence*, and it makes equal sense to realist and empiricist alike."

Both concepts are epistemic, and the demarcation line between the contingent and the necessary, too. Indeed, the distinction is relative and depends on theoretical background knowledge. Due to the reduction of contingency by accepted theories, the demarcation line between the contingent and the necessary is shifting with time in two ways. On the one hand, theory formation predicts events. In this way their occurrence becomes law-like, that is, necessary in relation to our knowledge. Newton's law of gravitation and its boundary conditions in the solar system enable us to predict the next eclipse of the sun with very high precision. No one who knows that would call this event contingent, in contradistinction, say, to the people before the times of ancient science who were afraid of this kind of event. On the other hand, theoretical background knowledge enters the data analysis of physics. This gives rise to theory-laden phenomena and data that appear contingent to the physicist, whereas a defender of social constructivism would regard them as theoretical constructions.

The distinction of the contingent and the necessary, and with it the kind of modal realism proposed here, is irreducibly epistemic. Compared with the positions discussed above in Sect. 1.2, it is close to critical realism as well as to moderate empiricism. Indeed it is a Kantian position, a version of *empirical realism* in a Kantian sense.[53] According to this view, any contingent phenomena, measurement results, experimental data, and phenomenological laws of physics count as part of the empirical world. In addition, theoretical structures such as symmetries which derive directly (*pace* safe background knowledge) from the phenomena or data may be considered to capture some structure in reality. It is a safe bet to rely on them even if one suspects that the fundamental laws of physics may lie.[54] The parts of empirical reality with which physical theories and experiments deal may then be called *physical reality*.

What exactly counts to empirical reality in the modal sense proposed here, i.e., to the actual world embracing its history and its inherent possibilities, has to be examined more closely in case studies. Things are getting very complicated due to the theoretical reduction of contingency mentioned above. According to the above criterion we may count every matter of fact which has ever been contingent among empirical reality, even if maybe later we became able to deduce it from a theory. Thus, in order to find the contingent content of a theory it is necessary to study the process of theory formation and not a completed theory after it has been accepted. In order to find out what a theory says about the actual world, it is necessary to study its

---

[53] A consequence is that all physical properties, objects, systems, and processes are considered to belong to the *phenomena*, not to the *noumena*, in a Kantian sense. A related position is Putnam's internal realism; see Putnam 1980, 1990. However, for the purposes of the philosophy of physics it is not helpful to put the concept of truth in question, as Putnam does.

[54] See Cartwright 1983.

original observational basis and the way in which it was transformed into safe background knowledge. It is tedious to study these details. In Chap. 3 it will be shown step by step which kinds of contingent data lie at the heart of experimental particle physics.

1.4 Realism and Quantum Theory

It is a brute fact that the contingent events and measurement results of subatomic physics are incompatible with the structure of classical physics. Hence, according to the epistemic reality criterion given above, we certainly know that (i) the subject matter of subatomic physics exists and (ii) it does not have a classical structure. Quantum theory fits the contingent experimental results and is incompatible with the structure of classical physics, too.[55]

The views of Planck and Einstein show, however, that scientific realism emerged from a metaphysics closely tied to classical assumptions about physical objects.[56] It is a realism of objects with well-defined spatio-temporal and causal properties, which *cannot* be attributed to the quantum states of subatomic particles. Let us call this kind of realism *classical realism*. In the famous Bohr–Einstein debate, Einstein defended it against Bohr's complementarity view of quantum phenomena.[57]

It is easy to show that quantum phenomena are at odds with classical realism. Quantum measurements show that subatomic particles have no trajectories in a classical sense. Their tracks on a bubble chamber photograph are due to subsequent position measurements but quantum theory does not entitle us to link them with a continuous path.[58] Their values of position and momentum cannot be measured at once in such a way as to result in sharp values. They obey Heisenberg's indeterminacy relation. According to this, the behavior of subatomic particles is not completely determined. Quantum theory describes their logical space of properties in Hilbert space. It is spanned by roughly one half of the spatio-temporal and dynamic properties of a classical particle. A quantum state in Hilbert space has *either* a sharp position *or* a sharp momentum, but not both at the same time. The same is true for each pair of non-commuting observables or operators in Hilbert space.

In view of quantum theory, the causal aspects of *critical realism* sketched above becomes highly problematic. Except in the specific case of repeatable

---

[55]Here, 'quantum theory' means non-relativistic and relativistic quantum mechanics, quantum electrodynamics, and any quantum field theory. (Old quantum theory is excluded. It rested on the attempt to save as much of classical physics and classical realism as possible.)

[56]Planck 1908; Einstein 1936, 1949; Scheibe 2001, 142–155.

[57]Bohr 1928; Bohr 1949.

[58]See Sect. 5.3. The corresponding change in the meaning of the particle concept is discussed Chap. 6.

measurements, quantum theory does not allow one to attribute dynamical causes to individual events. In terms of causal realism, quantum theory commits us to dispense with individual causes.[59] In addition, the conservation laws of a quantum theory give rise to the prediction of acausal law-like relations between individual events, namely EPR (Einstein–Podolsky–Rosen) correlations. EPR correlations are non-local. In view of Einstein's causality condition (according to which no signal can be transmitted over spacelike distances), they are acausal and they are incompatible with Bell's inequality.[60] In terms of structural realism, quantum theory commits us to non-local structures.

Quantum theory is much less disturbing for strict empiricism than for classical realism. If one asks neither for non-observable causes of quantum phenomena nor for the quantum domain beyond its empirical substructure, one has no problem at all with interpreting quantum theory. Indeed the non-local features of quantum theory belong to its empirical substructure. From an empiricist point of view, there is no need to restore a classical account of reality for a better understanding of quantum phenomena. Strict empiricism comes together with an instrumentalist interpretation of the quantum mechanical wave function, which is understood only as a useful tool for the explanation and prediction of measurement outcomes.

When Born gave his preliminary sketch of the probabilistic standard interpretation of quantum mechanics, in his famous first communication on the quantum mechanics of collision, he already asked whether the quantum domain exhibits a "pre-established harmony", according to which it is by no means accidental that theory and experiment agree in the lack of causal conditions of an individual quantum process.[61] Asking so, he obviously does not defend empiricism but draws our attention to the issue that quantum theory may be regarded as grist for the empiricist's mill. Indeed, from an empiricist point of view the whole debate on the interpretations of quantum theory is superfluous because it deals with the very question of what happens in the quantum domain *beyond* the empirical phenomena. Thus, strict empiricism simply eliminates the philosophical problems of quantum theory which had been the subject of the Bohr–Einstein debate for two decades.[62] Or to put it

---

[59]Indeed, Cassirer 1937 suggested that in quantum theory the principle of causality can be maintained only at the level of the Schrödinger equation, i.e., for the statistical ensemble. See also Chap. 7, in particular Sect. 7.6.

[60]Einstein et al. 1935; Bell 1964; Aspect et al. 1982. In recent experiments, EPR correlations were observed over a distance of 12 km.

[61]Born 1926a, 806.

[62]Surely van Fraassen is well aware of this fact. He sees his own 'modal' interpretation of quantum theory as a possible understanding of the theory which is by no means compelling; see van Fraassen 1991, 9: "But the interpretational demands of *What is really going on (according to this theory)?* or even the more modest *How could the world possibly be how this theory says it is?* will not disappear if science is to help us construct and revise our world-pictures."

in other terms, strict empiricism has no access to the interpretational problems of quantum theory. These problems only come into view if a stronger or a more complex account of physical reality is presupposed.

The account of reality, according to which quantum theory is at odds with realism, is *classical realism*. It stems from classical point mechanics and electrodynamics. The theories of classical physics have been developed in the spirit of rationalist metaphysics. According to this, physical reality is completely determined. On this basis, Einstein was convinced that the quantum mechanical description of physical reality is *not* complete.[63] Classical realism embraces much stronger assumptions about the structure of physical reality than the distinction of the qualitative and the modal features of empirical reality which I suggested in the preceding section (i.e., the distinction between the real and the actual). In particular, it ties its assumptions about the theoretical universe of discourse of physics and the properties of physical systems to two metaphysical presuppositions which are rooted in rational metaphysics. The first is the principle that physical objects are *substances*. A substance in Descartes' or Leibniz' sense is a thing-in-itself, which is explained as follows. It is independent of the rest of the world, it is a carrier of primary qualities, and it can be defined in terms of monadic predicates. (According to Descartes, physical substance, or matter, is made up of corpuscular matter constituents. For him, their primary qualities are extension, duration, and motion. Newton added to this list the dynamical properties of solidity and inertial mass.) The second metaphysical presupposition is the assumption that all physical events and processes are completely determined by the laws of Nature, in accordance with the principle of *causality*.

These metaphysical presuppositions were always criticized by empiricists. Locke criticized the non-empirical concept of substance,[64] Hume criticized the principle of causality. In taking up this criticism, Kant understood both principles as conditions *a priori* for the possibility of our experience. For him, empirical reality, or Nature, is the law-like connection of all empirical phenomena. His theory of Nature is modelled on the structure of Newton's mechanics.[65] In his theory of Nature, the principle of the conservation of substance and the principle of causality play a crucial role. In his *Metaphysical Foundations of Natural Science* they are related to the concepts of mass and force of Newton's mechanics. Later, Mach criticized Newton's concepts of mass and force in the spirit of Locke and Hume from an empiricist point of view.

The principles of substance and causality are residues of rationalist metaphysics. They enter classical physics in terms of physical objects and deter-

---

[63]Einstein et al. 1935; Einstein 1949.

[64]... which has the unclear meaning of a propertyless carrier of properties, in contradistinction to his concept of an empirical substance; see Locke 1689, Book II, beginning of Chap. XXIII.

[65]Friedman 1992; Falkenburg 2000.

ministic laws. In classical point mechanics, they come in terms of mass points and their trajectories. In classical electrodynamics, they come in terms of spacetime points as carriers of the electric and magnetic field strengths, and in terms of retarded potentials which according to the textbooks count as 'physical' (whereas the use of the 'unphysical' advanced potentials is usually suppressed according to the principle of causality).[66] Mass points are carriers of physical properties such as mass, momentum, energy and charge. For these properties, conservation laws are valid.[67]

A charged point mass has the following features in common with a substance in the traditional sense. It is regarded as a thing-in-itself which exists on its own and carries its physical properties independently of the rest of the world, according to the law of inertia. Forces cause deviations of the inertial trajectory. The forceless case is regarded as primary, and interactions are taken into account later. The same is true for field theory. It starts from free fields, while their interactions with each other and with matter come later. Where the interactions become too complicated, perturbation theory is used. It starts from the unperturbed case, i.e., free point masses and uncoupled fields, and approximates their interactions stepwise. Indeed, the metaphysical concept of a substance which is independent of the rest of the world underlies the most familiar idealizations of physics.[68] The law-like connection between the contingent properties of classical particles (or mass points) in a many-particle system is expressed in terms of a trajectory in phase space. If the position and momentum values of all particles at a given time are given as initial conditions, the trajectory is completely and unambiguously determined for all times.

As Cassirer has shown, quantum mechanics is neither compatible with the traditional concept of substance (that is, the principle of attributing properties to property carriers) nor with the principle of causality in its usual application to individual systems and processes.[69] These traditional princi-

---

[66]Indeed, the use of advanced potentials is controversial. In the quantum theory of radiation, they were introduced by Wheeler and Feynman 1945. Cramer 1986 suggests the 'transactional' interpretation of quantum theory on these lines.

[67]In non-relativistic mechanics, the concept of mass splits into inertial and gravitational mass, where gravitational mass is the charge of gravitation; and mass, momentum and energy are conserved separately. In special relativity, we have rest mass and velocity-dependent relativistic mass; and the non-relativistic conservation laws are replaced by the combined conservation laws for mass–energy and energy–momentum.

[68]See Falkenburg 1993a, 1993b, 2006b.

[69]Cassirer 1937; see in addition Falkenburg 1993b. Bohm's approach (see Bohm 1952) of a hidden variable theory is an attempt to save both principles for the quantum domain. However, any hidden variable theory pays the price of introducing actions-at-a-distance into physics, namely a non-local potential which violates Einstein causality and which is at odds with special relativity. See Bell 1964, 200: "Moreover, the signal involved must propagate instantaneously, so that such a the-

ples belong to the classical universe of discourse, but they are at odds with the structure of quantum phenomena. In terms of the distinction of the real and the actual introduced in the preceding section this means that in the subatomic domain our account of the real and our experience of the actual come into conflict. We conceive of the real in terms of carriers of physical properties and their complete determination, but Heisenberg's indeterminacy relations and quantum theory in its usual probabilistic interpretation tell us that the contingent outcomes of quantum measurements do *not* reduce to such a theoretical construal of physical reality. There is no simple way out of this struggle, as the never-ending story of realistic, instrumentalistic, holistic, quantum-logical, physical, metaphysical, etc., interpretations of quantum theory shows. This problem has the following reason, which is hardly recognized in recent debates on quantum theory. Our usual construal of reality may be useful and plausible for everyday experience. However, it is deeply rooted in 17th century metaphysics, and in the face of relativity and quantum theory it turned out to be too strong with regard to the concepts of space and time, substance, and causality.[70]

Thus, quantum theory is at odds with classical realism, since our classical construal of reality is at odds with the contingent quantum structure of the actual world. The real, as we got used to conceiving it with a traditional (Cartesian) metaphysical background, and the actual, as the experiments of subatomic physics exhibit it, do not agree. Our usual metaphysics and the contents of modern physics mostly agree for objects of everyday experience and classical physics. Classically, we conceive of the real in terms of a completely determined spatio-temporal and dynamical connection of things, events and processes. The actual, however, is contingent, and in the quantum domain it does not fit in with our classical construal of reality. Indeed, the classical principles of substance and causality overdetermine the quantum domain, as regards the concept of a physical object with completely determined properties.

If we do not adhere to some version of constructivism, and if we do not believe in the possibility of re-establishing a classical construal of the actual world, we are compelled to dispense with a thoroughly classical construal of the world. In the quantum domain, physicists like Planck, Einstein and Bohr had to dispense with crucial classical assumptions about matter and radiation, even though they were not willing to do so. Indeed, this very matter of

---

ory could not be Lorentz invariant." In the domain of relativistic quantum field theory this price becomes very high indeed. A Bohm-type relativistic quantum theory is beset with many difficulties, amongst them the problem that relativistic particles cannot be localized; see Clifton and Halvorson 2002 and Sect. 6.4.1.

[70]Mittelstaedt 2006 shows how special relativity and quantum theory criticize classical realism as an overdetermined construal of empirical reality in two respects. Special relativity teaches that the classical concepts of space, time and simultaneity are not general enough; quantum theory teaches that the classical concept of substance is too strong.

fact is a case against constructivism, and a case for a realistic interpretation of quantum structures. The classical construal of the subatomic domain failed. As far as classical particle trajectories and the classical theory of radiation are incompatible with quantum phenomena, the laws of classical physics are wrong and the laws of quantum theory should be considered to be true. But the debate on the interpretations of quantum theory shows that the terms in which a *structural realism* of quantum phenomena may be specified remain very unclear.

The terms in which a *causal realism* of subatomic particles may be spelled out as the causes of quantum phenomena are even less clear. Indeed, this is one of the main problems of the present book, which will be investigated in more detail in the subsequent chapters. Since the very concept of a subatomic particle has a causal aspect, in view of quantum theory it becomes unclear what exactly physicists mean when they talk about subatomic particles such as electrons, protons, photons and quarks.

One point, however, *is* clear with regard to the distinction between the real and the actual, or our theoretical construals and the contingent. There is no thoroughly classical construal of the quantum domain, even though there are partial and approximate models of quantum phenomena. A key feature of quantum phenomena is that many of them cannot be embedded in the theories and models of classical physics at all, whereas others only correspond to the classical case at a statistical level. In general, the results of quantum measurements are irreducibly contingent. Thus, in the quantum domain the classical construal of physical reality has to be replaced by the quantum structure of the contingent phenomena.

This quantum structure is minimal. It relies on the minimal interpretation of quantum theory, namely Born–von Neumann's probabilistic interpretation.[71] It reduces to quantum phenomena and to the measured and predicted values of physical quantities. But in the subsequent chapters we shall see that the phenomena and models of particle physics are *not* grist for the empiricist's mill. As far as particle physics contributes to our understanding of the actual world, the quantum structure of the phenomena does *not* reduce to an empiricist account of physical reality. Indeed, any quantum phenomenon and especially the measurements of particle physics presuppose not only the existence of a classical world but also the language of classical physics. Many of these measurements are explained in complicated semi-classical models which *cannot* be interpreted in terms of, say, van Fraassen's empiricism. Many of these models make tacit use of a generalized version of Bohr's principle of correspondence. In order to understand them it is useful to recall Bohr's complementarity interpretation of quantum mechanics.[72]

It is characteristic of the quantum domain that our theoretical account of the real and our experimental access to the actual can by no means be

---

[71]Born 1926a, 1926b; von Neumann 1932, 101–110.
[72]See Sects. 5.4.1–5.4.2 and 7.2.2.

made to coincide. The actual measurement results show up in a classical macroscopic world that quantum mechanics cannot explain. Conversely, the decoherence approach which explains the emergence of definite measurement results at least at a statistical level, on the basis of a quantum mechanics of open systems, has to presuppose a macroscopic local environment into which the quantum superpositions are damped away.[73] This is an important step towards closing the gap between the real and the actual, or between quantum theory (as a theory of probabilities) and (single) quantum events in a classical world. But in this way the gap is only partially closed. The intriguing question of how a classical world may emerge from subatomic reality still remains. As long as only *classical* realism is available, the unresolved quantum measurement problem provokes *anti*-realism. Hence, there seems to be no account of subatomic *reality*.

1.5 The Metaphysics of Physics

In view of quantum theory it becomes obvious that the debate on scientific realism deals with questions of traditional natural philosophy. The options of classical realism or strict empiricism stand in the philosophical traditions of rationalism and empiricism. Both options fall short of a satisfactory interpretation of quantum theory. The classical construal of physical reality is too strong and strict empiricism is too weak for understanding the quantum laws of Nature. The situation is similar to what Kant explained in his doctrine of the antinomy of pure reason. Rationalism tells us too much about the spatio-temporal structure of the world, whereas empiricism does not tell us enough. In saying this, Kant dealt with the traditional problems of the beginning of the world in space and time and the existence of atoms as ultimate constituent parts of matter.[74] His solution was to investigate the metaphysical foundations of natural science, that is, the *a priori* assumptions which underlie our theories of the spatio-temporal world and its constituents.

Indeed, Kant's project of investigating the metaphysical foundations of natural science is *not* obsolete, as the empiricist philosophy of science would have liked us to believe. Twentieth century empiricism failed in its attempts to find the demarcation between empirical science and metaphysics. Carnap's research program of reconstructing physical theories from concepts with empirical significance had to face Quine's insight that the relation between a physical theory and its data is holistic. However, the architectonics of physics is complicated. It is a many-dimensional network of hierarchical structures, the empirical and the non-empirical elements of which are in many ways interwoven. Closer examination of specific physical theories makes it possible to isolate the following kinds of non-empirical elements in physical theories:

---

[73]Giulini et al. 1996.
[74]See my analysis in Falkenburg 2000, Chap. 5.

(i) **Metrical Standards.** Measurement presupposes a metric. The choice
of a metric, however, presupposes a theory of the measurement devices.
Length and time may be measured by means of rigid rods and Newtonian clocks or by means of light signals, that is, with a Galilean or a
Lorentzian metric. The same is true of velocity, which may be measured
in a Galilean frame or a Lorentz frame. Mass may be measured in classical terms, i.e., with a mass spectrograph which obeys the Lorentz force
law, or in non-classical terms, e.g., in terms of the mean energy of the
resonance of the decay of an unstable subatomic particle and Einstein's
equivalence of mass and energy.

(ii) **Conventional Elements.** Many physical theories contain quantities
which are empirically underdetermined and which can be fixed in favor
of a mathematically simple and elegant structure of the theory. A famous example is the convention about the two-way speed of light, upon
which Einstein's famous operational definition of simultaneity is based.
In his seminal paper on special relativity, Einstein suggested that, when
we synchronize clocks by means of light signals, light travels in both
directions with the same speed.[75]

(iii) **Meta-Theoretical Principles.** These rely on traditional metaphysical assumptions about the structure of empirical reality and about the
laws of Nature. They fix classes of theories. In this sense they may be
called meta-theoretical.[76] Most meta-theoretical principles are inherited
from rationalism. Classical realism assumes that empirical reality is completely determined and local. The structure of quantum mechanics tells
us the contrary. And which kinds of symmetries do the laws of Nature
have? Classical mechanics is invariant under spatial reflection and time
reversal. Non-relativistic mechanics is Galilean invariant. Electrodynamics is Lorentz invariant. Thermodynamics introduces the arrow of time.
Weak interactions violate parity. The assumption that the fundamen-

---

[75]Einstein 1905, 894. Reichenbach 1957 and Grünbaum 1963 derived conventionalist positions from the existence of such conventional elements in physical theories.
Friedman 1983, 264–339, argues against conventionalism. He points out the holistic
structure of special relativity by illustrating "how firmly the standard simultaneity
relation is embedded in relativity theory. One cannot question the objectivity of this
relation without also questioning significant parts of the rest of the theory. In particular, one cannot maintain that distant simultaneity is conventional without also
maintaining that such basic quantities as the proper time metric are conventional
as well." (Friedman 1983, 317.)

[76]Here, 'meta-theoretical' is not used as a linguistic term but following Wigner.
Wigner 1964, 16, observed "a great similarity between the relation of the laws of
nature to the events on one hand, and the relation of symmetry principles to the
laws of nature on the other." See Falkenburg 1988. A meta-theoretical principle
fixes a certain structural feature of a theory. It can be expressed in the language of
the theory as well as in an informal meta-language.

tal laws of Nature must be CPT invariant[77] has survived up to today. These examples show that in 20th century physics some of these meta-theoretical principles have become testable. The respective experiments have the nice feature that they test not a specific theory but rather a *class* of theoretical structures.[78] Indeed the crucial experiments of Bell's inequality exclude all hidden variable theories with local coupling.[79]

These three kinds of non-empirical elements of physical theories are inter-woven. The spatio-temporal symmetry of a theory defines the choice of a Galilean or Lorentzian metric. The choice of the conventional elements of a spacetime theory affects the metric too. To reject the Einstein convention of special relativity means to opt for a very complicated metric.[80] All three kinds of assumptions affect what is called the ontology of a theory, that is, the kind of entities to which a theory commits one.[81] To believe in a preferred inertial frame means re-establishing an absolute, neo-Newtonian spacetime and to reject Lorentz invariance, and vice versa. To believe in a quantum theory with hidden variables means re-establishing action-at-a-distance, re-jecting Lorentz invariance, and so on.

The above classification of metaphysical assumptions still rests on the empiricist account of metaphysics, according to which empirical concepts are devoid of metaphysics and, conversely, all non-empirical concepts are metaphysical. Since this distinction of the empirical and the metaphysical had to be revised in view of the vain search for demarcation criteria, one should also re-examine the underlying concept of metaphysics. Indeed, the above three kinds of non-empirical elements in physical theories represent three *very* different kinds of metaphysical assumptions. I list them in reverse order:

(iii) Meta-theoretical principles that may become testable. The attempt at a demarcation between empirical concepts and metaphysical assump-tions fails in both directions. Some parts of metaphysics turned into

---

[77]See Streater and Wightman 1964, 142–146. P is the parity transformation, the subatomic analogue of mirroring; C is charge conjugation, the transformation of particles into antiparticles; T is time reversal.

[78]Franklin 1986, 35–38, investigates this feature for the experimental tests of parity violation.

[79]See, e.g., Scheibe 1986, 1991.

[80]See the discussion of nonstandard simultaneity relations in Friedman 1983, 165–176, and the related discussion of empirically equivalent theories, loc. cit. 266–339.

[81]To talk about ontology in this way means accepting the ontological commit-ments of a theory in Quine's sense. See the famous dictum: "To be is to be the value of a [bound] variable" (Quine 1953, 15), according to which the ontology of a theory is the domain of the individuals which the statements of the theory quantify. In view of empirical underdeterminacy and the debate on scientific realism, one may argue that this notion of ontology is not really helpful. However, it helps to clarify the metaphysical beliefs to which the specific choice of a metric, etc., commits one.

empirical science. That is to say, in the experiments of 20th century physics, several meta-theoretical assumptions of physics turned out to have empirical content and became falsified in the quantum domain. In particular, the principles of locality and parity conservation derive from traditional metaphysical assumptions about Nature, but they failed. In the quantum domain, locality was disproved by the violations of Bell's inequality. That parity is violated by the weak interactions was first confirmed in the ^{60}Co decay. Other meta-theoretical principles of physics, such as the dimensional invariance of physical quantities[82] and CPT invariance are considered to be testable. The principle of energy conservation was repeatedly put into question.[83] However, up to the present day these general principles of physics have escaped falsification and it is not obvious to what extent they are empirically underdetermined.

(ii) Assumptions that remain empirically underdetermined for principal reasons. The conventional assumptions of physical theories are a clear case of empirical underdeterminacy, but not the only one. Other examples are the interpretations of quantum mechanics *beyond* Born's and von Neumann's probabilistic interpretation, such as Bohm's theory of hidden parameters, the many-worlds account of the measurement process, or the recent 'transactional' interpretation of quantum mechanics.[84] As far as they only are empirically compatible with current knowledge about quantum systems, they may some day become testable. As far as they are thoroughly empirically equivalent to ordinary non-relativistic quantum mechanics (like the many-worlds approach), they are empirically underdetermined for reasons of principle. A further example is Newton's mechanics with or without absolute space.[85]

(i) Indispensable methodological principles. The most familiar metaphysical principles of physics are the most neglected in the current philosophy of physics. The metrical standards of physics seem so obvious that they have not attracted any attention at all in the philosophy of physics of the last decades. However, they are clear cases of indispensable methodological principles of physics. Without measurements, no quantitative empirical science is possible. Other such indispensable principles underlie the experimental method. Most important is the principle that physical phenomena and systems can be decomposed and recomposed, that is, analyzed into parts (*mereological analysis*) and causally relevant factors (*causal analysis*), which in turn are assumed to combine in a more or less obvious way to the objects of physical explanations. All

---

[82]See Appendix B and Sect. 4.2.

[83]For example, in the BKS theory (Bohr et al. 1924) or before the confirmation of the neutrino hypothesis.

[84]Cramer 1986.

[85]This example of empirical underdeterminacy and empirically equivalent theories is discussed in van Fraassen 1980, 44–47.

these methodological principles are *constitutive* of physics as an empirical science. They are transcendental principles of physics in Kant's sense of being conditions of the very possibility of physical experience. They are closely related to the traditional concepts of substance and causality. Their constructive features will be analyzed in the next chapter. There are also indispensable *regulative* principles of physics, such as the principles of the unity and simplicity of the laws of physics. Newton expressed them in the words: "Nature will be very conformable to her self and very simple."[86]

Hence, on the one hand physical theories may be understood as sophisticated theoretical models of a metaphysical theory of Nature which has a weaker formal structure. Such a metaphysical theory relies on pre-theoretical, informal concepts and principles. When these metaphysical assumptions become specified in order to develop physical theories, some of them turn out in the long run to have empirical content. On the other hand, physical theories and physical practice depend on indispensable methodological principles which are *not* testable since to abandon them would mean to dispense with physics as an empirical science. However, they are interwoven. Therefore, it cannot be known in advance whether one of these principles may some day obtain empirical content and become testable to a certain extent in a certain domain.

For example, the traditional metaphysical assumptions about the order of Nature gave rise to symmetry principles which Leibniz used as powerful tools against Newton's concepts of absolute space and time in the famous Leibniz–Clarke debate.[87] Some of these symmetry principles became testable and were falsified, as parity violation shows. But the same metaphysical assumptions about the rational order of phenomena give rise to the methodological principles of the *unity* and *simplicity* of the laws of physics. They lie at the very heart of modern physics, and they are *not* testable. The two principles are closely related. Unification is a central goal of physics. Planck emphasized that the goal of physics is to discover a constant mathematical unity, which is actually the real, behind the phenomena.[88] Stripped of Planck's metaphysical realism (in which structuralism and mathematical Platonism seem to go together), the quest for unification still remains an indispensable *methodological principle* of physics. Simplicity is an important principle for the construction of a unified theory. Modern philosophy of science emphasizes that scientific explanation is unification and that simplicity is a crucial criterion for the choice between empirically equivalent theories.[89] All the principles mentioned here stem from rationalism. They express the belief that the world is rational. They claim that we are able to decipher the book of Nature because it is written in mathematical letters that can be put in axiomatic

---

[86] Newton 1730, 397.

[87] Leibniz and Clarke 1715–1716, in particular Leibniz' third letter.

[88] Planck 1908, 49.

[89] See Friedman 1983, 266–271; Friedman 1988.

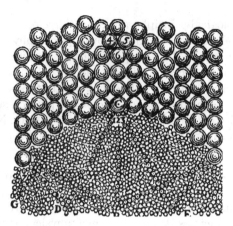

Fig. 1.1. Descartes' corpuscular theory of matter and impact theory of light propagation (Descartes 1644, Fig. 24)

terms. Even though Nature turned out to be much less symmetric, unified, and simple than the founders of modern physics believed, the belief in the principles of symmetry, unification, and simplicity remains indispensable for theory formation in physics.

As symmetry violations and non-local quantum correlations show, experiments and thought experiments are not only a touchstone of specific physical theories. They are also a touchstone of the underlying metaphysics. Experiments test the empirical consequences of a theory, while thought experiments test its logical consequences. Together with the empirical adequacy and internal consistency of a theory, the underlying metaphysical assumptions are tested. If certain metaphysical assumptions turn out to be too strong in view of the phenomena, they should be rejected. For the indispensable *methodological* principles of physics, this is obviously impossible. However, the suspicion may arise that due to the very methods of physics there are crucial limitations on physical knowledge. This suspicion seems to underlie Bohr's complementarity interpretation of quantum theory, since, according to Bohr, Planck's constant indicates the limitations of experimental analysis.[90]

In view of quantum theory, the traditional concepts of substance and causality are at stake. They have a very peculiar role. On the one hand, they belong to the informal metaphysical theory of nature that was specified when Newton developed the first physical dynamics. On the other hand, they are closely related to the indispensable methodological principles of physics,

---

[90]In the famous Como lecture, Bohr emphasized that "the so-called quantum postulate [...] attributes to any atomic process an essential discontinuity, or rather individuality, completely foreign to the classical theories and symbolized by Planck's quantum of action." (Bohr 1928, 580.) Since 'individuality' for Bohr means 'indivisibility', this seems to be closely related to the limitations of *analysis and synthesis of experience* he repeatedly stated; see Chevalley 1991, 373–378.

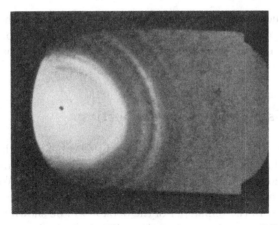

Fig. 1.2. Wave–particle duality. Magnetic influence on electron diffraction (Thomson 1927, after Grimsehl 1938, 221)

namely to the principles of mereological and causal analysis that underlie the experimental method. Modern physics rests on the assumption that phenomena consist of substances with causal powers. Based on this assumption, the experimental method pursues the mereological and causal analysis of phenomena with the goal of finding these substances and powers. By the way, these substances are called *particles* in the universe of discourse constructed by the founders of modern physics. This book will investigate in detail the fate of the mereological and causal particle concepts that underlie particle physics up to the present day.[91] The metaphysical assumptions about mechanical bodies, their least parts (called atoms), and other substances were constitutive in the development of classical physics. In just the same way, these assumptions gave rise to classical realism, the metaphysics of classical physics. Quantum theory, however, teaches that they are *too strong*.[92]

In a physical dynamics, the informal concepts of substance and causality are replaced by formal concepts of the state of a physical system and its dynamical development. In classical mechanics, the dynamical development of a state is described by a trajectory in phase space. Quantum mechanics replaces the trajectory in phase space by the unitary development of the state vector in Hilbert space according to the Schrödinger equation *plus* the non-unitary reduction of the quantum state by measurement. The development of the quantum state of an individual system cannot be translated back into the traditional terms of substance and causality. It has to be expressed in terms of waves and their probabilistic interpretation. The propagation of the waves or their probabilities violates classical locality and causality conditions. In the quantum domain, Nature is *not* as classical physics had expected. The

---

[91] These particle concepts are introduced in Sect. 3.1. See also Appendix E.
[92] See Cassirer 1937; Mittelstaedt 2006.

contingent structure of quantum phenomena is at odds with the classical construal of empirical reality which had been the universe of discourse of physics for more than two centuries. This is why it is so hard to find a satisfying interpretation of quantum theory.

1.6 Towards a Realism of Properties

Indeed quantum theory has shaken the traditional belief in the rationality, uniformity and simplicity of Nature. But to completely dispense with these principles would mean dispensing with physics as a science. Weakened versions of the principles of unity and simplicity have survived the transition to quantum theory. In the remainder of the book it will be shown in great detail that today, due to quantum theory, the unity of physics is a *semantic* rather than an ontological unity. Physics still has a unified language, namely the language of physical quantities, even though the unity of axiomatic theories and their objects has been lost.

The semantic unity of current physics is based on the construction of *scales of physical quantities*. The scales of length, time and mass are constructed from 0 to ∞ as if the classical and non-classical, non-relativistic and relativistic theories of physics were theories about the same natural kinds. Indeed, the construction of the scales of physical quantities expresses a strong belief in the reality and uniformity of corresponding properties in Nature. To this belief that length, time, and mass are of the same kind at a large scale or at a small scale corresponds the assumption that the units of these quantities can be chosen arbitrarily. Length, time, and mass may be defined in the cgs system as well as in cosmological units, atomic units, or the units of the Planck scale.

Indeed, the construction of the scales of physical quantities rests on a very weak axiomatic basis. The arbitrary choice of the units is legitimated by the Archimedean axiom of the theory of real numbers.[93] The theory of real numbers is surely shared by all known measurement theories of physics. Therefore, it is possible to extend the scales of physical quantities from the classical to the quantum domain, even though classical observables and quantum observables are formally distinct. We shall see that this extension makes tacit use of a bridge principle or correspondence rule, namely a generalized version of Bohr's famous principle of correspondence. By constructing the scales of physical quantities in this way, current physics is able to ignore the problem of *incommensurability* in Thomas S. Kuhn's sense. This problem, which has been discussed extensively for decades in the philosophy of science, is indeed only a philosophical problem which plays no role in physical

---

[93]This is emphasized in Hilbert 1918 as an essential part of the use of the axiomatic method in physics. See also Appendix A.

practice. The semantic details of how to overcome this problem, however, are tricky.[94]

The semantic unity of current physics, however, is still associated with metaphysical commitments. To construct scales of physical quantities means to believe that the *properties* of natural kinds are the same in the cosmological, meso-cosmic and subatomic realm. (Here, 'the same' means 'of the same kind', or 'similar' in such a way that they make up a continuum which maps to the positive real numbers.) This belief corresponds to a version of scientific realism which is weaker than the positions of global realism and causal realism sketched in Sect. 1.2 but stronger than structural realism, strict empiricism, or constructivism. It is a certain belief in the existence of physical *properties*,[95] which is not necessarily connected with belief in the existence of separate physical *objects* as the carriers of such properties. When Stoney created the name 'electron' in 1891, it denoted a "natural unit of electricity", i.e., a charge which had been measured in electrolytic processes.[96] A charge is just a physical quantity, that is, a physical property which is substantiated in a certain kind of physical process and which can be measured. In contradistinction to such a physical property, a massive charged particle is a *collection* of physical properties that come *constantly together* in measurements, i.e., an empirical substance in Locke's sense.[97]

The belief in physical properties which correspond to measured quantities has an operational basis. It corresponds to the position of moderate empiricism sketched Sect. 1.2. The construction of the scales of quantities, however, corresponds to a stronger metaphysical belief. It corresponds to the trust that there is a continuum of properties in Nature which covers the characteristic spatio-temporal and dynamic quantities of all real or possible processes of physics at all scales. This belief is stronger than operationalism in filling the gaps between the diverse operational definitions of physical quantities which are bound to specific measurement methods. And it is stronger than structural realism in its *qualitative* features. Structural realism is the belief that physical processes exhibit (at least approximately) certain formal structures. Property realism is the belief that length, time, mass, charge, etc., are qualities which can be attributed to systems and processes in Nature. Indeed,

---

[94]See Chap. 5.

[95]It should not be understood in the sense of Platonism, that is, as a belief in the existence of properties as universals. It is closer to the Aristotelian account of properties, according to which properties are always carried by individuals. In quantum physics, these individuals are the measuring devices in which a subatomic process happens, or the macroscopic environment in which a quantum system is embedded; see Falkenburg 1993b. This physical view of properties comes close to the concept of tropes in current ontology; see Seibt 2002.

[96]See Millikan 1917; Pais 1986, 73–74. See Sect. 3.2.1.

[97]According to Locke, an empirical substance is a collection of properties that come constantly together and the concept of a carrier of these properties is metaphysical. See Locke 1689, Book II, Chap. 23.

without such a realism of properties, physics as a science would not be possible. Without belief in physical properties that are common to all natural kinds of all scales (whatever these natural kinds may be), neither subatomic physics nor cosmology would be possible as empirical disciplines.

Obviously, the scales of physical quantities are in a Kantian sense *a priori* conditions of physics as an empirical science. In Sect. 1.5, I classified the metrical standards of physics among the indispensable methodological principles of physics which are *constitutive* of physical experience. This is true of all measurement methods separately. The scales of physical quantities, however, *unify* the different measurement methods of physics. Hence, they are not *a priori* in the sense of Kant's principle of causality or the other principles of pure understanding. They are *a priori* in the same way as the principle of the unity of the empirical laws of Nature is *a priori* according to the *Critique of Judgment*. In constructing the length, time, and mass scale we conceive of Nature *as if* there were a continuum of physical properties. No one grants that the construction does not break down, say, at the Planck scale. But *if* it breaks down somewhere, the semantic unity of physics breaks down at that scale, and with it, the very possibility of physics as an empirical science.

2 Extending Physical Reality

Modern science rests on a sophisticated interplay between theory and experiment. The way in which this interplay extends physical reality beyond the observable realm gave rise to the conflict between Galileo and his Aristotelian opponents and it has been nurturing the later debates on scientific realism up to today. In order to get a better understanding of the problems involved in the debate about subatomic reality, this interplay has to be analyzed in more detail. In particular, the constructive features of modern physics have to be investigated. The classical construal of subatomic reality turned out to be too strong. This is a point in favor of constructivism (even though as a philosophical position, constructivism cannot cope with the empirical successes of modern physics and the self-correction mechanisms inherent in theory formation).

The experimental and mathematical methods of modern physics are intimately related to the traditional metaphysical presuppositions of physical theories sketched in Chap. 1. They aim at the decomposition and composition of natural phenomena, and they rely on the separability of causal agents in nature. Traditional metaphysics conceived of these causal agents in terms of substance. We have already seen that this concept is made to cope with the domain of classical physics, whereas it becomes highly problematic in the quantum domain. Therefore the present chapter takes up the problems of the former by discussing them now in the context of methodological aspects. In its course, several objections against scientific realism will be taken into account in more detail. Some of them stem from empiricism, others from constructivist positions. Some of them are quite general, others are closely related to the specific nomological structure of the quantum domain. They have in common that they belong to the Aristotelian tradition in one way or another. They oppose the way in which modern physics extends physical reality.

Modern physics was called Galilean science. Galileo created the specific way in which physics extends empirical reality by means of experiment and mathematical techniques. Galilean science rests on three pillars:

1. the application of *mathematics*,
2. the systematic performance of *experiments*,
3. the use of *observation instruments*.

Mathematics had already been used in ancient natural science, namely in Ptolemy's astronomy or in Archimedean statics. Galileo was the first to apply it to terrestrial motions in order to investigate mechanical phenomena. He carried out measurements of mechanical motions and expressed their results in terms of kinematical quantities such as velocity and acceleration. Newton added dynamical quantities such as mass. Experiments are designed to standardize the conditions under which measurements are performed. They investigate the phenomena under well-defined artificial conditions in order to measure their physical quantities. Galileo and Newton developed new methods of experimental analysis which are based on mathematical models of the phenomena. Observation instruments are technological devices. Galileo's telescope made Jupiter's moons and the phases of Venus visible, but for his Aristotelian opponents the telescope was an artefact that did not serve the observation of natural phenomena. Hence, experiments make use of technological devices. They bring observation and measurement together and they supply modern science with new and highly sophisticated observation instruments such as the electron microscope or the particle beams and detectors of current high energy physics.

Galilean science aims at scientific objectivity but it pays its price. As we have seen in the last chapter, physics exhibits a semantic unity which is expressed in terms of physical quantities. It achieves objectivity by performing measurements and by expressing their results in terms of quantities such as length, time, mass, energy, charge, temperature and so on. These quantities replace the subjective qualities of our sensory perceptions. Max Planck emphasized that this replacement is a process of de-anthropomorphization, in which the loss of our immediate sensory qualities is compensated by the gain of a genuine, mathematical reality (Sect. 2.1). Indeed, Galileo's experimental method is tailored to the application of mathematics to the phenomena. The mathematization of the phenomena comes in several steps. It is based on what has been called Galilean idealization, and it results in standardized phenomena and their measurement. Experiment aims primarily at the decomposition and composition of phenomena, that is, at *experimental analysis*. Galileo called the decomposition and composition of phenomena the *resolutive–compositive method*.[1] The background of his experimental method is the traditional method of analysis and synthesis. On the one hand, experimental analysis aims at standardizing the phenomena. On the other hand, it aims at causal analysis (Sect. 2.2). The questions of whether the experimental phenomena are detected or constructed, and whether their causes are discovered or invented, have been discussed in various philosophical contexts. Some recent constructivist arguments against the reality of experimental re-

---

[1]See Losee 1993, 57–62. Experiment has its own purpose: it aims to investigate nature independently. Testing theoretical hypotheses is a secondary step in the interplay of theory and experiment.

sults and the existence of their causes outside experiment are taken up and countered in Sect. 2.3. Here, causal realism becomes crucial (Sect. 2.4).

Experiment is one way of extending empirical reality, observation by means of technical devices is the other. But the extension of empirical reality into the quantum domain carries the burden of bridging the conceptual gaps between classical physics and quantum theory. The observation instruments of subatomic physics are not only much more sophisticated than Galileo's telescope. They employ also highly developed technologies which are brought to work in accordance with quantum physics. Our concept of observation, however, stems from a classical world. What do we observe and in which sense with, say, an electron microscope or with a particle detector? To what extent can the traditional, empiricist account of observation be generalized for the quantum domain (Sect. 2.5)? However, the claim that a particle detector serves to observe particles is highly problematic. What kind of entity is a subatomic particle that might be observed? The problem is intimately related to the question of how it is possible to constitute objects in the quantum domain. Somehow the empiricist is right in refusing to talk about theoretical objects. However, this does not commit us to a strictly empiricist account of the empirical content of theories. In the crucial case of a particle track, the empirical content of a theoretical model is a classical construal that is incompatible with quantum theory (Sect. 2.6).

2.1 Introducing Physical Quantities

Objective knowledge must not depend on our subjective cognitive faculties and epistemic conditions. Hence, the claim that science aims at objectivity means that it aims to dispense with the subjective observer. It focuses on objects that are taken to be independent from the observer's subjective views. The mathematical and experimental procedures of physics lead us away from our ordinary understanding of reality, away from the usual qualitative experience of the events occurring around us, and away from the natural language in which we normally express our experiences and sensory perceptions.

Prior to any experiment of modern physics, measurement methods had been developed in order to find objective standards of length, time and weight. Originally, measurement was based on human standards but in the course of the scientific revolutions which gave rise to modern science enormous efforts were made to standardize them. However, some principal features of measurement have remained the same since the days of ancient geometry. Measurement is rooted in comparison.[2] To measure means to compare an object or process, or one of its physical properties, with a standard such as the standard meter. The standard is taken as the unit of a scale. The measured property is expressed in terms of this unit. It is crucial that *many*

---

[2]See Carnap 1966, Part II, and Appendix A.

objects or processes are compared to the *same* unit. Their physical properties correspond to multiples of the unit, that is, to quantities which are expressed by numbers in the scale. A number in the scale corresponds to a physical property, and indeed to *all* objects or processes which possess that property.

Thus, measurement serves to generate well-defined classes of physical objects or processes. Each class of objects or processes corresponds to a physical property, say, the property of having a length of 1 cm, lasting a time of 3 ns, or having a temperature of 3 K. A physical property corresponds to the class of objects or processes that share that property. All classes of experimental results are expressed in terms of real numbers and the dimension 'length', 'time', 'mass', 'temperature' of the respective quantity. Each real number expresses a physical property in terms of a multiple of the unit of the appropriate scale. The scale ranges from 0 to ∞. It represents one class of physical properties, for example, all possible mass values. It corresponds to all physical properties of this type.[3] The physical quantity 'mass' is represented by the mass scale. The mass scale is the abstract class of all classes of concrete bodies of equal weight.

Hence, physical quantities have an abstract and symbolic character, as was especially emphasized by Pierre Duhem.[4] They are symbolic since they are concepts that denote physical objects and processes. In addition, the meaning of a physical concept such as 'mass' is very abstract. It does not reside in concrete objects but in a class of physical properties with a whole scale of corresponding numerical values. Every measured value in turn corresponds to a class of concrete objects or processes that exhibit the same physical property. Following Frege's account of numbers as second-order concepts,[5] physical quantities turn out to be second-order concepts too:[6]

1. Every numerical value of a physical quantity corresponds to a class of concrete physical phenomena for which a given measurement method gives, within certain error margins, the same measurement result.
2. Formally, a physical quantity is a function which maps a class of properties onto the set of positive real numbers.[7] This set – the scale of a quantity – is defined only up to the choice of the unit.

---

[3]Physical properties are of the same type if and only if it is possible to add them, like the masses of two bodies. Physical properties of different types may be multiplied, making up new properties, such as velocity, momentum, energy, etc. This gives rise to the algebra of physical quantities and to the Π theorem of dimensional analysis; see Appendix B and Krantz et al. 1971, Chaps. 8 and 10.

[4]See Duhem 1906.

[5]See Frege 1884.

[6]See Appendix A.

[7]This applies only to classical physics with the inclusion of both theories of relativity. In a quantum theory, the quantities are not real-valued, but operator-valued.

Thus, every concept of a physical quantity refers to a class of classes of concrete physical phenomena—or to a class of physical properties constituting the scale of the quantity. Most physicists think that the formation of such classes in physics rests on fundamental physical properties which belong to natural kinds, to entities such as atoms, elementary particles, or black holes which are not immediately observable. This is an *essentialist* position, whereby the concept of a physical quantity refers to the essential properties (or primary qualities) of natural kinds, that is, the entities which ultimately cause the natural phenomena due to their respective properties. However, the class of properties corresponding to the concept of a physical quantity can also be defined operationally – without any essentialist metaphysics – by connecting it to measurement. This requires the definition of a chain of well-established measurement methods by means of which the scale of a quantity can be completely apprehended.[8]

In contrast, according to Bridgman's radical operationalism, every measurement method defines a distinct kind of quantity.[9] His position shows that the operational concept of a scale remains based on metaphysics. The very formation of a scale is a theoretical construal. One assumes that in different parts of the length, time, mass, or temperature scale distinct measurement methods measure the same kind of physical property. The scales are constructed from the Planck scale to the size of the universe as if there were no conceptual gap between the quantum domain and the classical world. The scales of physical quantities constitute the theoretical universe of discourse of modern physics as if there were no such gap. The construction of this theoretical universe of discourse is based on a crucial metaphysical assumption. It makes use of Newton's belief that "nature is very conformable to herself."[10] It is based on trust that constructing a *semantic unity* of physics does not give rise to contradiction.[11]

Duhem believed neither in natural kinds nor in the unity of nature. He defended an anti-metaphysical interpretation of the abstract and symbolic language of physical quantities. According to him, physical concepts, laws, and theories do not stand for things or events *in concreto*. Neither did he regard them as elements of a true description of a physical reality which would lie at the basis of observable phenomena in the form of essential properties and true causes.[12] From Duhem's perspective, the abstract concepts of a physical theory are mere *instruments*, they are tools of unification and prediction. He emphasized that their definition always includes a certain degree of arbitrari-

---

[8]See Falkenburg, 1997, and Appendix A.

[9]See Bridgman 1927.

[10]Newton 1730, *Query 31*.

[11]I will come back to this problem in Chap. 5, after discussing the typical measurement methods of particle physics in Chaps. 3 and 4.

[12]In the spirit of Newton's first rule of reasoning in philosophy (Newton 1729, 398 and 1687, 794).

ness and that they serve primarily to bundle together, in a most economical way, as many qualitatively distinct phenomena as possible. On the first issue, he tended towards Poincaré's conventionalist view of physics. On the second, he was close to Ernst Mach's empiricist view of physical theories. According to Mach, physical theories only aim at an economical representation of experimental phenomena and the assumption of atoms does *not* belong to such an economical representation.

In contrast, the founders of 20th century physics defended scientific realism. They were convinced that physical quantities and theories aim at the essential properties and structural features of things and causal processes in nature. Boltzmann, Planck, Einstein, Rutherford and Bohr were atomists, personally participating decisively in the investigation of atoms.[13] In his lecture about the unity of a physical world view, 1908, Max Planck distanced himself decidedly from Mach's empiricist and phenomenalist conception of physics and presented a realist and essentialist, if not Platonist, view of the abstract language of physical theory. According to Planck, the formation of physical concepts aims at liberating our understanding of nature more and more from anthropomorphic conceptions. Thus, the development of the classical concept of force of classical mechanics emancipated us from the idea of physical power which we must apply in order to do work, for example, when we want to lift an object.[14] Unlike Mach, Planck does not regard the increasing distance of physical theory from immediate sense experience as a loss, but rather as a gain – or, stated more adequately, the gain, in his eyes, by far exceeds the associated loss:

> If we look back upon the past, we can briefly summarize by saying that the signature of the past development of theoretical physics is a unification of its system that is achieved through a certain emancipation from the anthropomorphic elements, especially sensations. [...] Indeed, the advantages must be invaluable, if they deserve such fundamental self-sacrifice![15]

The "invaluable advantages" lie, as Planck subsequently explains, in the increasing unity of the physical view of the world. For him, this means much more than an economy of thought in Mach's sense. He emphasizes that the unification of theories leads to physical *universalism*. The conceptual unity of a comprehensive theory – a theory which rests only on a few principles and which is as free as possible from the specific circumstances under which we perceive natural phenomena – makes the results of physical investigation independent of place and time, of the individuality of the investigator, of nation and of culture.[16] According to Planck, a unified world view is infinitely

---

[13] See Scheibe 2001.
[14] Planck 1965, 30.
[15] Planck 1965, 31. My translation.
[16] Planck 1965, 45.

superior to the idea of an adjustment of our theories to the facts, as demanded by Mach.[17] Physical universalism frees our scientific knowledge from the contingencies of human existence and leads to a constant reality behind the variable and manifold phenomena of the senses:

> As I have tried to show, the constant unified world view is precisely the fixed goal which the real natural science, in all its transformations, constantly approaches. [...] This constant, independent of any human, or rather, of any intellectual individuality, is just what we call the real.[18]

According to Duhem, the formal language of physics consists of mere symbols and leads away from reality. According to Mach, this language achieves the highest possible adjustment of our ideas to the facts only at the price of abstraction, simplification, schematisation and idealization.[19] According to Planck, on the other hand, it is this language which first leads to the recognition of a constant reality.

Here, we face obviously diametrically opposed conceptions of what is real. For Duhem or Mach, reality lies in immediately observable phenomena. Duhem, the experimental physicist, identifies them with the observational results gained in the experiments of physics. Mach, the phenomenalist, sees them in fundamental elements of our sensations – in sense-atoms, so to speak, instead of the physical atoms of Boltzmann, Planck, Einstein, Rutherford or Bohr.[20] For Planck, however, there is a constant reality to be found behind the changing play of the phenomena of our sensory perception. This play of sensory phenomena is conditioned, on the one hand, by our cognitive faculties from which physics ought to emancipate itself as much as possible and, on the other hand, by the unchangeable laws of an underlying reality, the constitution of which only becomes comprehensible through this emancipation.

The opposition thus sketched between Planck's scientific realism and Mach's empiricist or Duhem's instrumentalist position has its origin in the methods and the goal of modern physics. The abstract language of physical quantities aims at constituting scientific objectivity. But what are the achievements of this scientific objectivity? Does it reveal an independent physical reality which underlies the phenomena? Or does it replace the phenomena by mathematical construals?

---

[17]See Mach 1926, 164.

[18]Planck 1965, 49.

[19]See Mach 1926, 455.

[20]Mach's phenomenalism is obviously more radical than Duhem's instrumentalism or modern versions of scientific anti-realism. It was adopted by Carnap in *Der logische Aufbau der Welt* but later abandoned by 20th century empiricism.

2.2 Idealization and the Experimental Method

Measurement is not the only way of mathematizing phenomena. It goes hand in hand with various other kinds of idealization. In the debate on scientific realism, all of them are under the suspicion of leading away from empirical reality rather than revealing the true laws of nature. Closer analysis shows that some kinds of idealization are typical of Galilean and Newtonian science, whereas others had been there before. McMullin's typology of Galilean idealization suggests the following distinctions:[21]

1. **Mathematical Idealization.** Imposing a mathematical formalism on the phenomena:
   - measurement;
   - mathematical description.
2. **Model Construction.** Replacing a complex object by a simple mathematical model:
   - formal idealization, or idealization proper;
   - material idealization, or abstraction.
3. **Causal Analysis.** Replacing a complex process by a simple causal model:
   - experimental analysis;
   - thought experiments.

Mathematical idealization is very old. Measurements have been made and mathematical descriptions of certain phenomena have been given since the days of ancient geometry and astronomy. Both kinds of mathematical idealization can come with greater or lesser accuracy. Mathematical descriptions are often based on approximation and correction procedures. A process of approximation had already been the basic idea behind Ptolemy's complicated astronomical system. Any observed deviation of the predicted circular planetary motions was captured in terms of epicycles, excenters and so on. As in modern Fourier analysis, any observation could be taken into account as precisely as desired. This example shows that mathematical idealization *per se* does not necessarily make construals out of the phenomena. McMullin emphasizes that the "extent to which it *is* an idealization has steadily diminished as the mathematical language itself has become progressively adapted to the purposes" of the natural sciences.[22] Measurement also aims at saving the phenomena with mathematical rigor. But the realm of the observable

---

[21] McMullin 1985. I extend his typology by taking measurement into account, and I change the terminology slightly. See also Hüttemann 1997, 86–104. McMullin's "formal" and "material" idealization corresponds to Hüttemann's "idealization in a narrow sense" (which I call "idealization proper" below), and "abstraction". None of these distinctions are sharp. They give only a rough sketch of the various ways in which mathematical structures do *not* match the empirical phenomena they replace.

[22] Ibid., 254.

is not yet exceeded by measuring some empirical phenomena approximately in terms of some empirically accessible unit (such as the standard meter). However, it is obviously exceeded when we construct the scales of physical quantities from the Planck scale to the size of the universe in modern physics.

In Galilean science, however, mathematical descriptions and measurement come together with new kinds of idealization. Model construction, causal analysis and the experimental method exceed the realm of the observable.

Model construction replaces a complex object by a model with a simple structure. Model construction is typical of mathematical physics. A model is not designed to give a true description. Deliberately, it has a structure that differs from the modelled object. It is constructed in order to obtain a simple mathematical description of an object under investigation. *Formal idealizations*, or idealizations proper, make formal descriptions feasible. Some typical examples are: classical mechanics describes bodies as mass points and two-body systems in terms of the reduced mass. Bohr's atomic model makes use of an infinite-mass approximation of the atomic nucleus and neglects relativistic effects. Classical electrodynamics deals with infinite wires or capacitor plates for making the boundary conditions of the Maxwell equations simple and the solutions calculable. In such cases the model is formally simpler than the modelled objects. In contradistinction to formal idealization, *material idealization*, or abstraction, is a way of neglecting certain physical properties, material aspects or internal structures of the objects under investigation. Kinetic theory does not specify the internal structure of the molecules of a gas. In a similar way, Rutherford's model of the backward scattering of α rays at a thin gold foil did not take into account the size of the atomic nucleus. In both cases this can be corrected by means of more complicated models of the objects under investigation. In the case of Rutherford scattering this is done be introducing form factors into the simple Coulomb potential.[23]

In model construction, too crude idealizations can be refined by *de-idealization*. One starts with the most simplified structure or system description and considers to what extent it explains the relevant phenomena. A model is good if it can be made better by means of corrections. This means that specifications of its unspecified features are added. If the model is good, these additions are not ad hoc, and the specified model results in improved predictions. In this way, idealization is followed by putting back some of the omitted features of the object under investigation, that is, by de-idealization. The procedure is justified by empirical success. McMullin argues that model construction by means of idealization and de-idealization is "truth-bearing in a very strong sense"; improvements show that a model has "the means for self-correction which is the best testimony of truth."[24] Thus, he sees model construction on the same footing as mathematical description. The process of adding back some of the omitted features is like a process of approximation.

---

[23]See Chap. 4.
[24]Ibid., 264.

De-idealization makes it possible to capture the phenomena more and more accurately in a step-by-step manner.

According to McMullin, *causal analysis* is different. It aims at the decomposition of the single causal agents or factors which give rise to a complex causal process. Causal analysis rests on idealization as well. It is based on the following assumptions:

1. different causal factors act independently;
2. some of them are relevant and others can be neglected;
3. the relevant causal factors can be isolated;
4. their effects add up linearly.

Causal analysis is done in real experiments as well as in thought experiments. Real experiments investigate different causal agents and the combination of their effects under different experimental conditions. The variation of the conditions under which an experiment is performed is crucial; no experiment comes on its own.[25] Thought experiments probe the internal consistency of causal explanations. They may either be performed in reality or not. In the second case they are counterfactual. McMullin discusses these cases for Galileo's investigations of the fall of mechanical bodies. Causal analysis results in the distinction of free fall and the influence of air resistance. *Experimental analysis* consists of the famous experiments with an inclined plane and a pendulum. Variation of the conditions under which the experiment is performed confirm the hypothesis that air resistance hinders free fall and that free fall obeys the law of constant acceleration. *Thought experiments* are designed to show that Aristotle's assumptions about the fall of bodies are absurd and that diminishing the influence of air resistance up to the limiting case of the fall in a vacuum *would* make it physically possible to isolate free fall as a single causal agent.[26]

Is it possible to interpret the results of causal idealization in terms of (approximate) truth and the results of stepwise causal de-idealization in terms of increasingly better approximations? In Galileo's or Newton's view, causal analysis aims at recognising the *true causes* of natural phenomena, in the sense of Newton's *first rule of reasoning*.[27] In McMullin's view, causal realism is far from being obvious. He notes that

the issue hinges around the technique long known as 'composition of causes', a technique which both Galileo and Newton utilized but which gave both of them trouble. [...] I would conclude that the manifest successes of the natural sciences since Galileo's day furnish

---

[25] See Falkenburg and Ihmig 2004.

[26] McMullin 1985, 266–237, and Galileo 1638.

[27] Newton 1729, 398 and 1687, 794. Newton's search for the *true causes* of the phenomena has quite often been understood in terms of causal realism, but it is indeed more cautious; see below.

adequate testimony to the worth of the technique of causal idealization [...]

This conclusion does indeed not support scientific realism. It expresses an
instrumentalist account of the causal factors stemming from causal analysis
and Galileo's experimental method. McMullin's arguments defend structural
realism with regard to mathematical idealization and model construction, but
they do not defend causal realism with regard to the causes of experimental
effects. Is he right?

Like measurement, Galileo's experimental method aims at constituting
scientific objectivity. Like the introduction of physical quantities, it aims at
mathematization of the phenomena. Let us have a closer look at the way in
which it works. It aims at isolating certain partial aspects of natural phenomena and analyzing them under the most ideal conditions possible. The
experiments of physics are designed to produce isolated, regularized and reproducible phenomena and to vary them in a controllable manner. Only the
disassembly of phenomena into isolated standardized components makes the
application of mathematics to natural phenomena possible. Only an observable effect in an experimental setup, one that can be arbitrarily reproduced
in a controlled manner, can be grasped as an element of a well-defined class
of homogeneous physical phenomena which then become accessible to mathematical description.

Thus, the experimental method serves the generation of well-defined
classes of phenomena and the investigation of the systematic relations of their
elements. The results of experiments serve measurement, they correspond to
the abstract and symbolic character of physical quantities. In addition, experiments investigate the functional interdependence of physical quantities.
Every experiment is designed to decompose complex natural processes into
specific components, the composition of which is then systematically varied
in order to measure numerical values and functional relations of physical
quantities. In this way, experiments establish stable phenomena and phenomenological laws, and in doing so, they constitute scientific objectivity.

The constructive features of the experimental method cannot be overlooked. The experimental method makes the phenomena fit the application
of mathematics, and not vice versa. And it results in the conceptual and
factual construction of separate causal factors which are relevant for the generation of a phenomenon. Once these causal factors are considered to act as
independent causal agents, they can be conceived in metaphysical terms of
substance and causality.

We have already seen that varying the conditions under which an experiment is performed aims at causal analysis. Some experiments also aim at
the recomposition of these components. A simple qualitative measurement of
this kind is found in Newton's *Opticks*. White light is decomposed by sending it through a prism. The resulting color spectrum is superposed on the
spectrum resulting from the passage of white light through a second, parallel

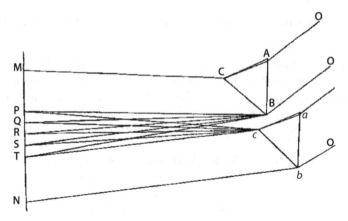

Fig. 2.1. Spectral decomposition and recomposition of white light (Newton 1730, 147)

prism. The result is the recomposition of white light from the two spectra.[28] In Newton's view, the causal agents which are investigated in this experiment are the components of white light. In his causal analysis of the experiment, he identifies these components with light rays of different colors but *not* with light particles. He put all metaphysical questions concerning the nature of light into the *Queries* part of *Opticks*. In his physical theories he disregarded the causal agents of nature *as long as* their physical quantities were not subject to experimental analysis and measurement. In this way, he reconciled a realism of physical quantities with his famous verdict *hypotheses non fingo*.[29]

In 17th century natural philosophy, causal decomposition and recomposition was called the method of *analysis* and *synthesis*. Galileo's experimental method has also been called the compositive–resolutive method.[30] Indeed, 'composition' and 'resolution' are just the Latin words for the Greek expressions 'analysis' and 'synthesis'. The latter terms have survived until today in a literal sense in the language of chemistry. In the 17th century, several variants of the method of analysis and synthesis coexisted. Newton's account of it was very complicated. It was a complex procedure that stepped back and forth between experimental phenomena and mathematical models, by repeatedly varying the experimental conditions and adjusting the mathematical principles.[31] It was close to what physicists still do today when they design experiments and construct theoretical models in order to investigate a new physical domain.

---

[28]Newton 1730.

[29]See Falkenburg and Ihmig 2004.

[30]See Losee 1993, 57–62.

[31]Newton's method of analysis and synthesis has its origin in ancient geometry; see Ihmig 2004. It is closely related to Francis Bacon's account of induction which has the same complexity; see Falkenburg and Ihmig 2004.

Thus saying that experiments aim at causal analysis amounts to saying that they aim at the isolation of causal agents which can be separated from each other. Causal realism considers these causal agents to act independently, as entities in their own right. Once they are regarded as independent carriers of physical properties, they count as substances in a traditional sense. The step from causal *factors* to causal *agents*, from Newton's realism of physical quantities to modern causal realism, does not have to be made. However, it is quite a small step. Galileo and Newton's experimental method is not metaphysically neutral. It aims at the generation of standardized phenomena and at the isolation of causal agents with the structural properties of metaphysical substances.

2.3 Discovery or Manufacture?

The typical causal agents of modern physics are forces, fields, atoms and subatomic matter constituents. The question of whether the experimental method discovers or construes such entities has a long tradition. It ranges from British empiricism and German idealism to modern philosophy of science. In an Aristotelian tradition, Berkeley and Hegel opposed the reality of atoms or forces.[32] In neo-Kantianism, Cohen suggested that the physicist's hidden reality is a theoretical construal, thus opposing Mach's empiricism for quite other reasons than Planck.[33] Amongst those who insisted on the constructive features of modern physics one of the most prominent is the physicist Eddington. He compared the body of physical knowledge to a net which catches only fish of a certain minimum size.[34] This is a constructivist point that in the following aspects comes close to a Kantian epistemology: our physical knowledge depends on certain epistemic conditions; physical theory and the experimental method impose an *a priori* structure on phenomena. According to Eddington, scientific objectivity is based on selective subjectivism. In the same thrust he raised doubts about the results of the experimental method:

> The question I am going to raise is – how much do we discover and how much do we manufacture by our experiments? When the late Lord Rutherford showed us the atomic nucleus, did he *find* it or did he *make* it? It will not affect our admiration of his achievement either way – only we should rather like to know what he did. The question is one that scarcely admits of a definite answer.[35]

---

[32] See Berkeley 1721 and Breidert 1969; Hegel 1830, § 266 and § 276; Falkenburg 1993c, 1998a.
[33] Cohen 1896 and Falkenburg 2000, 316–317.
[34] Eddington 1939, 16.
[35] Eddington 1939, 109.

In order to oppose naive realism, Eddington advocates anti-realism. His objections against the reality of forces, fields and subatomic matter constituents are sophisticated and substantial, they are based on intimate knowledge of relativity and quantum theory. In recent philosophy and history of science, related objections have been raised, most often missing Eddington's delicacy. Three kinds of constructivist ways of understanding – or misunderstanding – modern physics became influential:[36]

1. social constructivism,
2. interventionalist causality,
3. constructive empiricism.

1. Social Constructivism

This version of constructivism was stimulated by Thomas S. Kuhn's thesis that rival theories are *incommensurable*. Kuhn's incommensurability thesis embraces the semantic claim that the vocabulary of rival theories is so distinct that an unambiguous translation between them is impossible, and the ontological claim that rival theories are associated with incompatible world views.[37] Sociological constructivism is based on the view that such different ontologies or scientific world views depend on scientific cultures, that is, on the specific development of schools within the scientific community. In accordance with this view, sociologists focus only on the external, socio-cultural factors under which scientific theories are developed. So far, so good. No one will deny that modern physics developed under very specific social conditions which were *necessary* for the rise of classical mechanics, field theory, thermodynamics, relativity, quantum mechanics and so on. Modern physics is crudely misunderstood, however, if such external factors are confused with *sufficient* conditions of theory development. Recent sociology and history of science has had strong tendencies to neglect the internal factors of theory development, namely scientific methodology and the internal criteria of accepting scientific theories.[38]

Typical of a radically sociological view of theory development in physics is Pickering's book *Constructing Quarks*.[39] It comes with the authority of a theoretician of physics and it defends the view that the quark model of current particle physics is a social construct of the scientific community. The book has

---

[36]They make up sub-positions of the positions (5), respectively (4), of the list given in Sect. 1.2. Scheibe 1997, 9–29, gives a more exhaustive list of recent constructivist positions.

[37]Kuhn 1970.

[38]See Knorr-Cetina 1981, 1999; Latour and Woolgar 1979. Daston 2000, 2–3, emphasizes that historical studies of the coming into being and passing away of scientific objects are 'orthogonal' to the debate on scientific realism since "scientific objects can be simultaneously real *and* historical" (ibid., 3).

[39]Pickering 1984.

been influential in sociology of science. It puts together a most interesting and fascinating history of the schools of theoretical particle physics in the 1960s and the genesis of the current standard model and a total disdain for the very scientific contents of the quark model. Pickering writes:

> The quark–gauge theory picture of elementary particles should be seen as a culturally specific product. The theoretical entities of the new physics, and the natural phenomena which pointed to their existence, were the joint products of a historical process. [...] On the view advocated in this chapter, there is no obligation upon anyone framing a view of the world to take account of what twentieth-century science has to say. [...] There is no reason for outsiders to show the present HEP (high energy physics) world-view any more respect.[40]

The background of this view is a historical analysis of the analysis and re-analysis of the data confirming the neutral weak currents (which were decisive for the validity of the early quark model). In the Gargamelle bubble chamber data taken in 1963, *no* weak neutral currents had been found. They were re-analyzed a decade later, when the quark model had been confirmed. Now, evidence of neutral weak currents was found in the data. Pickering claims that a strong symbiosis between theory and experiment was the reason for this empirical success:

> The neutrino experiments did reappraise their interpretative procedure in the early 1970s and [...] they succeeded in finding a new set of practices which made the neutral current manifest.[41]

However, the authority of a theoretician is *not* an experimenter's authority. Measurement and data analysis are theory-laden, but the theories involved must *not* be identical with the theories under test. The experimenter's job is to keep the theories under test and the theories of measurement or data analysis strictly apart. If they cannot be separated, the measurement goes astray and the experimental results are not regarded as reliable. For an understanding of this point the distinction between safe background knowledge and theoretical uncertainty is crucial.[42] Indeed, the re-analysis of the old neutral current data in the early 1970s relied on a better understanding of the radiative corrections of quantum electrodynamics and *not* on the quark model under test. In contradistinction to Pickering, the historian Galison

---

[40] Pickering 1984, 413. Pickering continues: "In certain contexts, such as foundational studies in philosophy of science, it may be profitable to pay close attention to contemporary scientific beliefs. In other contexts, to listen too closely to scientists may be simply to stifle imagination." According to this view, philosophical investigations like this book which aim at making the contents of modern physics more comprehensible, are also nothing but a socio-cultural phenomenon.

[41] Pickering 1984, 194.

[42] Shapere 1982; see Sect. 2.5.

emphasizes the autonomy of experimental practice against theory construction.[43] The moral is that theory-ladenness *per se* does not prove that data are sociological construals.

2. Interventionalism

This variant of constructivism makes a pragmatist turn. Pragmatism focuses on science as a human activity. Experiments and measurement are obviously specific kinds of human actions. In a controlled way, they intervene in natural processes, with the goal of generating standardized phenomena and imposing mathematical structures on them. The Erlangen school of constructivism attempted to reconstruct the mathematical structure of physical theories from the concrete operations which give rise to approximately ideal objects and processes.[44] In this tradition, Tetens proposed to interpret the results of experiments in terms of von Wright's concept of experimental or interventional causality.[45] Physicists make experiments in order to realize ideal conditions which do not exist in nature. Under these artificial conditions a process is initiated that would obviously not occur exactly like this in 'untouched' nature. From this undebatable matter of fact Tetens draws anti-realistic conclusions which are indeed reminiscent of the ancient distinction between $\varphi\upsilon\sigma\iota\varsigma$ and $\theta\varepsilon\chi\nu\varepsilon$: the regularities exhibited in an experiment are the goal of the experiment. They belong to an intentional structure which is typical of human activities; they are expressed in equations of prognostic value. These equations do not correspond to objective laws of nature but only to our subjective expectation of what will happen. The conclusion is radical instrumentalism: an experiment is a teleological activity, and the main goal of an experimental theory is to deliver tools of prediction which serve our technological purposes. Tetens emphasizes that our view of nature cannot remain unaffected by the experimental method:

> Experimental experience is therefore not an observation of something which is already present in nature, but it is the experience of the changeability of the natural which is actively made. In experimental experience, nature becomes visible mainly as an insurmountable limitation of technological manipulation.[46]

---

[43]See the detailed analysis of the neutral current data analysis and re-analysis in Galison 1987, 135–241.

[44]See Lorenzen and Janich's program of protophysics (Butts and Brown 1989). The most advanced theory which could be reconstructed in this way was special relativity. For quantum theory, the project was bound to fail. Concrete human actions operate on macro-objects and not in the subatomic domain.

[45]Tetens 1987; von Wright 1971.

[46]Tetens 1987, 12; my translation. It should be added that Tetens no longer defends such a radical instrumentalism today. See for example his splendid book on the mind–body debate (Tetens 1994), which takes the results of modern cognitive science for granted in a way that is *incompatible* with instrumentalism.

I do not think that this claim is justified. It is true that the use of technology does not leave our concept of nature unaffected. Due to modern technologies, the borderline between what is extra-human nature and what is due to human activities has indeed become fuzzy. But does this striking observation make the laws according to which an experiment works – or does not work – a result of human action? In Galileo's days, Francis Bacon emphasized that even if science has technological purposes, any use of technology requires the laws of nature to be obeyed.[47] Interventionalism may be *necessary* for understanding experiments and the phenomenological laws in which they result, but it is not *sufficient*.

Like theory-ladenness, technology-ladenness is not fatal for scientific objectivity. Tetens' interventional view of experimental physics provokes several objections:

1. The very activity of experimenting is not relevant for understanding the laws of physics. Experimental results depend much more on the interplay of technological skills and measurement theory than on the goals and ambitions of the human beings at work. (Even the recent cases of cheating in biology or physics and in particular their discovery show that science aims primarily at truth rather than at human purposes.)
2. The borderline between observation and experiment is fuzzy too. The domains of the two kinds of epistemic activity overlap. For example, the phenomena of natural and man-made radioactivity overlap. In particle physics, the data of cosmic ray photographs and the data taken in the beams of particle accelerators overlap. A particle accelerator is fed by a natural radiative source. And a particle detector in a scattering experiment of high energy physics still records tracks of cosmic rays when the beam is switched off.
3. Human actions have intended and unintended consequences. In an experiment, only well-known laws of nature can give rise to intended consequences of the experimenter's activities. Usually, an experiment is performed because some specific laws of nature are *not* yet well known but still under investigation. If the results depended *exclusively* on the experimenter's activities, *why* should she make the effort to perform the experiment?

Experimental activity and technological use should not be confused. Technology aims at intended consequences of our actions. Experiments aim mainly at investigating unintended consequences. The unintended consequences of an experiment are *contingent*. They may be as expected or just the opposite. They exactly match our reality criterion developed in Sect. 1.3: they belong to the actual. *If and only if* an experiment works as expected can it be transformed into technology and become a useful technological device

---

[47]Bacon 1620, First Book, III: "Natura enim non nisi parendo vincitur; et quod in contemplatione instar causae est, id in operatione instar regulae est."

for the performance of *other* experiments. Indeed, the arguments of interventionalism may also serve to support scientific realism.[48]

3. Constructive Empiricism

Social constructivism and interventionalism have in common that they confuse necessary and sufficient conditions of the experimental method. In doing so, both positions arrive at the same conclusion: the experimental method yields *only* social or technological construals. They miss the point that an experimenter cannot deliberately generate arbitrary phenomena. They do not recognize the contingency of experimental results as evidence of an independent physical reality, of laws of nature which are *not* at our disposal. In both views, physics becomes a cultural product and natural science collapses into technology. However, van Fraassen's constructive empiricism shows that a constructivist view of the experimental method is compatible with an empiricist account of physics. Van Fraassen claims that theories are mathematical construals, and experiments result in empirical construals. The two kinds of construal come hand in hand. Our theoretical construals of the world are incomplete and experiments are designed to fill the blanks in our theories. They aim at determining the contingent boundary conditions of differential equations or at measuring universal constants of nature such as the speed of light or the electron charge.[49] Thus, the interplay of theory and experiment has constructive features. Van Fraassen illuminates this point with his *Clausewitz doctrine of experimentation*. According to this, performing an experiment means replacing the diplomacy of theory construction by a technological struggle with nature:

> Experimentation is the continuation of theory construction by other means. [...] I should like to call this view the 'Clausewitz doctrine of experimentation'. It makes the language of construction, rather than of discovery, appropriate for experimentation as much as for theorizing.[50]

This passage clearly recommends a constructivist view of theory formation in physics. In addition, however, it suggests that performing an experiment means dealing with an independent instance which is the touchstone of the empirical adequacy of our construals.[51]

---

[48]See Hacking 1983 and Sect. 1.4.

[49]Van Fraassen 1987, 119–120

[50]Ibid., 120. In van Fraassen 1980, 5, he emphasizes the constructive features of physics by calling his epistemological position *constructive empiricism*: "I use the adjective 'constructive' to indicate my view that scientific activity is one of construction rather than discovery: construction of models that must be adequate to the phenomena, and not discovery of truth concerning the unobservable."

[51]See Wind 1934 and my distinctions in Sect. 1.3.

Amongst the three constructivist views sketched here, van Fraassen's comes closest to Eddington's delicacy.[52] Indeed, Eddington's question of *discovery or manufacture* requires a complex answer. One of the two options will not do. The interplay of theory and experiment rests on a complicated interplay of human activities and nature's causal agents.

Here a genuine Kantian point has to be made. If an experiment is performed correctly, nature acts like an independent and non-corrupt witness. (The performance is correct if the theoretical background knowledge of the measurement is well split off from the subject under investigation, if the apparatus is well-known, and if the experimenter is not cheating.) If the experiment has an unambiguous result, it decides a binary choice. If the theory of the subject under investigation gives a prediction, the experimental result is either in agreement with the prediction or not. If a blank in the theory is filled, one value of a physical quantity is measured within certain error margins, and all other values are excluded. In a famous passage, Kant puts the interplay between our epistemic activities and nature's testimony as follows:

> When Galileo rolled balls of a weight chosen by himself down an inclined plane, or when Torricelli made the air bear a weight that he had previously thought to be equal to that of a column of water, or when in later time Stahl changed metals into calx and then changed the latter back into metal by first removing something and then putting it back again, a light dawned on all those who study nature. They comprehended that reason has insight only into what it itself produces according to its own design; that it must take the lead with principles for its judgments according to constant laws and compel nature to answer its questions. [...] Reason, in order to be taught by nature, must approach nature with its principles in one hand [...] and, in the other hand, the experiments thought out in accordance with these principles [...][53]

In the time of Kant or Eddington, the constructive features of the experimental method had to be advanced against empiricism. Today, it seems to be the other way round: the empirical features of experimental results have to be advanced against several variants of constructivism. Good witnesses are independent. To the questions put in good experiments, nature does not answer in an unstable or arbitrary way. Experiments exhibit law-like regularities which are *not* thoroughly at the experimenter's disposal. Like an incorrupt witness, the experiments may resist our theoretical preoccupations. This did indeed happen in subatomic physics when the classical construal of experimental results failed.

---

[52]The empiricist features of his *constructive empiricism* are less convincing; see Sect. 2.6.

[53]Kant 1787, B XIII.

2.4 Phenomena and Their Causes

It is crucial for Galilean or Newtonian science to ask for the causes of phenomena. However, phenomena do not reduce to what is observable in a strict empiricist sense, and neither are their causes observable. The experimental method does more than filling the blanks in theoretical descriptions. It aims at causal analysis. According to van Fraassen, the causes of the phenomena do *not* belong to the empirical domain of a physical theory but to the realm of the unobservable. Thus, in the 'empiricism' part, his position of *constructive empiricism* also falls short of the structure and methods of modern physics. Some of his arguments against causal realism are substantial. They will be taken up later.[54] But first, it has to be clarified what the phenomena of modern physics are, what their causes are, and what can be said in favor of the reality of the latter *against* empiricism and constructivism.

In current philosophy of science it is controversial what exactly the *phenomena* of modern physics may be. Indeed, views about phenomena depend on the respective position in the debate on scientific realism. For scientific realists, phenomena are the matters of fact in nature explained by physical theories. According to this view, which was defended by Bogen and Woodward in an influential paper,[55] phenomena are what physicists call *effects*: the Einstein–de Haas effect, the Bohm–Aharanov effect, the quantum Hall effect, etc. For empiricists like van Fraassen, the phenomena of physics are the appearances, that is, what can be observed or perceived by sensory experience. For constructivists, the phenomena of physics will be the structures generated by experiments and mathematical methods. It is indeed impossible to define an unambiguous concept of the phenomena of modern physics in accordance with physical practice. What phenomena are depends on the theoretical and experimental context. Even Newton's use of the concept is not stable. Sometimes he considers phenomena to be what is observable, whereas the phenomena of book I and III of the *Principia* are the planetary motions described by Kepler's laws.[56]

In his book *Representing and Intervening*, Hacking emphasizes that phenomena in the sense of modern physics occur quite rarely in nature.[57] Phenomena are regular. They occur according to law-like rules. They are opposed to the miraculous. They exhibit reliable laws of nature. They are standardized and can be expressed in terms of physical quantities. In all domains except astronomy, they have to be generated in experiment. In experiments, they may be half discovered and half manufactured. Only in astronomy and modern astrophysics did they simply occur. In any case, they are based on mathematical idealization. It has to be concluded that most (if not all) phenomena of physics result from mathematical idealization and the experimental method.

---

[54]Van Fraassen 1980, 17. See Sect. 2.6.
[55]Bogen and Woodward 1988.
[56]Newton 1689, 797–801. See also Matthews 2003.
[57]Hacking 1983, 227–232.

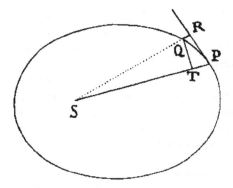

Fig. 2.2. The phenomenon of Keplerian motion (Newton 1687, 467)

Prior to Galileo's experimental method, the phenomena of natural science were confined to astronomical observations. In Ptolemy's or Kepler's laws they were subject to mathematical idealization. With Galileo and Newton, the mathematization of the phenomena reached a new level. For Galileo, phenomena were the standardized results of his experiments. For Newton, the mechanical phenomena which he wanted to explain were indeed phenomenological laws, namely Kepler's laws and Galileo's law of free fall. Similarly, the phenomena of his optical experiments resulted from geometrical models of light rays. Thus, from the very beginning of modern science the term 'phenomenon' was used in such a way that it meant a *theory-laden* mathematical description of what had been called phenomena *before*. The phenomena of modern physics are theory-laden empirical structures.

Scientific realism and any moderate empiricist position agree that theory-laden empirical structures belong to empirical reality. The empirically accessible parts of physical reality embrace more than the raw sense data admitted by strict empiricism. The empirical basis of modern physics consists neither of observable phenomena in a strict sense nor of empirical data given in terms of pointer values of measuring devices. Whereas phenomena are empirical structures, experimental data are numbers. Measurement gives numerical values of physical quantities. The numerical data are obtained from reading out the measuring devices. They result in mathematical data models, i.e., mathematical minimum structures which fit the data. Data models are empirical structures, too. To a certain extent, the distinction between phenomena and data is blurred in modern physics.[58]

But the experimental method goes one crucial step further. It generates standardized phenomena in order to investigate their causes. Varying the conditions of a given experiment aims at recombining and reconstructing independent causal agents. Newton's decomposition and composition of white light by means of a prism aimed at *qualitative* causal analysis.[59] Varying

---

[58]See Matthews 2003 and my remarks above about the various ways of understanding the phenomena of physics.

[59]See end of Sect. 2.2.

the physical quantities and analyzing their functional dependencies aims at *quantitative* causal analysis. Galileo studied the free fall under two different kinds of constraint, namely with the inclined plane and the pendulum, varying the inclination of the plane and measuring the rolling times of the ball, or the amplitude of the pendulum. Newton varied the arrangement of rays of sunlight and prisms in order to study the invariant features of the resulting spectra and to measure the refraction coefficient of light rays of different colors. Modern high energy physicists vary the energy of the incoming particle beam and measure the cross-section of a certain kind of scattering with several kinds of particle detectors.

To measure the quantities means recording their numerical values. In modern experiments the recording is called *data taking*. The data are values of length, time, mass, charge, electric current, voltage, electric or magnetic field strength, temperature, etc., in a given system of units. In former times the data were protocolled in a logbook. Today they are recorded by means of electronic devices which are read out and analyzed by computers. The data of modern physics are also theory-laden empirical structures. They are fed into data models[60] and analyzed in order to determine the functional dependencies between physical quantities. Usually these functional dependencies are expressed in terms of differential equations, as the mathematical correlate of causal explanations.

Regarding the results of causal analysis, van Fraassen's empiricism and Galilean science split. Galileo or Newton explain phenomena in terms of causes. Newton's first *rule of reasoning* demands a search for the "true causes" of the phenomena. Even though he identifies them primarily as physical quantities rather than unobservable causal agents,[61] a tendency towards causal realism is certainly inherent in modern physics. Newton excluded atoms, the ether, and light particles or waves only because for him their physical properties were beyond experimental reach. Today, the physical quantities of subatomic particles are measured, as well as the properties of the vacuum states of a quantum field (which are currently the best candidates for *some* kind of physical ether). Most particle physicists would surely defend causal realism with respect to subatomic particles. In complete contrast, van Fraassen requires them to stay agnostic about the unobservable causes of observable phenomena or empirical structures. According to him, the causal agents of modern physics are metaphysical. Unobservable entities such as forces, atoms or subatomic particles are hypothetical in the sense of being fictitious. For him, to draw conclusions about unobservable causes means exceeding empirical reality.

Hacking has an opposite view of the causal agents of modern physics. He shares van Fraassen's or Tetens' constructivist views of the experimental method but he defends causal realism. In his view, causal realism is sup-

---

[60]Suppes 1962.

[61]Newton 1729, 398; Falkenburg and Ihmig 2004. See Sect. 2.2.

ported rather than rejected by the constructive features of experimentation. According to him, the intentional or teleological features of the experimental method give rise to the following reality criterion: If it is possible to use unobservable causal agents as a regular tool in a successful experiment, they must exist:

> We are completely convinced of the reality of electrons when we regularly set out to build – and often enough succeed in building – new kinds of device that use various well-understood causal properties of electrons to interfere in other more hypothetical parts of nature.[62]

He demonstrates his reality criterion by discussing a measurement of parity violation which was made in 1978. Since an electron gun constituted an essential part of the experimental device, the conception, the performance, and the data analysis of the measurement presupposed the existence of electrons.[63] Here, Hacking grasps an essential feature of the way in which physicists extend their measurement theories. If a theory about a certain kind of entities or causal agents is held to be well-confirmed and accepted, it becomes part of well-known background knowledge and is no longer subject to experimental tests. Now it may give rise to new measurement theories and to new technologies used for the experimental testing of other theories. In this way, the background knowledge about electrons entered into the construction of electron guns and particle accelerators. From an instrumentalist point of view, the way in which an electron gun works cannot be explained. The effects of the pulsed electron beam in the measurement of parity violation discussed by Hacking can only be understood if one assumes that linearly polarized electron pulses exist and cause the observed left–right asymmetry in the scattering events.

Hacking's argument advances the physicist's pragmatic intuition that entities which you can manipulate in order to produce some desired effect must be real. It may be that their physical properties and their structure are not thoroughly known, but they serve a given purpose and they must therefore exist. To deny their existence means to claim that you can knock a nail into the wall with a non-existent hammer. This philosophical *hammer-and-nail argument* is indeed close to the reality criteria used within the scientific community of physics itself. In 1939, Eddington made the same point about neutrinos. When he wrote his book *The Philosophy of Physical Science*, the neutrino hypothesis served to guarantee energy conservation in β-decays, but there was no evidence for neutrinos. He emphasized that he would not take them seriously before they could be generated in experiment and used as a technological tool:

> I am not much impressed by the neutrino theory. In an ordinary way I might say that I do not believe in neutrinos. [...] Dare I say

---

[62] Hacking 1983, 265.
[63] Loc. cit., 266–273.

that experimental physicists will not have sufficient ingenuity to make neutrinos? [...] If they succeed in making neutrinos, perhaps even in developing industrial applications of them, I suppose I shall have to believe – though I may feel that they have not been playing quite fair.[64]

Neutrinos were first detected indirectly in 1956, in inverse β-decay. Since their discovery and manufacture they have belonged to the toolbox of high energy physics. Today, they are on the same footing as electron beams. Several neutrino scattering experiments have investigated the structure of the proton and neutron.[65] In Eddington's and Hacking's reality criterion three principles come together:

1. the principle of *causality*, according to which an observable effect must have a cause;
2. a principle of *adequacy*, according to which we must have adequate knowledge of this cause in order to use it as a tool; and
3. an *intentional* principle, according to which technological intentions aim at realizing human purposes by means of given tools.

In order to make an experiment succeed, none of these three principles must be violated. Without causality, no effect would be generated at all. Without adequate knowledge of the cause, no effect could be got under control. Without the tool being given, no goal of a technological intention would be realized at all. All three principles show that the instrumental character of physical theories is crudely misunderstood by holding that the prognostic value of a theory does not imply the existence of corresponding causal agents. For the technological applications of a theory, quite the contrary is true.[66]

The causal aspect of physical theories has also been emphasized by Nancy Cartwright. Following Mill's inductive logic, she suggests adding causal modelling to an empiricist view of science.[67] According to her, the causal agents of physics are real dispositions. They are *nature's capacities* to evoke cer-

---

[64]Eddington 1939, 112.

[65]See any recent book on high energy physics, e.g., Perkins 2000, and Sect. 4.4. See also Franklin 2001.

[66]Carrier 1991, 1993 shows, however, that the prognostic potential of a theory does not sufficiently constrain the *kinds* of causal agents involved. You may explain certain caloric phenomena with the assumption of phlogiston in place of molecular motion. And you may drive a nail into the wall with a shoe in place of a hammer.

[67]Cartwright 1989, 163–182, following Mill 1872. A related approach is Mackie 1980. It differs from Cartwright's views in one crucial point. To incorporate causality into empiricism, a non-nomological account of causality is needed, such as Hume's regularity view or Mackie's conditional analysis. In an anti-Humean thrust, Cartwright (1989, 25) emphasizes that Mill's (and her own) views do not reduce to a mere regularity theory of causation. See also Falkenburg and Schnepf 1998.

tain effects under given circumstances.[68] Cartwright's empiricist position is close to causal realism. She suggests explaining physical objects, forces, fields, etc., in terms of their causal powers. Nature's causal powers give rise to the observable phenomena and experimental results of physics. Thus, an electromagnetic wave is the capacity to propagate a signal with the speed of light; subatomic particles are the capacities to make black spots on a photographic plate or particle tracks in a bubble chamber, etc. Exactly what kind of entity such a capacity is and which physical laws it obeys are left open.

Causal realism emphasizes the *actual* features of empirical reality that Kant emphasized. The unobservable causes of observable effects do not belong to empirical reality in a strict empiricist sense. But if a causal story connects them with empirical observations, they belong to empirical reality in a wider sense. The criterion reminds us of Kant's postulates of empirical thinking. However, it is more general. Kant demands a series of possible sensory perceptions[69] but we can have no sensory perception of subatomic causes. Nevertheless, the unobservable causes of the phenomena belong to actuality or to *contingent* reality in a quite literal sense. They *act*, and they do so neither due to our intentions, nor due to our theories. A cautious, Kantian causal realist should conceive of them in terms of entities of which we have limited empirical knowledge. Here, this empirical knowledge has been spelled out in the spirit of Kant's postulates of empirical thinking. Indeed, it comes in terms of a generalized account of observation not yet developed by Kant.

2.5 Observation Generalized

From its very beginnings, Galilean science was based on observation instruments such as the telescope and the microscope. In the recent debates on scientific realism it is still debated whether they make empirical reality accessible to us at a new scale. Against recent empiricist or constructivist views in the Aristotelian tradition, generalized accounts of observation have been developed. Like Hacking's hammer-and-nail argument discussed in the preceding section, they come together with causal realism. Experiments take data, and observational devices take pictures. Performing an experiment is intervening. Taking a picture at a new scale is representing. Intervening makes use of real objects which function as the tools of data taking. Representing makes analogous use of real objects which cause the pictures taken by a microscope or telescope.

---

[68] Cartwright 1989. Hüttemann 1997, 145–151, also defends the thesis that physics aims to describe dispositions.

[69] See Kant 1781/87, A 225–226/B 273: "[...] with the guidance of the analogies we can get from our actual perceptions to the thing in the series of possible perceptions. Thus we cognise the existence of a magnetic matter penetrating all bodies from the perception of attracted iron filings [...]."

In 1962, Grover Maxwell argued that observation instruments make unobservable objects, or "theoretical entities", visible.[70] According to him, it is crucial that they open a continuum of domains. The available observation instruments give us access to a spectrum of objects at all scales. Binoculars, telescopes and radio telescopes make objects visible at larger and larger distances. Similarly, glasses, magnifiers, microscopes, electron microscopes, cloud chambers, and particle accelerators of increasing energy make smaller and smaller objects visible. Due to this continuum of observation devices, the existence of physical objects of all scales from remote galaxies to subatomic particles is granted. Against this kind of reasoning, van Fraassen has objected that any inference from the existence of observation devices to the existence of observed objects is a non sequitur.[71] He contrasts the vapour trail of an airplane in the sky with the particle track in a cloud chamber. We are able to see the airplane which causes the vapour trail at the airport, and we are able to follow its trace with the naked eye. In contradistinction to the airplane, we are in principle not able to see an electron with the naked eye. He emphasizes the distinction between "detection" and "observation":[72]

> So while the particle is detected by means of the cloud chamber, and the detection is based on observation, it is clearly not a case of the particle's being observed.

This criticism is based on van Fraassen's empiricism. What we cannot perceive directly in any way is not subject to observation. This may be too narrow a concept of observation. But in the given example, quantum theory supports van Fraassen's criticism. An electron is no object in the usual, macroscopic sense. Its track is no classical trajectory. Closer analysis shows indeed that it does not *cause* a particle track in the usual sense according to which cause and effect are separable.[73] Thus, Maxwell's generalized concept of observed objects belongs to a naive, classical realism which cannot come to grips with the structure of the quantum domain.

In 1982, Dudley Shapere proposed a more sophisticated account of generalized observation. His account of observation does not focus on the observation devices but on the theories involved in their use. In focusing on the laws of physics involved in an observation, it is in Max Planck's spirit of physical universalism. It aims at liberating our understanding of observation from anthropomorphic conceptions. Consequently, it is based on a naturalistic epistemology which considers the eye to be an observation instrument and which explains any human perception as a result of physical and neurophysiological processes.

---

[70]Maxwell 1962, 7.

[71]Van Fraassen 1980, 14–18.

[72]Van Fraassen 1980, 17.

[73]See Sect. 6.6.

Shapere suggests disentangling the use of the term 'observation' in empirical science from the traditional epistemic concept of observation, which is tied to perception. Surely a concept of observation based on scientific practice and cut off from perception will disappoint philosophers in important respects. But such a concept is perfectly suited to take into account the theory dependence of scientific experience. It helps to determine in which cases such theory-ladenness of scientific experience is fatal. This is precisely Shapere's main goal when he gives the following criterion for direct observation (observability) of an entity x:[74]

Shapere's Criterion: x is directly observed (observable) if:
- information is received (can be received) by an appropriate receptor; and
- this information is (can be) transmitted directly, i.e., without interference, to the receptor from the entity x (which is the source of the information).

The criterion is illustrated by an example from recent astroparticle physics. Shapere analyzes the first experiments which measured the neutrino flux of the sun.[75] According to the current theory of solar physics, the neutrinos leave the sun with a given probability and propagate to the earth, where they may induce with a given probability processes of inverse β-decay in a suitable liquid like perchlorethylene. All these probabilities are precisely calculable from the current theories of the sun and the weak interaction. Thus the detection of the solar neutrinos and the measurement of their relative frequencies, the solar neutrino flux, should make it possible to investigate the processes which occur inside the sun. The measurement of the neutrino flux is an experimental test of the physical theory of the sun. (By the way, neutrinos are used here as tools of an experimental investigation *without* being manufactured in Eddington's sense.)

According to Shapere's criterion, the solar neutrinos serve the direct observation of the interior of the sun. As far as the neutrinos have practically no interactions between the sun and the neutrino detector below the earth's surface, they satisfy condition (2). In contradistinction to light (photons) or massive particles, they have very few interactions. Therefore, they get from the source of information, the interior of the sun, to the receptor, the neutrino detector which is an underground perchlorethylene tank,[76] without being disturbed.

The resulting concept of observation generalizes the corresponding empiricist concept held, e.g., by van Fraassen. The generalization comes in two

---

[74]Shapere 1982, 492.

[75]... in the days when the possibility of neutrino oscillations was still a vague theoretical speculation.

[76]However, in 1999 the Kamiokande experiment showed that between sun and earth a certain percentage of the solar neutrinos is definitely lost, due to neutrino oscillations. See below and Perkins 2000, 294.

steps: first, by disregarding the specific mind-dependent features of sensory perception (i.e., the distinction between primary and secondary qualities, the nature of ideas, of consciousness, etc.); and secondly, by drawing an analogy between the physiological process underlying an act of perception, and the transmission of information. According to this analogy, the eye is only one kind of information receptor among many, and light is only one medium of transmission among many. We may take a reliable picture of our surroundings by means of a camera (and we know that the eye functions like a camera). Analogously, the solar neutrino flux may give us a reliable picture of the interior of the sun. Similarly, an electron microscope takes a reliable picture of a microscopic structure. And a particle beam takes a reliable picture of subatomic structures.[77] In all these cases, it is not sufficient to know how the receptor works. When Maxwell claimed that there is a continuum of observation, he overlooked the fact that there is an additional condition: the transmission of information must be reliable. Shapere's condition (2) that the information be transmitted directly aims to guarantee this requirement. Let me call this requirement the *reliability condition* and take a closer look at it.

Shapere's criterion aims at giving a *sufficient* condition for direct observation, or observability, of an entity x. According to this, three theories may be involved in the analysis of how to directly observe an entity x: "the theory of the source, the theory of transmission and the theory of the receptor."[78] But it implies a *necessary* condition for what makes an observation *direct*: the act of transmission must be direct, that is, *without loss of information*. [In the quotation above, Shapere says: "without interference", indeed taking into account the fact that neutrino oscillations or other ways in which quantum states interfere may violate his condition (2). Since we need a general definition of 'direct transmission', I prefer 'without loss of information'.] This necessary condition relies on the theory of transmission of information. Thus, the reliability condition for direct observation in Shapere's sense is:

> **Reliability Condition:** The observation of an entity x is *direct* only if the transmission from x to a receptor of information *preserves* the original information.

The reliability of some kind of transmission of information obviously depends on our knowledge of the physical laws which describe a concrete act of transmission. Our assumptions about these laws are only reliable if they belong to the body of well-confirmed physical background knowledge. In the course of de-anthropomorphizing our knowledge, the observation of some entity may be generalized in such a way that it is uncoupled from sensory perception. The gain is scientific objectivity. The price is the theory-ladenness of what counts as direct observation. Here, Shapere's distinction of "background information" and "what is uncertain" becomes crucial. An observation generalized

---

[77]See Chap. 5.
[78]Shapere 1982, 492.

in Shapere's sense counts as direct only if safe background knowledge tells us that the transmission of information is reliable. Indeed, data analysis in experimental physics is usually based on the distinction between safe background knowledge and uncertain theories, and much effort is expended in making the distinction itself as safe as possible. However, *if* the transmission process becomes uncertain, our information about the source of information is crucially affected. Then Shapere's criteria for direct observability are no longer fulfilled.

To take Shapere's own example: the propagation of the neutrinos, which are emitted from the center of the sun, is a process in which the neutrinos either oscillate or do not. If they oscillate significantly, the reliability condition is violated. If we suspect that they do (but we do not know whether they do or not) our theory of transmission is uncertain. In this case, Shapere's reliability condition is violated.

Indeed, in the 1990s it turned out that the measured solar neutrino flux was much lower than expected. It was unclear whether the theory of the sun or the neutrino theory was wrong. In 1999, the existence of neutrino oscillations was confirmed.[79] Now it was clear that the information had *not* been transmitted directly, due to quantum interferences of two kinds of neutrinos. The moral is that the theory of transmission involved in Shapere's analysis had *not* been based on safe background knowledge.

Thus, we see that Shapere's criterion for direct observation relies on epistemic grounds. It depends on the reliability of our knowledge of the transmission of information. It yields no sharp distinction between *direct observation* and its opposite, that is, *inference based on observation* – Shapere's expression for what an experimenter would call 'indirect evidence'. The distinction depends on our background knowledge. The line between what counts as an observation or as an inference is shifting with the development of physics. Shapere emphasizes that

> specification of what counts as directly observed (observable), and therefore of what counts as an observation, is a function of the current state of physical knowledge.[80]

However, the dividing line between direct observation and theoretical inference cannot be shifted arbitrarily in such a way that, finally, *any* experimental result may count as a direct observation of some entity x. To treat the distinction of observation and non-observation as completely arbitrary and vague (as did, for example, van Fraassen for purely epistemic reasons, when

---

[79]See Perkins 2000, 294.

[80]See Shapere 1982, 492. See also 517: "The *epistemically* important line between the non-inferential and the inferential is drawn in terms of the distinction between that which we have specific reasons to doubt [...] and that upon which we can build confidently."

he criticized Maxwell's famous defence of theoretical entities[81]), is therefore not convincing.

The above considerations suggest the question: are there any further necessary conditions for the direct observation of an entity x, on the basis of Shapere's analysis? And, might these conditions help us to determine what has been observed in a generalized sense and based on well-confirmed background knowledge, in accordance with the reliability condition above? Indeed, this becomes possible if we focus on measurement and translate Shapere's criterion of observation into the language of physical quantities as follows:

1. the *source x* is a *physical system* which is characterized by certain physical properties or quantities;
2. the *receptor* is a *measuring device*;
3. the *transmission* is a *sequence of interactions* which results in a measurement;
4. the *information* consists in *measured values of physical quantities*;
5. the measured values of physical quantities can be reliably attributed to the physical system under investigation.

In terms of physical quantities, a direct observation in Shapere's sense is a measurement of values of physical quantities which can be unambiguously attributed to a physical system. The transmission of information is a sequence of interactions which results in a measurement of the physical properties of the system under investigation. In this way, observation is generalized to being a kind of measurement.

In the language of physical quantities the reliability condition of observation must be expressed in two parts. (i) A *general description* of a physical system and its interaction with a measuring device must be given in terms of physical quantities. (ii) It must be granted that these quantities, as they were measured in an actual experiment with a definite numerical result, can be unambiguously attributed to an actual *individual system*. Together, these conditions should guarantee that the measured values of physical quantities correspond to the physical properties of an individual system. Obviously, these conditions are closely related to the possibility of *constituting physical objects* in a Kantian sense and to the problems of constituting objects in the *quantum domain* which will be taken up in later chapters.[82] Here, only one aspect of these problems needs to be emphasized. Condition (ii) is closely related to *causal realism*. It requires that the outcome of a measurement can be traced back to an individual system which causes a certain effect in the measuring device. Therefore, I suggest finally the following general (necessary and sufficient) criterion for the observation of a physical system:

---

[81]See van Fraassen 1980, 14–19, and Maxwell 1962.
[82]See Chaps. 5–8, and Mittelstaedt 1998b.

General Observation Criterion: A physical system x is observed if and only if

- the measurement results correspond to values of certain characteristic physical quantities of the system x, and
- an *individual causal story* can be told which shows that the measured quantities can indeed be attributed to an individual system located in a certain spacetime region, accessible to the actual measuring device.

This general criterion no longer distinguishes between 'observation' and 'direct observation'. When observation is generalized to measurement, the crucial issue is the question of whether it is possible to trace the measurement results back to an individual system which counts as their cause. This causal story may be as indirect as our most highly sophisticated measurement theories admit, as long as these measurement theories rest on well-confirmed, safe background knowledge; and as long as they admit of singling out an individual system to which certain measured values of physical quantities can be unambiguously attributed.

In subsequent chapters it will be shown in detail to what extent this is the case for the particles of current particle physics. As we shall see, the subatomic matter constituents investigated in the scattering experiments of high energy physics can indeed be observed in this sense – whereas an electron which makes a particle track *cannot* be observed as the individual cause of a track. Thus, van Fraassen's empiricist's claim that the particle track does not give rise to the observation of an electron is right. However, van Fraassen argues in favor of it for the wrong reasons. We shall see that the general criterion given above does *not* support the analogous empiricist claim about the observation of an individual quark system in a jet event![83]

2.6 The Empirical Reality of Physics

What is, after all, the empirical reality of physics, and what is empirical reality in the subatomic domain? In this chapter I have sketched how Galilean science constitutes scientific objectivity. It does so by *going beyond* experience in an empiricist sense, namely sensory perception, in several ways. Max Planck emphasized the achievements of this emancipation from the anthropomorphic elements in our view of reality.[84] The very scientific methods of modern physics are at odds with an Aristotelian view of empirical reality. Therefore, the less one knows about the interplay of theoretical background knowledge and technological devices, the more one is inclined to defend a simplified view of what may count as empirical in view of modern science. Since the

---

[83] See Sect. 6.5.2 and Falkenburg 2000a.
[84] See Planck 1965, 31, and Sect. 2.1.

subject matter is very complex, such a simplified view belongs to what Hegel called *lazy reason*. Current physics is at odds with *any* account of empirical reality which is *not* based on detailed physical knowledge. And what counts as empirical evidence in modern physics changes with time. It depends on well-confirmed background knowledge and on highly developed technologies. Due to theoretical and experimental progress, the empirical basis of physical theories becomes more and more theory-laden and technology-laden.

In Sect. 1.3, I suggested distinguishing between the real and the actual, and taking the actuality of contingent events as an independent epistemic criterion of empirical reality:

> Closest to the criterion of happening contingently and not being at our disposal is the notion of an *event*. An event is something that occurs locally in such a way that it marks an observable difference from its environment: a light flash, a 'click' in a Geiger counter, a spot on a photographic plate.[85]

Such events are indeed the empirical basis of modern particle physics. An event is the empirical basis of a position measurement; a sequence of adjacent events is a particle track; a coincidence of two or more particle tracks is a scattering event; and a collection of many scattering events goes into the measurement of a cross-section, the quantity in which the scattering experiments of high energy physics meet quantum field theory. The increasing theory-ladenness of these measurements is investigated in detail in the next chapter.

To the contingent events one must add contingent measurement results, data, and phenomena. Whatever is not at the theoretician's and the experimenter's disposal actually belongs to empirical reality. In particular, the transition from classical mechanics and electrodynamics to a series of quantum theories was *not* at the physicist's disposal. In view of Planck's law of black-body radiation, Rutherford's scattering experiment, and the structure of atomic spectra, it was no longer possible to maintain a classical construal of empirical reality. The empirical structures of subatomic physics are *quantum* structures. Particle tracks are *not* actually caused by *classical* particles. For example, the classical model of the energy loss of charged particles in matter fails.[86] The generation of particle tracks can only be explained in terms of quantum processes.

Identifying empirical reality with the actual and taking the contingent events, phenomena, and data as the empirical evidence for it, we may draw some preliminary general conclusions concerning the debate on scientific realism. (More detailed conclusions can only be given at the end of the book.) Quantum theory makes things a little bit easier. The usual probabilistic interpretation of quantum theory rules out the extreme positions of the spec-

---

[85] See Sects. 1.3 and 3.3.
[86] See Sect. 5.3.

trum sketched above in Sect. 1.2, namely global realism and constructivism. Both positions struggle with the transition from classical physics to quantum physics. Global realism is too strong. It is indebted to classical realism. It tells us that electrons exist as objectively as stones or trees. Heisenberg's indeterminacy relations tell us that they do *not*. Electrons are not objects in a classical sense.[87] Constructivism is the other extreme. It is too weak. It tells us that electrons do not belong to nature like stones or trees but to the artifacts of a physics laboratory. Nevertheless, photographs taken from cosmic rays or from a particle detector when the beam is switched off demonstrate that electrons also make particle tracks without human interaction. Their tracks are as real and objective as a thing or person in a photo. (Do we *manufacture* a thing or a person we photograph?) And their weird quantum properties are not manufactured. They are not at the honest experimenter's disposal, they are contingent matters of fact.

What about the other positions? The options of critical realism, moderate empiricism and strict empiricism are left open. Critical realism comes in terms of laws of nature (structural realism) and/or causal agents (causal realism). In view of quantum structures, structural realism seems tenable whereas causal realism becomes highly problematic. If we maintain a generalized concept of observation like the one developed in Sect. 2.5, we cannot dispense with causal realism in search of subatomic reality. Therefore, we have to carefully investigate the possibilities of individuating subatomic causes and constituting subatomic objects. Structural realism comes in terms of abstract mathematical structures. They need a physical interpretation which comes in terms of physical quantities. Therefore, structural realism cannot be sharply distinguished from modest empiricism. The latter restricts our empirical knowledge of physical reality to measurement results, that is, to values of physical quantities. Since they in turn make up certain algebraic structures (such as a Boolean or a non-Boolean lattice), moderate empiricism collapses into structural realism as soon as the algebra of physical quantities is admitted. In addition, the measurement results may be highly theory-laden and technology-laden, as long as they are based on safe background knowledge.

Strict empiricism explains empirical reality (respectively our knowledge of it) exclusively in terms of observables in a literal sense, that is, in terms of what *we* observe. In quantum physics we observe the following: at a scintillator screen, light flashes occur at given angles with a relative frequency that corresponds to the Rutherford scattering cross-section; particle tracks appear

---

[87]For electrons and other subatomic particles, classical realism commits us to untestable ontological speculations such as hidden parameters or many worlds. Bohmian hidden parameters are at odds with relativistic quantum field theory, whereas an Everett-type many-worlds interpretation of quantum theory declares all *possible* measurement results for *actual*, even though only one of them is actual *for us*. Both interpretations of quantum theory attempt to re-establish what Putnam calls the *God's Eye View* of empirical reality; see Putnam 1990, 10–14.

on bubble chamber photographs; interference fringes are photographed behind a crystal; the famous diffraction pattern of the double-slit experiment is detected by a particle detector that runs a long time and counts many individual detections over the whole screen area; and so on. At first glance, quantum structures look like grist for the mills of strict empiricism. Nevertheless, quantum physics is no more compatible with strict empiricism than any other branch of modern physics. It demands at least a version of *moderate* empiricism admitting of quantitative measurement results, mathematical data models fitting the data, and the use of mathematical statistics.[88] Any traditional (Aristotelian) or recent version of empiricism is untenable in view of Galilean science.

This becomes obvious in van Fraassen's constructive empiricism. It admits use of the experimental method but it restricts empirical reality to the observable phenomena of experiments. For van Fraassen, causal explanations and observations based on theoretical background knowledge do not contribute to our empirical knowledge. They have the status of metaphysical hypotheses. He claims that "what is observable is a theory-independent question."[89] To what extent does his empiricism admit of mathematical idealization? It includes values of physical quantities.[90] Once measurement results are involved, however, measurement theories are employed and the crucial step is made from mere observations toward theoretical background knowledge. In quantum physics this step seems innocent. The empirical content of a quantum theory lies in the prediction of measurement results which give rise to the algebraic structure of quantum observables. For van Fraassen, this algebraic structure is empirical.[91] But the expression 'observable' is misleading. The quantum observables are given in terms of expectation values of physical quantities. (Formally, the observables correspond to operators in Hilbert space. Empirically, they correspond to relative frequencies of measurement outcomes.)

Due to this uncontroversial view of the empirical content of a quantum theory, van Fraassen makes no distinction between observable quantum phenomena and the algebraic structure of quantum observables. At this point, his constructive empiricism merges with structural realism. For him, the empirical substructure of a theory is crucial. It lies in "structures definable from measurement data", and it is based on spatio-temporal quantities.[92] However, in quantum physics the step of embedding isolated empirical data into an empirical structure is *not* innocent. Think of a particle track in a cloud chamber. The quantum phenomenon is a sequence of observed droplets. This sequence is discontinuous. It consists of single subsequent position measure-

---

[88]See Suppes 1962.

[89]Van Fraassen 1980, 57.

[90]See van Fraassen 1991, 42.

[91]See van Fraassen 1980, 64–68.

[92]Van Fraassen 1980, 59–60.

ments. The track is the continuous path which connects the measurement results. The empirical structure of the track is a spatio-temporal trajectory. But the quantum mechanics of the track is at odds with the existence of a classical trajectory.[93] The trajectory is a classical construal. Van Fraassen does not accept that one can meaningfully talk about the observation of a particle in view of the track, whereas he accepts that one speak of the track in view of the discrete position measurements. He is not aware that, according to his own empiricist principles, this is also metaphysics.

Strict empiricism is heroic. It has to dispense with constructing particle tracks from the observed droplets in the cloud chamber or from the spots on the bubble chamber photograph. In other words, it must be so strict that it avoids making data models out of data and measurement results out of phenomena. But this means missing *any* substructure of a quantum theory. In focusing on algebraic quantum structures, van Fraassen's account of the empirical substructure of a theory turns out to be less heroic. Indeed, in the age of quantum computers and nanotechnology an empiricist world view on the lines of Mach or Aristotle is definitely obsolete.

One conclusion is inevitable: *all* philosophical positions in the current debate on scientific realism fall short of the quantum structure of subatomic physics. Global realism and constructivism are ruled out by quantum theory. Strict empiricism is at odds with the methods of Galilean science. The remaining positions are indebted to classical realism in one way or another. Causal realism is based on a classical account of causes, structural realism and moderate empiricism presuppose the classical scales of physical quantities. However, subatomic reality is *not* classical, it exhibits typical quantum phenomena such as the famous double-slit experiment with a low intensity laser or electron beam.

Thus, the crucial question is: in terms of which concepts can we conceive of subatomic empirical reality? This is a Kantian question, even though quantum structures compel us to weaken Kant's original *a priori*. Niels Bohr always insisted that classical concepts remain indispensable for interpreting the observed quantum phenomena. He has been proved right up to now with regard to the physical concepts involved in measurement. To make the subatomic domain empirically accessible, the classical scales of length, time, mass and energy are needed. Any experiment in the quantum domain makes use of macroscopic measuring devices and quasi-classical measurement theories. The reality of the subatomic domain can only be explained in terms of the familiar classical physical quantities, and in terms of mathematical structures such as Hilbert spaces or Lie groups. As far as these mathematical structures do not correspond to familiar classical models,[94] the empirical quantum structures teach how the classical construal of reality breaks down

---

[93]See Falkenburg 1996 and Sect. 5.3.

[94]In the sense of a generalized version of Bohr's principle of correspondence; see Heisenberg 1930a,b and Sect. 5.4.2.

in the subatomic domain. It will turn out in the subsequent chapters of the book that it is only possible to reconstruct the way in which quantum structures refer to physical *objects* as far as they correspond to *some* classical model.

To what extent is it justified to speak about *subatomic reality* at all? In doing so, certain assumptions of *causal* realism are involved. Subatomic reality is the kind of reality which exists at a small scale, at a length scale of about 10^{-13} cm and below; and it causes quantum phenomena and quantum structures. Quantum theory is a consistent theoretical construal of these phenomena and structures. It shows that subatomic reality does *not* fit in with the strong assumptions of classical realism. I will explain this subatomic reality from a 'middle' position which is minimum in view of the methods of Galilean science. It is based on moderate empiricism, and associated with a *weakened version of classical realism*. It can be spelled out in terms of physical properties, as explained in Sect. 1.6. This starting point avoids metaphysical *a priori* assumptions about the subatomic carriers of physical properties, namely subatomic particles. The only classical elements involved in the subsequent investigations are classical and semi-classical measurement theories. These measurement theories are based on constructing the scales of physical quantities quasi-classically, that is, *as if* there were physical objects with a given size, mass, charge, momentum and energy. (Indeed, today the scales are constructed far beyond experimental accessibility, namely down to the Planck scale.) However, one must ask to what extent we can attribute these measured quantities to *subatomic particles* if we conceive of subatomic particles in terms of *causal agents* which underlie the quantum phenomena.

The answer to this crucial question should tell us what causes a particle track and what the subatomic constituents of matter and light may be. We will see that, compared with the uniform grand world view of classical realism, the details of the answer are complicated and disillusioning. In order to understand the relation between subatomic reality and the phenomena of particle physics, the detailed investigations of Chaps. 4–6 are indispensable. What remains of subatomic *particles* will be summarized in a preliminary way at the end of Chap. 6. In order to explain what finally remains of subatomic *reality*, however, the additional investigation of wave–particle duality in Chap. 7 is needed.

3 Particle Observation and Measurement

In order to understand the empirical basis of current particle physics, the actual structure of the subatomic domain has to be reconstructed step by step. The task is tedious. Several structural layers of physical phenomena have been discovered and investigated during a century of experimental particle physics. Some of these layers are classical, others are not, and others have a bridging function. They must be analyzed carefully. It must be made transparent step by step what physicists themselves consider to be the empirical basis for current knowledge of particle physics. And it must be made transparent what they mean in detail when they talk about subatomic *particles* and *fields*. The continued use of these terms in quantum physics gives rise to serious semantic problems. Modern particle physics is indeed the hardest case for incommensurability in Kuhn's sense.[1] Let us now study the clever ways in which physicists overcome these problems, creating the semantics of quantum physics.

In this chapter the observational basis of particle physics will be considered in detail. It has a layered structure. It consists of empirical observations, experimental data, measured values of physical quantities, and phenomenological laws. At all levels the interpretation of these ingredients is far from obvious. Even for experimenters it is complicated to figure out what belongs to the contingent structures of subatomic physics. From a philosophical point of view, things are even more complicated. The dividing line between actual measurement results, experimental phenomena, phenomenological laws and observations on the one hand and theory construction on the other hand is shifting with time.

Indeed, it is shifting in two opposite directions. What was once mere theoretical speculation is well-confirmed experimental knowledge today, like the existence and the quantum dynamics of atoms and electrons. Conversely, what once counted as empirical in a narrow sense is intimately interwoven with theory today. Like the weight of bodies, which has been measured for thousands of years by means of the balance. Prior to Newton's *Principia*, it was empirical. Today, it is a theoretical concept of classical mechanics. Or like the particle track discussed in Sect. 2.6. After all, theory-ladenness is a

---

[1]Kuhn 1962, 1970.

bad criterion for making the distinction between safe background knowledge and uncertain assumptions or hypotheses.[2]

How are observation, measurement, and theory interwoven in experimental particle physics? Which empirical phenomena and theoretical assumptions are crucial for the claim that subatomic particles exist? Like any physical theory, the current particle concept has a twofold empirical basis. Firstly, it has an *observational basis* in a strict empiricist sense which consists of the observed phenomena and their comparison. In particle physics, this observational basis consists of Geiger counter clicks, particle tracks in a cloud chamber (or bubble chamber) photograph and so on, that is, in local events detected by particle detectors and in the spatio-temporal shape of event sequences. Secondly, this observational basis is extended by theory-laden measurement results and empirical structures. This *theory-laden extension* consists of the measured numerical values of physical quantities and their data models. In particle physics, this extension of the empirical basis is threefold. It consists of:

(i)   *dynamic quantities* like mass and charge in terms of which distinct kinds of subatomic particles are discriminated,

(ii) *kinematic quantities* like momentum or energy attributed to the observed particle tracks (called 'kinematic' because they belong to the relativistic kinematics which describes the propagation and collisions of particles or waves of high energy, as in the Compton effect),

(iii) *further quantities* which are obtained by analyzing particle tracks in terms of the kinematic and dynamic quantities and which correspond to those of a quantum field theory.

The determination of these quantities of particle tracks is called *data analysis*. Measurement theories are involved in this data analysis, and these are based on sophisticated theoretical background knowledge, where the background knowledge has increased substantially over a century of particle physics. For the last few decades, particle tracks have no longer been observed in an empiricist sense. They are recorded electronically and reconstructed by means of computer programs, giving rise to a further theory-laden extension of the observational basis:

(iv) *computer-reconstructed particle tracks* obtained by analyzing the electronic data recorded by spark chambers, drift chambers, scintillation counters, or other kinds of electronic particle detectors.

---

[2]In this regard, structuralistic philosophy of science is in danger of missing the point. In the context of the debate on scientific realism, Sneed's account of T-theoretical concepts (Sneed 1971) becomes misleading. According to Sneed's approach to the structure of empirical theories, weight is a T-theoretical concept of classical mechanics. But it became so only through the rise of classical mechanics. Prior to this, it was empirical. For two millennia, it has been measured in the same way using a balance. The moral is that Sneed's distinction between pre-theoretical and T-theoretical concepts itself depends on theory.

In what sense does particle physics deal with *particles*? When the theoretical foundations of the current particle concept(s) are investigated, it becomes obvious that an *operational particle concept* is crucial.[3] Particle detectors detect local events and particle tracks. When local events and their sequences were first observed in the experimental devices of subatomic physics, they were associated with particles in the traditional sense of small constituent parts of matter. The observed phenomena were embedded into the mathematical structures of classical mechanics, according to the dynamic and kinematic quantities (i) and (ii). In any empirical evidence for a subatomic particle the two parts of the empirical basis of the particle concept come together, the observational basis and its theory-laden extension (i)–(iv). During a century of particle physics the theoretical background knowledge of subatomic particles increased enormously. The observational basis and the measurement theories of particle physics became more and more intertwined. What has been observed and to what kind of evidence it gives rise became more and more theory-laden. That is, the observational basis was *theorized*. With the extensions (iii) and (iv), the phenomena of particle physics have turned into mathematical data and computer constructs. In the following I will analyze some crucial stages of this process by explaining the key theoretical concepts, the observed phenomena, and the measurement theories involved in particle identification. The systematic structure of the relation between empirical observations and theoretical constructs is illustrated by means of several historical case studies.

The conceptual starting point lies in *two informal particle concepts*. The first is defined in terms of part and whole. The second is defined in terms of cause and effect. They have gone hand in hand since the beginnings of modern particle physics. They rest upon the indispensable methodological principles of modern physics (Sect. 3.1). The way in which they are combined into a *law of classical mechanics* is decisive for what counts as *empirical evidence of a particle*. The Lorentz force was crucial for the discovery of the electron in 1897. In a similar way, the relativistic principle of momentum–energy conservation was decisive for the acceptance of Einstein's light quantum hypothesis within the scientific community (Sect. 3.2). On the basis of these case studies the kinds of theoretical assumptions involved in particle identification are analyzed. They are related to the possibilities of localizing a particle and of applying the classical dynamics of a mass point to it. They give rise to several kinds of empirical evidence for particles which in the development of particle physics became *increasingly theory-laden* and complicated: position measurements, particle tracks, scattering events, resonances of unstable particles. In the well-confirmed background knowledge of these kinds of evidence, quantum theory got increasingly involved (Sect. 3.3). After this systematic account of particle identification and its empirical and theoretical basis, the data analysis of particle tracks is illustrated by two further case studies. The

---

[3]See Chap. 6, in particular Sect. 6.3.

next crucial historical step identifying a new kind of particle from its track was the discovery of the positron by Anderson in 1932. It resulted from a tricky experiment. However, it was based on immediate observation and on a simple measurement theory which was absolutely independent of the Dirac equation (Sect. 3.4). Further crucial systematic steps concerning particle identification were the puzzle around the 'mesotron' and quantum electrodynamics; and the recent possibilities of identifying individual quark events from the scattering events called jet events (Sect. 3.5). Finally, the philosophical question of whether subatomic particles exist is discussed in terms of the debate on scientific realism; and the empiricist and constructivist opponents to a moderate scientific realism are countered.

3.1 Two Particle Concepts

The current particle concept has two informal pre-theoretical meanings. They are *not* pre-theoretical in the sense of empiricist philosophy of science. Both are philosophical. They derive from the metaphysical presuppositions of modern physics and Galilean science, respectively. They do indeed rest upon the principles of mereological and causal analysis of phenomena, which belong to the indispensable methodological principles of physics discussed in Sect. 1.5. They are not *based* on empirical observations but they *make them possible*. In this sense, they are *a priori* in a Kantian sense. The mereological particle concept is traditional. It stems from ancient natural philosophy. Newton introduced it in the remark to his third rule of philosophy, which entitles one to generalize the extensive quantities of all macroscopic bodies to the atoms.[4] Since the beginnings of modern physics, it has gone hand in hand with the causal particle concept. Newton suspected that atoms of matter and light cause the interactions of matter and light.[5] A century ago, the causal particle concept became crucial. The local events observed in the experimental phenomena of subatomic physics were traced back to particles as their causes.

> **Mereological Particle Concept:**[6] Particles are the constituent parts of matter or light.

> This mereological particle concept is based on a *part–whole relation*. Particles are the microscopic parts of a macroscopic whole. The concept dates back to ancient atomism. It was taken up in 17th century theories of matter (Galileo, Descartes) and of light (Newton). Prior to the experimental discovery of electrons, α-particles, and the atomic nucleus, this traditional particle concept had no empirical content.

---

[4] Newton 1687, 795–796.

[5] Newton 1730, 370–374.

[6] In modern logic, mereology is the formal logic of the part–whole relation; see Simons 1987 and Appendix E.

Causal Particle Concept: Particles are the causes of local events in detectors.

This causal particle concept is based on the *relation of cause and effect* associated with a dynamic process. It applies to the local events of subatomic physics that are explained in causal terms. By way of this causal concept the mereological particle concept gets experimental meaning. The causal particle concept explains the successive local events in particle detectors. Particles from natural radioactivity, cosmic rays, a nuclear reactor, or a particle accelerator make a Geiger counter click. They produce particle tracks in the cloud chamber, in a nuclear emulsion or in the bubble chamber. They give rise to measurable currents and light flashes in drift chambers, scintillators, Čerenkov counters, and other detectors of current particle physics.

Both informal particle concepts are rooted in traditional metaphysics. From a Kantian point of view, they are conditions for the possibility of the experience of particles. The first is based on the concept of *substance*. According to this, particles are parts of matter, or material substances at a small scale. (If light is conceived as being made up of particles, it is also conceived as made up of substances.) The second is based on the *principle of causality* which in the philosophical tradition was associated with determinism. For physicists, the local events and event sequences (tracks) measured by particle detectors are the *fingerprints* of subatomic particles. They indicate that *something* had once or repeatedly a local interaction with the matter of the particle detector. This *something* is called a *particle*. In this way, the two meanings are connected.

A *theory of particles* makes this connection explicit. It aims at referring to unobservable entities which are microscopic parts of matter causing observable macroscopic phenomena in particle detectors. The particle theory corresponding to the metaphysical concepts of substance and causality is *classical point mechanics*. However, it turned out that this theory and its underlying metaphysical presuppositions are *too strong*.[7] It is an irony of the history of physics that the very causal particle concept which established experimental particle physics and thus made metaphysics turn into physics finally turned out to be untenable.[8]

The assumption that particles exist has its observational confirmation in observable local events in particle detectors. The *empirical* evidence or signature of a particle is tied to the *causal* particle concept. The effect is observable, while the cause is inferred according to the principle of causality. The causal particle concept provides the local phenomena observed in a particle detector with a causal story. In this way it gives rise to the observation of particles,

---

[7]See Mittelstaedt 2006.

[8]Quantum theory teaches that the individuation of the cause of a particle track is highly problematic; see Sects. 5.3 and 7.6.

in the generalized sense of observation which has been explained in Sect. 2.5. In a physical theory, the objects to which the causal particle concept refers are described in terms of physical quantities such as mass, charge, or energy. Crucial for the identification of a *kind* of particle is its characterization by values of *dynamic* quantities. Today, the kinds of particles are distinguished in terms of mass or energy, spin, parity, and several kinds of charges (electric charge, flavor, color). These charges determine the coupling constants of subatomic interactions (electromagnetism, weak interaction, strong interaction). Particle identification presupposes a measurement theory of these quantities. Ascribing their values to a kind of subatomic particle relies on the conservation laws of the associated physical dynamics and on the *mereological* particle concept. Mass or charge are ascribed to the constituent parts of matter (or light) *as if* the mass and charge of macroscopic bodies were summed from the mass and charge of electrons, protons, and neutrons.

To be more exact, 'constituent' parts of matter means that certain sum rules are valid for the quantities of a compound system of subatomic particles. Such sum rules are in general more complicated than a simple sum running over the corresponding quantities of the constituent parts of chemical atoms. The whole may be more or less than the sum of its parts. For example, the mass of an atomic nucleus is smaller than the sum of the masses of the corresponding number of protons and neutrons. The sum rule for its mass is based on the principle of energy conservation. It takes the binding energy and the relativistic mass–energy equivalence into account. Due to the binding energy, the nucleus as a bound system is more stable than a mere collection of its constituent parts. It has less energy. Due to the equivalence of mass and energy, its mass is therefore also lower. The mass difference between the bound system as a whole and the sum of its parts is called the *mass defect*. The sum rules of the quark model of current particle physics are even more complicated.

The mereological particle concept and the sum rules provide the subatomic models of matter with a physical semantics. The familiar meaning of quantities such as mass, energy and charge, which is well-known for macroscopic physical objects, is carried over to the microscopic constituent parts of matter. It is assumed that the physical quantities of macroscopic objects sum up from those of subatomic particles according to the sum rules given by the corresponding physical dynamics.

Around 1900, when the subatomic constituent parts of matter finally became accessible to experimental investigation, it seemed obvious to think that they obeyed the well-known *classical* laws of nature. In particular the massive charged particles which were discovered in cathode rays and radioactivity, as well as the constituents of Rutherford's or Bohr's atomic model, were described in terms of *classical point mechanics*. Prior to Bohr's atomic model, their dynamics was conceived in terms of *classical electrodynamics*. With the rise of quantum theory, classical electrodynamics was abandoned first, and classical point mechanics later. However, for an approximate description of particle tracks and for the kinematics of subatomic scattering processes, classical mechanics has remained empirically adequate up to the present day. The measurement theories are still based on the laws of classical physics. Even in view of quantum theory, experimenters working in the field of current particle physics remain inclined to think that a particle track is caused by a *local* physical object. However, according to quantum theory the track is only a series of subsequent position measurements, and the particle which causes the track does not remain in a localized quantum state after a measurement of its position. Position measurements repeatedly localize a certain portion of mass, charge, or energy in a particle detector. But the subsequent position measurements that make up a track are *not* connected by a classical trajectory.[9]

According to the values of mass, spin, parity, and charge(s), which have actually been measured from the data of particle tracks, distinct kinds of subatomic particles have been distinguished. Most popular are the massive charged electron, the much heavier nucleons, proton and neutron, which make up the atomic nucleus, the photon as the massless particle or field quantum of light (and electromagnetic radiation in general), and the neutrino which is generated in β-decays of atomic nuclei. According to recent experimental evidence, it has a non-vanishing, very small mass.[10]

In addition, there are less obvious particle states which are also associated with mass or energy, spin, parity and charge(s). It is assumed that in scattering experiments of increasing energies unstable quantum states give rise to resonances which show up at a given scattering energy. And there are the quarks which figure as pointlike scattering centers of lepton–nucleon and neutrino scattering as well as the causes of jet events. The measurement theories of resonances and quarks are highly sophisticated. Their principles will be explained in Sects. 3.3 and 6.5.

3.2 Evidence for a Particle: Two Case Studies

The experimental evidence for a particle has always had two necessary ingredients. Even after quantum theory made the classical particle concept doubtful these ingredients remain closely related to the metaphysical particle concepts discussed in the preceding section: (1) The empirical observation of a *localized effect* is tied to the *causal* particle concept. (2) The successful application of a measuring law which is typical of a particle and which is based on the *classical mechanics of a mass point* is tied to the *mereological* particle concept in terms of sum rules for conserved dynamic quantities.

How these conditions come together will now be shown in two case studies. Experimental particle physics began with the discovery of the *electron* as a

---

[9]See Sects. 3.3.2 and 5.3.
[10]See Perkins 2003, 3 and 170–179.

charged mass. Twenty years later, Einstein's light quantum hypothesis was extended to the quanta of electromagnetic radiation in general, the particle nature of light and X rays was accepted, and the new particle was called the *photon*. In both cases, experimental physicists did not content themselves until the particle in question was proved to be a localizable quantum of the quantities ascribed to it.

3.2.1 The Electron

The so-called discovery of the electron is based on the measurement of its charge and mass. It is usually ascribed to J.J. Thomson and dated to the year 1897 when the debate about the waves or particle nature of cathode rays was still ongoing. Cathode rays are generated when a high voltage is applied to two electrodes in an evacuated tube. Like X rays, cathode rays cannot be seen, but their effects can be observed (e.g., on a television screen or on a computer display). In a textbook of experimental physics from the year 1900, they are phenomenologically described as follows:

> Strange effects which propagate rectilinearly leave the cathode B and are therefore called cathode rays. These beams are invisible. However, when they meet glass, ruby, fluorite, corals and many other substances, they make them shine intensely. This shining fails to appear behind a metal or a mica plate, etc.; so the obstacle is pictured as a shade. Little fan wheels, etc., are set in motion by the cathode rays.[11]

Cathode rays are described here as the rectilinear propagation of an effect whose locality or non-locality does not become evident from the observations. In 1897, Thomson determined the ratio e/m of charge and mass for cathode rays by measuring the deflection of the cathode rays in a magnetic field and then the compensation of this deflection by an electrical field. He determined the value for e/m by calculating the two deflections from the classical *Lorentz force* $F = (q/c)(E + v \times B)$, which acts on a charged particle with charge q and velocity v in an electric field E and a magnetic field B. Lorentz was, like Thomson, a defender of the atomistic paradigm. He had introduced the expression for this force to theoretical physics only two years before.[12] Thomson interpreted his measurement result for e/m, which differed from zero, as empirical evidence for the particle nature of the cathode rays because the measured charge was not compatible with the alternative wave hypothesis. According to Maxwell's electrodynamics, electromagnetic waves do not carry a charge.

---

[11] Jochmann et al. 1900, 417. My translation.

[12] Thomson 1897; see Pais 1986, 85. Lorentz 1895; see Pais 1986, 76.

Thomson's measurement theory involved only the expression for the Lorentz force. His use of this law connected the particle concept of classical physics with an atomistic theory of electricity. For Thomson, the electrons were mechanical objects at a small scale which have on the one hand a characteristic inertial mass m, and on the other hand carry the elementary charge unit e and obey the Lorentz force $\boldsymbol{F} = (q/c)(\boldsymbol{E} + \boldsymbol{v} \times \boldsymbol{B})$ therefore. As Millikan later stressed in his retrospective account of the discovery of the electron, the concept of the electric elementary charge unit was originally completely independent of the classical particle concept. In 1891, J.G. Stoney had introduced the term 'electron' as a name for the fundamental unit of charge tied to electrolytic processes. Only in Thomson's concept of the electron did the electric charge e become associated with an inertial mass m.[13] Thomson ascribed four properties or attributes to the electron which characterize it as a charged particle:

(i)   an *inertial mass m*,
(ii) an *electric charge e*,
(iii) *pointlikeness* or *locality regarding the effect of external forces*, and
(iv) a *spacetime trajectory* which is subject to the classical force law.

According to (i), (iii), and (iv), the electron is a *mass point* in the sense of classical point mechanics, i.e., an inertial mass which is localized in a space point and moves along a classical trajectory. According to (ii), this mass point has an *electric charge* which is identical with the elementary charge unit e, the original 'electron' from electrolytic processes. This charge qualifies the force law which determines a spacetime trajectory (iv) to the expression for the classical Lorentz force. (i) and (ii) are the dynamic characteristics, while (iii) and (iv) are the spatio-temporal characteristics of the electron. They are connected to each other by classical dynamics. None of them is subject to a measurement on its own. What is measured is the ratio e/m. It is determined from the spatial deflections of the cathode rays according to the Lorentz force. In particular, neither the locality of the mass nor the charge associated with the mass point is measured. Whoever did not yet believe in 1897 that a massive fundamental charge unit existed, did not have to be convinced by Thomson's measurement result either. The measurement did not in any way test the hypothesis that cathode rays consist of single massive charged particles. It only confirmed a consequence of this hypothesis, namely that cathode rays are subject to the Lorentz force. Although the successful application of this law was a strong indication for the existence of the electron, it was not regarded as sufficient by all physicists.

Thomson's measurement is a good example of the way in which a measurement theory imposes a dynamic structure on an observed experimental phenomenon. This dynamic structure makes it possible to proceed from a qualitative explanation to a quantitative description of the phenomenon or

---

[13]Millikan 1917, 25–26. See also Pais 1986, 73–74.

its cause. Given that the particle hypothesis regarding the nature of the cathode rays could not be confirmed by direct observation, the measurement provided it with the value of a physical quantity which could not be interpreted in other terms. The e/m measurement was carried out under the presupposition that cathode rays consist of charged massive particles of charge e and mass m. The measurement delivered a constant of nature e/m, the theoretical interpretation of which makes use of the particle hypothesis. The Lorentz force applies to a carrier of charge and mass. Thus its application rests on the assumption that there are massive charges in the cathode rays. This assumption itself, however, remained untested in Thomson's measurement. To check the meaning of the measured e/m value *independently* of it, measurements were needed which identified *single* carriers of charge and mass rather than measuring only *rays* carrying charge and mass.

Townsend made the first step in this direction in Thomson's laboratory in the same year. He succeeded in using the principle of the cloud chamber (developed by Wilson in 1895) for localizing single charge carriers unsharply, and he determined their charges independently of mass.[14] If saturated steam, in which charged gas was produced by electrolysis, was oversaturated by expansion, it was observed that condensation droplets formed in it. This observation presupposed a complex theory of the way in which charged particles are localized in a cloud chamber. According to this, the electrons are charged particles, and they are released by electrolytic processes from the molecular or atomic constituents of matter which in turn is ionized, i.e., gets a positive charge. To this theory, the following experimental knowledge was added. One had to know that saturated steam can be taken by expansion to the state of supersaturation, and one had to have quantitative knowledge of the conditions under which supersaturated vapor forms little water drops at neutral condensation cores, as well as at ions of negative or positive charge. It is decisive that the theory of the cloud chamber only gives rise to the construction of the measuring device in which the observed phenomenon can take place. However, the occurrence of the phenomenon is not based on it, and neither is its quantitative analysis. To perform and to analyze position measurements with a cloud chamber requires substantially less theoretical knowledge than to design the experimental equipment which makes it possible to carry these measurements out.

Townsend determined the charge of the single condensation droplets in the cloud chamber from the total charge of the steam and the number of droplets per volume. The total charge of the steam was measured electrostatically, and the number of droplets was calculated from the weight of the cloud and the mean weight of the single droplets.[15] As opposed to Thomson's e/m

---

[14]For the invention of the cloud chamber by C.T.R. Wilson, see Millikan 1917, 46, and Pais 1986, 86.

[15]Townsend 1897. Millikan 1917, 43–47, discusses Townsend's experimental work in detail. In Pais 1986, however, it is not mentioned.

measurement, the force law of classical mechanics or its specification is not related here to invisible particles but to the macroscopic properties of the droplet cloud. After 1898, Thomson also carried out measurements based on this principle.[16]

It is not the e/m determination from cathode rays but only such a charge measurement from condensation droplets that assigns a characteristic value of charge to entities which are identified one by one in a detector. Only in the latter case is the measurement a good candidate for an observation in the generalized sense discussed in Sect. 2.5. However, the premise that the charge portions localized in the individual charge carriers were the same remained untested. All these measurements still allowed doubt as to whether every single condensation droplet carries an elementary charge unit (or at least an integer multiple of it). Only after 1909 were these doubts largely dispelled by Millikan's oil droplet experiments. In lengthy and difficult measurements, Millikan determined the force which an electric field exerts on a single charged oil droplet.[17] When he published his value for e, he emphasized that his measuring method was the first method for measuring the charge of individual charge carriers.[18]

Millikan himself no longer saw the principal benefit of his measuring method in the experimental validation of isolated elementary charges, but rather in the precision measurement for e which made this method possible. At this time, the existence of the electron and the atomistic constitution of

---

[16]Thomson 1899. Millikan 1917, 47–52, stresses that Thomson's measuring method was even less direct than that of Townsend: "Instead of measuring the weight of this cloud directly, as Townsend had done, Thomson computed it by a theoretical consideration. [...] The careful examination of Thomson's experiment shows that it contains all the theoretical uncertainties involved in Townsend's work, while it adds considerably to the experimental uncertainties." See, however, Pais 1986, 86, who ascribes the first direct measurement of e to Thomson (and does not mention the later oil droplet experiments by Millikan at all): "Thomson's measurement of e is one of the earliest applications of this cloud chamber technique. He determined the number of charged particles by droplet counting, and their overall charge by electrometric methods, arriving at $e \simeq 6.8 \times 10^{-10}$ esu, a very respectable result in view of the novelty of the method. And that is why Thomson is the discoverer of the electron." Establishing the truth seems to be a difficult task in the historiography written by physicists.

[17]Millikan 1911; see Shamos 1959, 238–249.

[18]See Millikan 1911, 249, by reference to a previous publication: "In a preceding paper a method of measuring the elementary electric charge was presented which differed essentially from methods which had been used by earlier observers only in that all of the measurements from which the charge was deduced were made upon one individual charged carrier. This modification eliminated the chief sources of uncertainty which inhered in preceding determination by similar methods such as those made by Sir Joseph Thomson, H.A. Wilson, Ehrenhaft and Broglie, all of whom had deduced the elementary charge from the average behavior in electrical and gravitational fields of swarms of charged particles."

matter were for him well-established due to an abundance of experimental evidence from several other areas of physics.

3.2.2 The Photon

For many years the physics community did not accept Einstein's light quantum hypothesis of 1905. However, after the observation made in 1919 that light is bent by gravity, as general relativity predicted, Einstein's public reputation increased so enormously that in 1921 he was awarded the Nobel prize. His light quantum hypothesis was chosen because it made the bridge to Bohr's atomic model.[19] The Nobel prizes of 1921 and 1922 were given to Einstein and Bohr at the same ceremony in 1922. Only afterwards was the photon hypothesis definitely confirmed by experimental proof of the Compton effect.[20] For the experimental confirmation the energy–momentum conservation law of relativistic kinematics was decisive. It had similar significance as the Lorentz force had had for Thomson's e/m measurement. It explained the decrease in frequency of scattered X rays in terms of a measurement law of particles.

From Einstein's light quantum hypothesis[21] up to the experimental validation of the photon by Bothe and Geiger[22] almost two decades had gone by. The light quantum hypothesis of 1905 was very speculative. It was a theoretical assumption reminiscent of Newton's emission theory of light, long disproved by Young's interference experiments and Fresnel's wave theory of light.[23] In 1905, it could not be tested because it was not connected with a measurement law for the spacetime trajectory of a particle. Up to the 1920s, only a few believed in this.[24] In Einstein's paper of 1905, the light quantum hypothesis had primarily a heuristic value, namely its unifying power. It explained in a uniform way the photoelectric effect and several other experimental phenomena which show up in the interaction of light and matter.

---

[19]See Wheaton 1983, 279–281.

[20]Wheaton 1983, 282–286.

[21]Einstein 1905.

[22]Bothe and Geiger 1925.

[23]See Einstein 1905, 133: "According to the assumption to be envisaged here, in the propagation of a light beam from a point the energy is not distributed continuously spreading over greater and greater spaces, but it consists of a finite number of photons localized in space points which move without being divided and can be absorbed and produced only as a whole." Compare this with the words in the *Queries* at the end of book III in Newton's *Opticks*: "Are not the Rays of Light very small Bodies emitted from shining Substances?" Newton 1730, 370 (*Query 29*). And: "Are not gross Bodies and Light convertible into one another, and may not Bodies receive much of their Activity from the Particles of Light which enter their Composition?" Ibid., 374 (*Query 30*).

[24]The breakthrough for the photon hypothesis finally came with the Compton effect, that is, one year after Einstein got the Nobel prize for the explanation of the photoelectric effect (at that time still regarded as a phenomenological law); see Wheaton 1983, 279–286.

That light can cause a photocurrent had been known since the end of the nineteenth century. However, the way in which the photocurrent depends on the frequency and intensity of the incoming light could not be explained in the context of classical physics. According to Einstein's light quantum hypothesis, the photocurrent arises as follows. Light is absorbed in energy portions $h\nu$ and electrons are released with kinetic energy $h\nu - K$, where K is the releasing energy of an electron. In 1914, Millikan checked this theoretical explanation of the photoelectric effect experimentally and confirmed it quantitatively with high precision.[25] Since the light quantum hypothesis was not connected with other measuring laws, and moreover was in conflict with the wave theory of light, Millikan's results were not taken as an experimental confirmation of it. The light quantum hypothesis lacked both ingredients for indicating a particle. It was neither measured by a law of nature typical of particles and nor did it have an observational basis connected with such a measuring law. Therefore, Einstein's description of the photoelectric effect by the formula $E = h\nu - K$ was split off from the particle model of light which formed its heuristic background. In the following years, it was only considered to be a well-confirmed phenomenological law which lacked theoretical understanding.

In 1916 Einstein extended his light quantum hypothesis of 1905 to a first, statistically well-founded theory of the absorption and emission of light.[26] This theory laid the grounds for the acceptance of the photon hypothesis by enlarging its empirical content in two decisive respects: on the one hand, it connected the light quantum hypothesis to the quantization conditions of Bohr's atomic model. It made it possible to derive the frequency of the radiative transitions in the hydrogen atom.[27] Suddenly, the unifying power of the light quantum hypothesis extended to the area of experimental phenomena in which the failure of classical physics had become obvious. On the other hand, Einstein attributed to the photon in addition to the energy $E = h\nu$ a momentum $\boldsymbol{p} = \hbar\boldsymbol{k}$, where $\boldsymbol{k}$ is the wave vector of an electromagnetic wave with wave number $k = 2\pi\nu/c$. In this way, the energy–momentum relation $E^2 = \boldsymbol{p}^2 c^2$ of relativistic kinematics became applicable to it.[28] Hence the photon hypothesis was connected with the measurement theory of a *relativistic particle of zero rest mass*. This particle propagates with the speed of light. According to the relativistic principle of equivalence of mass and energy, it has an *inertial mass* $m = h\nu/c^2$. And its interactions with other particles obey the relativistic law of energy–momentum conservation. According to energy–momentum conservation, a particle gives a *recoil* to another particle.

---

[25] See Wheaton 1983, 238–241; Trigg 1971, Chap. 7.

[26] Einstein 1917.

[27] Ibid., 124, eq. (9).

[28] According to the terminology of current high energy physics, relativistic kinematics includes the laws which describe the propagation of (free, i.e., unbound) particles with relativistic velocities, as well as the energy–momentum conservation of their reactions.

The recoil corresponds to the momentum transferred in the interaction. Due to the momentum transfer the particle loses energy. According to the photon hypothesis, this energy loss does not have any effect on the intensity but it makes the frequency of the electromagnetic radiation lower.

With the integration of the photon hypothesis into relativistic kinematics, a similar step was taken to the one made in 1895 when Thomson applied the Lorentz force law to the electrons he expected to be the constituents of cathode rays. Now the photon had become subject to a measurement theory typical of classical relativistic particles. The spatio-temporal propagation of light was connected with a conservation law which is valid for the energy exchange between the momentum of a massive particle and the frequency of a photon. The light quantum hypothesis was now connected with a testable prediction for the momentum associated with a wavelength and for the light frequencies before and after an interaction between light and matter. Unlike the bare photon hypothesis of 1905 which did not admit of a direct experimental confirmation, the empirical adequacy of this measuring law could be directly tested.

The breakthrough for the photon hypothesis came in 1922, when Compton applied relativistic kinematics to the scattering of gamma radiation and electrons. His quantitative prediction of the effect fitted surprisingly well with the already existing experimental data.[29] As in the case of the discovery of the electron, the applicability of a measurement theory typical of a particle was decisive for the acceptance of the photon hypothesis within the scientific community. However, this was not yet regarded as sufficient. Classical electrodynamics was at stake, and Bohr, Kramers, and Slater invented the BKS theory in order to save it.[30] It predicted violations of energy–momentum conservation for individual subatomic processes and saved the classical picture only at the probabilistic level (thus paving the way for the probabilistic interpretation of quantum mechanics). To establish the photon hypothesis against the BKS theory, the localization of single photons of energy h/ν was needed. Unlike the measuring of single elementary electric charge units by Millikan, it was not long in coming. The experiment of Bothe and Geiger proved in 1925 that relativistic energy–momentum conservation is not only valid in the time average for the scattering of light at electrons but also for the single scattering process. Bothe and Geiger checked energy–momentum conservation in the individual case, showing by means of a coincidence counter that in the Compton effect every single photon is actually correlated with a recoil electron.[31] The coincidences were accepted empirical evidence for the effects of individual photons.

---

[29]Compton 1923 and Debye 1923. See Wheaton 1983, 283–286, as well as Mehra and Rechenberg 1982, 512–514.

[30]Bohr et al. 1924.

[31]Bothe and Geiger 1925.

But this is not the whole story. There is a parallel with Millikan's e/m measurement about fifteen years after the electron hypothesis was accepted; and another parallel with the confirmation of the neutrino hypothesis by indirect β-decay, about two and a half decades after establishing the neutrino hypothesis. Since the early days of quantum theory, several semi-classical theories of the interaction of light and matter have been developed. They quantize the atom but keep the classical description of electromagnetic radiation. They do indeed explain the experimental phenomena that were taken as evidence for the photon, including the photoelectric effect and the Compton effect:[32]

> In 1922, Einstein received the Nobel Prize – not for his relativity theory, but for his interpretation of the photoelectric effect as being due to particle-like photons striking the surfaces of metals and ejecting electrons. Textbooks regularly repeat Einstein's arguments as proof that light possesses a particle nature. And yet, ironically, it has been cogently argued that Einstein's conclusions were not fully justified.[33] It is ironic that Albert Einstein, arguably the greatest physicist since Newton, received the Nobel Prize for work that subsequently turned out to be flawed. And it is doubly ironic that this work, which was instrumental in placing before us the concept of wave–particle duality, turned out to be correct even though flawed.[34]

The decisive proof of single photons only came by the coincidence experiments of modern quantum optics, about six decades after Einstein got the Nobel Prize![35] The moral of this history of the photon is that it is not easy to establish empirical evidence for single particles. The criteria discussed in the above two case studies are *necessary* rather than sufficient conditions for the detection of a particle. The only decisive proof of particles is apparently to *make* them and to use them as tools in *other* experiments.[36]

---

[32] See Greenstein and Zajonc 1997, 23–26, and the literature quoted there. They discuss in some detail (ibid., 23–26) the semi-classical theory of the photoelectric effect developed by James, Lamb, and Scully (1969). The most prominent defender of a semi-classical explanation of the photoelectric effect was Planck. A semi-classical explanation of the Compton effect was given by Schrödinger (1927).

[33] Greenstein and Zajonc 1997, 23.

[34] Greenstein and Zajonc 1997, 34.

[35] Wave–particle duality is analyzed in Chap. 7. Single photons were first generated experimentally when it became possible to *prepare* them as particles (see Sect. 7.3) by detecting the *other* photon of an entangled photon pair; see Greenstein and Zajonc 1997, 33–35, referring to Clauser 1974 and Grangier et al. 1986.

[36] See the remarks on Eddington's reservation concerning the neutrino hypothesis and Hacking's reality criterion, discussed in Sects. 2.3 and 2.4.

3.3 Theorizing the Observations

The discovery of the electron was definitively completed only by Millikan's oil droplet experiments. Similarly, the validation of the photon was only completed by the proof that the Compton effect takes place at the level of individual scattering processes. Both historical case studies show paradigmatically that experimental evidence is based on two types of theoretical assumptions about particles, referring to the localizability and to the dynamics of a classical mass point:

(i)   A *theory of position measurement*. This contains assumptions about the possibilities of localizing a particle by a suitable measuring device. It is a theory about how to make the local effects of a particle observable or visible.

(ii) A *dynamic measurement theory*. This is part of a physical dynamics and it makes it possible to measure the dynamic quantities of a particle, like mass and charge.

As we know today, the two theories cannot be separated. According to quantum theory, the localization of a particle *is* nothing else but the detection of a certain quantum of mass, energy, or charge by means of a particle detector. A sharp conceptual distinction between the observational basis and the measurement theories is therefore at the most possible for the very beginnings of particle physics. However, taken together the two theories serve to establish the conditions for a *generalized concept of observation*, as explained in Sect. 2.5. The theory of position measurement makes it possible to observe a local effect in a particle detector, which can be traced back to an individual cause located in a certain spacetime region. And the dynamic measurement theory makes it possible to attribute the characteristic physical quantities of a subatomic particle to this individual cause, i.e., mass and charge to the electron, and momentum and energy to the photon. Two reservations have to be expressed. First, the above conditions are necessary rather than sufficient. Second, according to quantum theory the cause itself *cannot* be individuated like a classical particle.[37]

In the decades following the confirmation of the electron and the photon, the observational basis of particle physics was substantially *theorized*. It was assumed that the position measurements and the generation of particle tracks are due to the interaction of subatomic particles with the atoms of a particle detector and that these interactions obey the laws of quantum theory. Based on quantum theory, the background knowledge of the generation of particle tracks and particle reactions got more and more sophisticated. This sophisticated background knowledge entered in turn into the measurement theories of particle physics.

This is obvious for the analysis of particle tracks. The dynamic quantities which characterize a track or the underlying particle are *mass* and *charge*.

---

[37]See Sect. 6.6.

This is true for the classical as well as for the quantum theoretical description of a particle or track. Originally, the measurement theory of mass and charge was purely classical. In the 1930s and 1940s it was refined, partly on the basis of empirical knowledge and partly on the basis of quantum theory. An improvement in the methods of quantitative analysis was urgently needed for understanding the observational material from cosmic rays. It contained new kinds of particle tracks which fitted neither with previous measurement theories nor with assumptions regarding the existing kinds of particles. Therefore, in the 1930s and 1940s an improved measurement theory for the analysis of particle tracks was developed which is still in use today. Its development made use of quantum electrodynamics and it is a prime example of the relationship between the measurement theories used in experimental physics and the formation of a fundamental theory of the phenomenological domain which one investigates experimentally. At first they were *strictly separated*. The discovery of the positron was made at this stage of the measurement theory of particle tracks (Sect. 3.4). The fact that both theories had the same intended applications created serious *difficulties* later. Then major efforts were made to *resolve* them by combining independent measurement methods with some already empirically approved applications of the new theory (Sect. 3.5). Finally, as soon as the new theory was consolidated, it became *interwoven* with its measurement theory and it served as an independent measurement theory for a new domain.

The *increasing theoreticity* of the measurement theories also concerned the *phenomena* of particle physics. Recall the explanations given in Sect. 2.4 concerning the phenomena of Galilean science! The empirical evidence for particles was allowed to involve more and more theory. In addition, new technologies gave rise to new kinds of measuring devices, making observation more and more theory-laden. Scientific practice in experimental particle physics embraces very complicated technologies and measurement theories. Indeed, today data are electronic and they can only be observed on a computer display. In current high energy scattering experiments, data are recorded electronically and analyzed by computers. The measurement theories are encoded in the computer programs.

In current particle physics, position measurements, particle tracks, scattering events and resonances count as empirical evidence for elementary particles. The theorizing of the observational basis takes place in the following systematic steps. *Position measurements* are the basis. In some crucial cases they can even be directly observed in a strict empiricist sense. *Tracks* are obtained by connecting adjacent position measurements. *Events* are obtained by connecting adjacent particle tracks. *Resonances* are obtained from statistical ensembles of scattering events. The following discussion will elucidate the theoretical aspects of this increasing theoreticity and explain in what sense one is still justified in talking about an *observational basis* of the particle concepts involved.

3.3.1 Position Measurement

As in other fields of physics, the measurement of spatio-temporal quantities is fundamental in particle physics. First come position measurements.
In the first decades of particle physics they were immediately based on the
observation of empirical phenomena such as the 'click' in a Geiger counter
or the visible scintillations caused by α-rays. They were supplemented by
observing the phenomenological features of a sequence of position measurements, such as the shape and distance of the single spots in the sequence
and the temporal order of the position measurements established by time
measurements. These characteristics constituted the observational basis for
attributing mass and charge to subatomic particles. (In the late 1920s, however, these classical features of a massive charged particle were joined by the
visible interference fringes produced by the passage of electrons through a
crystal. This phenomenon confirmed the wave–particle duality of quantum
mechanics.) Thomson's e/m measurement was trivially based on position
measurements, namely on measurement of the spatial deflections of cathode
rays in magnetic and electric fields. A decade later Millikan's position measurement confirmed that the measured e/m value may actually be assigned
to single charge carriers.

However, the localization of particles by a detector is also possible *without* any sophisticated theory of the particles and their interactions with a
detector.

A few years after Thomson's e/m measurement, the first experimental
proof of the particle nature of α-rays gave the particle hypothesis a pre-
theoretical observational basis in a strict empiricist sense. This basis was
decisively extended and standardized by Perrin's measurement of Brownian
motion a few years later.[38]

From 1906, Rutherford and his assistants carried out their famous scattering experiments with α-rays. Their constituents were the first particles
which could be localized and observed independently of any measurement
theory of their dynamics and their interaction with a measuring device. As
Crookes and others had already discovered in 1903, a screen laminated with
zinc sulfide starts to phosphoresce in total darkness when it is exposed to
α-rays. Observed with a magnifying glass, this glow could be resolved into a
variety of single light flashes. Rutherford, Chadwick and Ellis commented on
this phenomenon as follows in their 1930 radioactivity textbook:

> On viewing the surface of the screen with a magnifying glass, the light
> from the screen is seen not to be distributed uniformly but to consist
> of a number of scintillating points of light scattered over the surface
> and of short duration. Crookes devised a simple apparatus called a

---

[38] Due to Perrin's experiments which are described precisely in Trigg 1971,
Chap. 4, the atomistic hypothesis was approved even by defenders of energetism
like Ostwald; see MacKinnon 1982, 126–127.

'spinthariscope' to show the scintillations. A small point coated with a trace of radium is placed several millimeters away from a zinc sulphide screen which is fixed at one end of a short tube and viewed through a lens at the other end. In a dark room the surface of the screen is seen as a dark background dotted with brilliant points of light which come and go with great rapidity. This beautiful experiment brings vividly before the observer the idea that the radium is shooting out a stream of projectiles each of which causes a flash of light on striking the screen.[39]

The description of the observation is not like the usual textbook presentation of a physical experiment. It is more like the description given by Duhem of what a layman perceives if he looks over the shoulder of an experimental physicist.[40] The step from the empirical observation towards measurement and to what Duhem calls the 'symbolic' interpretation of the phenomena in terms of physical quantities[41] is not yet performed here at all. Neither a theory of the α-rays emitted by radium nor a theory of their interaction with the zinc sulfide enters into the description of the observation and its occurrence. What is described is the qualitative representation of an empirical phenomenon rather than its standardization on the basis of a theoretical explanation and a measurement method. The proof of the particle nature of α-rays is only based on a causal inference from the observed light flashes to the local character of the radiation which gives rise to single light flashes on the scintillation screen.

The dynamic quantities of the α-particles were determined in later experiments. These were based on the principles of Thomson's e/m measurement. Rutherford used the ratio E/M of charge and mass of the α-rays in them for the first time.[42] After 1906, Rutherford and his assistants used the scintillation method to measure the scattering angle of single α-particles. The experimental check of Rutherford's famous scattering formula was later based on this, too.[43]

Most methods of position measurement are much more theory-laden than the localization of particles on a scintillation screen. Some of them are based on making the macroscopic effects of the interaction of a particle with matter directly visible, according to the relevant theoretical background knowledge. Others measure them by amplifying the electric pulses they produce. Owing to the lack of background knowledge about subatomic processes around

[39]Rutherford et al. 1930, 54–55.

[40]See the oft-quoted description of the observations which a layman makes if he watches the activities of a physicist measuring the electric resistance of a spool, Duhem 1906, Chap. 8.

[41]Ibid. and Sect. 2.1.

[42]Rutherford et al. 1930, 41–46. It turned out that the α-particles are helium nuclei.

[43]Geiger and Marsden 1913; see also Trigg 1971, Chap. 5.

1900, the theory of position measurement used for the experimental confirmation of particles was at that time rudimentary. It was also independent of the measurement laws based on classical dynamics. Today, the theory of position measurement is interlocked with many dynamic laws which form a complicated piecemeal assemblage of classical and quantum theoretical assumptions. In comparison with the scintillation method, the localization of charged particles by means of the cloud chamber was substantially theory-laden. It was based on a theory of ionization conceived in the context of classical physics. However, it was qualitative rather than quantitative. Many later particle detectors from the Geiger counter to the bubble chamber and the drift chamber are based on a much more detailed theory of the ionization processes caused by the passage of charged particles through matter.

The theory of the interaction of charged particles with matter, however, remained considerably underdeveloped for decades as compared to the position measurement by particle detectors based on the principle of ionization. It was a long time before a useful quantum theoretical description of ionization processes became available.[44] This shows how little the theoretical assumptions needed to construct measurement devices will depend on the theory to be established, during the emergence phase of a theory. In the historiography of physics, this methodological aspect of scientific practice in particle physics has been called the *autonomy of experimental traditions*.[45] The autonomy is indeed methodological. It is one of the reasons why position measurements which are theory-laden may count as part of the observational basis of particle physics. Another reason lies in the contingency of the measurement points which are interpreted as the local effects of particles. A third is that these measurement points are observable in a strict empiricist sense. Despite all theory-ladenness of position measurements they can be made visible, either on a photo plate or on a computer display.

3.3.2 Particle Tracks

As the theory of the position measurement of particles was refined, the observational basis of particle physics extended. In 1912 particle tracks could be observed and photographed for the first time from α- and β-radiation in Wilson's cloud chamber. The particle tracks visible on these photographs are characterized by the very small distance between individual condensation droplets, if it is possible to resolve at all the individual measurement points. For α-rays one sees only the tracks and no condensation droplets on the cloud chamber photographs, while for β-rays (which consist of electrons) the individual measurement points of the tracks can be clearly distinguished. The tracks of α- and β-particles are described phenomenologically in the text-

---

[44] See Sect. 3.4.
[45] See Galison 1987, 6–13 and 244–255.

Fig. 3.1. Particle tracks from α-rays observed in the Wilson chamber. (Meitner, after Grimsehl 1938, 16)

book by Rutherford, Chadwick and Ellis on radioactivity which appeared in 1930:

> Owing to the density of the ionization, the path of the α-particle shows as a continuous line of water drops. A swift β-particle, on the other hand, gives so much smaller ionization that the individual ions formed along its track can be counted." [46]

The first particle tracks stemmed from radioactive radiation sources in the laboratory. They were joined by the particle tracks from cosmic rays. Their photographic pictures provided particle physics with an immense observation material for the 1930s. For the 1940s, nuclear emulsions became available. They made it possible to record the tracks of charged particles and to develop their pictures photographically with a very high spatial resolution (of 1 μm).[47] The bubble chamber was developed in the 1950s by Glaser,[48] for the scattering experiments of the beginning era of particle accelerators. It worked the other way round to the cloud chamber: gas bubbles in liquid hydrogen instead of condensation droplets in steam. It made it possible to detect and photograph a variety of particle tracks at the same time. In the particle detectors of current high energy physics, the particle tracks are no longer observable on a photograph. They are recorded electronically and reconstructed by computer programs. Now the observable phenomena of the first decades of particle physics have been replaced by electronic data and

---

[46]Rutherford et al. 1930, 57.

[47]Rossi 1952, 127–142; Powell et al. 1959, 26–32; Perkins 1989.

[48]See Pais 1986, 491.

their reconstruction. Only after a lengthy process of data analysis by means of the reconstruction programs do they become visible on a computer display. At this current stage of theorizing the observational basis is *constructed*, based on the complete theoretical background knowledge which is available up to the present day with regard to the interactions of charged particles with matter.

As compared to the observation of condensation droplets in the cloud chamber or of single light flashes on the scintillation screen, the empirical basis of particle physics extended substantially with the particle tracks observed in nuclear emulsions, in the bubble chamber, and on computer displays. In the sense of our independent criterion of empirical reality given in Sect. 1.3, their tracks belong to empirical reality. All these tracks are actual evidence of the dynamic structure of subatomic reality. Even a non-physicist recognizes the tracks on a bubble chamber photograph or on the computer display as an effect of a series of events connected with each other, which is radically different from an irregular collection of single measurement points. Particle tracks are indeed the strongest empirical evidence for the existence of single micro-objects which behave like classical particles under suitable experimental conditions. They indicate effects which are spatio-temporally individuated within the limits established by the measuring method and which can be grasped in a first approximation with adequate precision by means of a classical measurement theory. According to the well-confirmed background knowledge of subatomic physics, the measurement points of a track are due to ionization processes. Adjacent position measurements can obviously be connected by a spacetime trajectory. With the imprecision of a macroscopic measurement, it looks like a *classical* trajectory. Therefore particle tracks are described in the context of a classical particle model, as long as this proves to be empirically adequate.

The observational basis of this classical particle model is as follows. Since the first observation of particle tracks in the cloud chamber up to the present day, the charge and mass of the particle underlying a track are inferred from the spatio-temporal characteristics of the track. The most important empirical features of a track are the density of the individual measurement points, the curvature and length of the track, and the temporal order of single position measurements which is determined by means of suitable measuring devices. These characteristics are analyzed by means of the relevant safe background knowledge of particle physics.

Thomson's e/m measurement from cathode rays was based on a classical measurement theory which still gives systematic grounds for the data analysis of particle tracks today. It is based on laws of non-relativistic as well as relativistic classical point mechanics. The core of this measurement theory is the term for the *Lorentz force* $\boldsymbol{F} = (q/c)(\boldsymbol{E} + \boldsymbol{v} \times \boldsymbol{B})$. This connects the charge q and mass m of a classical particle with a momentum $\boldsymbol{p} = m\boldsymbol{v}$, and it describes the momentum change of a massive charged particle in external electric or

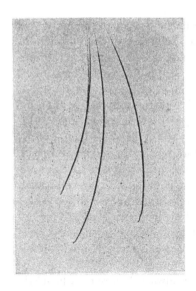

Fig. 3.2. Magnetic deflection of α-particles observed in the Wilson chamber. (Rutherford et al. 1930, after Grimsehl 1938, 15)

magnetic fields. The *momentum p* can be determined from a curved particle track according to the Lorentz force. Operationally, this gives a classical quantity. From a quantum-mechanical point of view, it is *quasi-classical*, i.e., it is a non-classical observable which has classical correspondence. Strictly speaking, quantum theory does not allow for a momentum value of an individual particle track determined from repeated position measurements. Nevertheless, it is empirically adequate to ascribe such a quasi-classical momentum value to a track the generation of which is only explained by the quantum theory of scattering.[49]

The flight direction of a particle is given empirically by the temporal sequence of the adjacent position measurements. (In quantum theory, this corresponds to the propagation of a quantum state over subsequent scattering processes and measurements.) If the flight direction is known, the *charge* and its sign are also immediately obvious from the *track curvature* of a charged particle in a magnetic field and the Lorentz force. The *mass* of a particle can be determined from its momentum and velocity, according to the classical measurement theory. The momentum is measured by means of the Lorentz force, and the velocity from two subsequent measurements of position and time, that is, from a measurement of distance and time-of-flight. For example, the mass spectrograph invented by Aston worked according to this principle in the early days of particle physics.

This measurement theory of a classical particle is based on specifications of the classical force law. It is joined by comparative knowledge about particle tracks of known origin. This knowledge is directly based on the observable features of a track. It enters into the track analysis also on the basis of a

---

[49]See Sect. 5.3.

classical particle model. So one infers, e.g., from the *density* of the individual measurement points on a particle track to the frequency of the interactions of the particle with the detector. This frequency is a measure for the *ionization degree* caused by the particle which in turn indicates the mass of an unidentified particle in comparison with the mass of known particles. For α-particles or protons, the ionization degree is substantially larger than for electrons, as was known from the first photos of particle tracks since 1912. In turn, the *length* of a particle track is related to the *energy* which a particle loses on its way through the detector before being stopped or absorbed. All this knowledge about the dynamic quantities of particles is half empirical, half based on classical models. As far as it is quantitatively confirmed, it is expressed in phenomenological laws. A half-empirical law, the so-called *energy–range* relation, was already formulated on the basis of scattering experiments with particles from radioactive radiation sources in the early days of particle physics. It connects the kinetic energy of a massive charged particle to its range (or track length) in different materials.[50]

According to the causal assumptions which always represent a heuristic background of theory formation in physics, the macroscopic spacetime trajectory in the cloud chamber or bubble chamber is explained in terms of repeated interactions of one and the same microscopic entity with a detector. To infer from the particle track to the particle is a *causal inference*. It is supported by the fact that the classical measurement laws (such as the term for the classical Lorentz force) used for the analysis of particle tracks are empirically adequate. With particle tracks one can nevertheless study how misleading the empirical adequacy of measurement laws can be if it concerns the truth of the fundamental laws of nature. Or to put it in Nancy Cartwright's terms:[51] the data analysis of particle tracks shows *how the laws of physics lie*. The interactions of charged particles with matter are subject to the quantum theory of scattering. This theory shows in contradistinction to the claims of classical point mechanics that there is no deterministic connection between the individual measurement points of a particle track at all.[52]

The way in which the observational basis of particle physics is theorized indicates two different issues. The empirical basis of the particle concept can on the one hand be extended immensely by means of familiar measurement laws. On the other hand exactly those kinds of evidence which speak for the existence of microscopic particles are put into question by the quantum theoretical description of the way in which particle tracks arise. Theorizing the observational basis is double-edged: it consists not only of the *extension of the empirical basis of particle physics* by contingent phenomena which get to be accessible only by theoretical data analysis, but also of the *uncovering of tacit theoretical presuppositions* already contained in the *pre-theoretical inter-*

---

[50]Rutherford et al. 1930, 294.

[51]Cartwright 1983.

[52]However, see Mott 1929, Falkenburg 1996 and Sect. 5.3.

pretation of the observation material. It shows the aspect that particle tracks only become visible by computer reconstruction today; and it has the aspect that the continuous spacetime trajectory connecting the discrete measurement points of a track is today considered to be extremely theory-laden. The continuous spacetime trajectory belongs to the classical measurement model of a particle track. As van Fraassen's criteria for the empirical substructure of a theory show, even a strict empiricist would like to regard it as an empirical data model.[53] It has, however, lost its innocence due to quantum theory. The track is not classical, it is only a *quasi*-classical phenomenon. Its classical model has limited validity. According to the quantum theory of scattering, the quasi-classical track only belongs in a weakened sense to the observational basis of the current particle concept. It is not due to a local cause but to a constant dynamic effect that underlies conservation laws and is repeatedly observable for a certain time at adjacent positions in a macroscopic environment. One nevertheless talks about the observation of this effect and thus also about the observation of the particle track, of course.

3.3.3 Scattering Events

Only *charged particles* can be detected directly from particle tracks and other macroscopic phenomena. They give rise to ionization processes at their interactions with the atoms of a particle detector. In contradistinction, neutral particles (e.g., the photon, neutron, neutrino or the neutral pion) are only indirectly observable. Here, 'indirectly observable' means that they do not give rise to observable tracks but can be identified from their interactions with charged particles on the basis of conservation laws. This distinction between direct and indirect observation is based on the criterion of whether a particle is identified from a single observed particle track, or whether it is identified from several adjacent tracks which are interpreted in terms of scattering events and conservation laws. The distinction differs from Shapere's distinction discussed in Sect. 2.5 with regard to the kind of causal analysis involved. According to Shapere's concept of observation, an observation is

---

[53]The decisive passages which show that he thinks that empirical data models must be formulated in the spatio-temporal quantities of classical mechanics is found on 59 ff.: "Hence let us designate as basic observables all quantities which are functions of time and position alone. These include velocity and acceleration, relative distances and angles of separation – all the quantities used, for example, in reporting the data astronomy provides for celestial mechanics." And 64: "To present a theory, is to specify a family of its structures, its *models*; and secondly, to specify certain parts of those models (the empirical substructures) as candidates for the direct representation of observable phenomena. The structures which can be described in experimental and measurement reports we can call appearances: the theory is empirically adequate if it has some model such that all appearances are isomorphic to empirical substructures of that model." Here, "appearance" more or less corresponds to what Suppes calls a data model; see Suppes 1962.

Fig. 3.3. Photon–electron scattering (Compton 1927, after Grimsehl 1938, 211)

direct if and only if an unambiguous causal story can be told which traces an observable effect back to its unobservable cause. In our case, the unambiguous causal story which can be told makes use of conservation laws and it traces an observed complex pattern of particle tracks back to a *common cause* and a *missing effect*. The common cause is a subatomic particle reaction which gives rise to what is called a scattering event, and the missing effect is a hitherto unknown particle which is assumed to carry away a missing portion of some conserved quantity.

The concept of a scattering event and the use of conservation laws in their data analysis have to be explained in more detail. Let us look again at the history of particle physics! In the experiment for confirming the photon (i.e., the energy–momentum conservation of the Compton effect at the individual level), Bothe and Geiger measured the coincidences of photons and recoil electrons.[54] They took correlated particle detections for the empirical signature of subatomic reactions in which particles interact. Such particle reactions are called scattering processes. They are described by the quantum theory of scattering for which Born laid the grounds.[55] The most striking empirical indication of subatomic reactions comes in the form of particle tracks of different curvature, direction and length which meet at one and the same spacetime point (or region) of a particle detector. On the photographs obtained from the bubble chamber or from nuclear emulsions, the connection of several particle tracks to such scattering events can often be recognized just as well as the individual particle tracks. In the theoretical analysis of the data recorded by a particle detector, adjacent tracks which start or end at the same time in the same place are combined for a scattering event. Here,

---

[54]Bothe and Geiger 1925.

[55]Born 1926a, 1926b. It is often forgotten that his seminal papers on the probabilistic interpretation of quantum mechanics actually provided quantum mechanics with a non-classical scattering theory.

'at the same time in the same place' means a finite spacetime region which is defined by the spatial and temporal resolution of the particle detector.

The examination of particle reactions is based on careful reconstruction of the scattering events for which particle tracks can be combined. This is the main business of data analysis in experimental particle physics today. The combination of particle tracks into scattering events is already based on a lot of background knowledge, which is required in addition to the bare observation of a maze of particle tracks on a bubble chamber photo. To decide whether or not to relate the given particle tracks to scattering events, one often needs background knowledge of the dynamic quantities to be assigned to the individual tracks. In particular, assumptions about the mass of a particle are needed, e.g., when a particle track with a kink is explained in terms of a scattering process in which the particle mass has changed.[56] Therefore the classification of a maze of particle tracks into scattering events always presupposes the complete available background knowledge for the analysis of particle tracks and the measurement of the corresponding values of mass and charge. The scattering events nevertheless belong to the observational basis of particle physics. Once strict empiricism is abandoned, there is no reason for making a *sharp* distinction between direct and indirect observation. At this point, defenders of scientific realism like Maxwell and Shapere are absolutely right.[57] The only crucial question is: how safe is the background knowledge about subatomic particles? Whenever it is well-confirmed, any causal inference based on it which traces a maze of particle tracks back to individual subatomic reactions may count as an observation in the sense of the generalized account of observation suggested in Sect. 2.5.[58]

Once the particle tracks are identified, the theoretical analysis of the scattering events is primarily based on conservation laws. They serve to infer the quantities of all particles involved in the reaction. The conservation laws of particle physics are based partly on relativistic kinematics (energy–momentum conservation) and partly on a quantum dynamics (conservation of spin, parity, charge, etc.). At first, 'traceless' particles without tracks like the neutron or the neutrino were conceived only on the basis of conservation laws. One ascribes to them an amount of energy (or another quantity) which is missing in the balance of all conserved quantities of the tracks observed in a scattering event. One assumes that the missing quantity is carried away by a hitherto unknown particle.

The principle of energy–momentum conservation plays a special role. It has been confirmed again and again, in contrast, say, to parity conservation. The prediction of the neutrino and the discovery of the neutron were also

---

[56]The development of the measurement theory which underlies the identification of charged particles is explained in Sects. 3.4–3.5.

[57]Maxwell 1962, Shapere 1982; see Sect. 2.5.

[58]See end of Sect. 2.5; and bear in mind that my suggestion aims at being as general *and* precise as possible.

based on this, as was the experimental proof of the photon. In 1930, to preserve the conservation of energy, Pauli suggested the neutrino hypothesis. It claimed that the continuous energy spectrum of the electrons sent out in β-decay was caused by a non-observable uncharged particle.[59] In 1932 Chadwick identified the neutron as a massive uncharged particle in the atomic nucleus. He showed that the radiation which arose in the reaction of α-particles with beryllium gave such a big recoil to the atoms of hydrogen or nitrogen that it could not be due to photons but only to massive particles with a mass of the order of the proton mass.[60] The neutrino hypothesis remained a *hypothesis* as long as no effects caused by neutrinos had been observed at the individual level. (The emission of electrons with a continuous energy spectrum in β-decay is a probabilistic process.) It was finally confirmed by an experiment in 1956 which generated the process of neutrino-induced inverse β-decay.[61] Only by the experimental evidence for particle reactions which can only be triggered by the particle in question is the hypothesis of such a 'traceless' neutral particle confirmed and given an observational basis. The cases of the neutron and the neutrino show once again that, for the empirical evidence of a particle, two conditions have to be fulfilled: the application of a measurement law typical of a particle, namely the principle of energy–momentum conservation; and a local effect associated with the measured dynamic quantities which is observable at the individual level. In the case of the neutron the local effect consists of the recoil of atoms. In the case of the neutrino it is the neutrino-induced β-decay.

Energy conservation has the character of a measurement law in these examples. It serves to standardize the observable phenomena and to infer from them the existence of uncharged particles of a certain dynamic type. Similarly, dynamic conservation laws provide the empirical basis for specific particle hypotheses in terms of quantized quantities like spin and parity, isospin and strangeness. Like the energy conservation law, they contribute to theoretical structuring, and by this to the refinement of the observational basis of particle physics. They make it possible to classify the empirical observation material into classes of particle tracks and scattering events, and to class them for particles and their reactions. The classification of the observation material by well-confirmed conservation laws leads to a new stage of theorizing the observational basis. Now it becomes 'sorted' into particle tracks and scattering events which are characterized by different values of conserved dynamic quantities.

---

[59] Pais 1986, 309–320.

[60] Shamos 1959, 266–280.

[61] Pais 1986, 569–570. Recall Eddington's scepticism about the possibilities of manufacturing neutrinos, expressed in Eddington 1939; see Sect. 2.3.

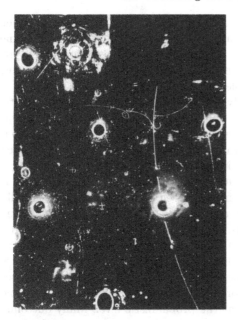

Fig. 3.4. Neutrino scattering event. First observation of a weak neutral current process in a bubble chamber (CERN, after Perkins 2000, 163)

3.3.4 Resonances

Since the 1950s, high energy scattering experiments have been performed in particle accelerators. There are two kinds of scattering experiments: *fixed-target* and *collider* experiments. A *fixed-target* experiment is performed as follows. A beam is generated in a particle accelerator. (It may be an electron beam, muon beam, proton beam, neutrino beam, etc.) Particle reactions are generated by sending the beam to a kind of matter block which functions at the same time as a particle detector. The matter block is called the fixed target. In the 1950s, the liquid hydrogen of the bubble chamber served as the fixed target. In later fixed target experiments the matter block consisted for example of iron plates which were sandwiched by scintillators equipped with photomultipliers, spark chambers, drift chambers or other kinds of particle detectors. In a *collider* experiment, two particle beams are crossed and scatter on each other within a small interaction zone surrounded by several kinds of particle detectors which detect the outgoing particles as completely as possible, depending on their kinematic and dynamic quantities. The era of the collider experiments began in the early 1980s.

The data of a high energy scattering experiment is analyzed as follows. One always collects *many* scattering events of certain dynamic types into statistical ensembles. By looking at these, the familiar grounds of individual particle tracks and their data analysis in terms of classical measurement laws are abandoned. Now the particle tracks and scattering experiments are interpreted in terms of the quantum theory of scattering. Like quantum mechanics, the quantum theory of scattering gives only probabilistic predictions in terms of expectation values of quantum observables. In high energy physics, these quantum observables belong to a quantum field theory.

Correspondingly, the data analysis of high energy scattering experiments makes use of statistical methods. Typically, a high energy scattering experiment in a particle accelerator has a run time of several months. It deals with huge amounts of statistical data which are analyzed in stochastic terms. The data are collected into *statistical ensembles* of scattering events. Theoretically, such an ensemble corresponds to the measurement results of a given kind of measurement in a given quantum state. Empirically, an ensemble of scattering events is a subset of the empirical event space on which the statistical data analysis of the respective experiment operates. It consists of events which are generated under the same experimental conditions and it splits into several sub-ensembles (or sub-subsets of the event space) which are defined in terms of the kinematic and dynamic quantities of the scattered particles. The quantum theory of scattering determines the outcome of a subatomic scattering process at the probabilistic level. Many outcomes are possible, in accordance with the respective quantum dynamics of subatomic interactions. The possible scattering outcomes are measurement results in the usual sense of a quantum mechanical measurement. In the quantum theory of scattering, their relative probabilities are given in terms of the S-matrix of the particle reactions. The S-matrix elements give the quantum theoretical *expectation values* for the possible kinds of particle reaction, which in turn correspond empirically to the *relative frequencies* of particle reactions of a given type which are measured.

In a high energy scattering experiment one determines the *effective cross-section* of a given particle reaction from such statistical ensembles of scattering events.[62] The measured effective cross-section is the relative frequency of scattering events of a given type. It is determined from the ratio of two measured numbers, namely the frequency of outgoing particles of given quantities relative to the frequency of incoming particles. The frequency of the incoming particles is given in terms of the *flux* of particles of a given type and energy which are generated by a particle accelerator. The particle flux is a magnitude of the particle accelerator or, to be more precise, a magnitude of the particle beam of a scattering experiment. It is measured independently of the measurement of the scattering results. The frequencies of the outgoing particles are obtained by counting the numbers of particles of a given type within given intervals of the kinematic quantities momentum or energy and the scattering angle. Thus, the effective cross-section of a certain kind of particle reaction is obtained straightforwardly by measuring two numbers: the number of incoming particles which is given in terms of the measured flux of

---

[62]See Appendix C.

the particle beam; and the number of scattered particles which is obtained by counting the scattering events of a given type. The latter number is in turn obtained by counting the corresponding particle tracks.

Thus, the *effective cross-section* is the physical magnitude with which the current quantum field theories come down to earth. As a *theoretical quantity*, the cross-section is calculated from the S-matrix of a quantum dynamics. It also expresses the relative coupling strength of the various kinds of physical interactions, namely the coupling constants of the electrodynamic, weak and strong interactions (which are analogous to the gravitational constant of classical mechanics). As an *empirical quantity*, it is the measured relative frequency of scattering events of a given type. Here, the scattering events are recognized using the methods described in Sect. 3.3, and counted in terms of the kinematic and dynamic quantities attributed to subatomic particles.

A *resonance* is the following phenomenon which shows up under the following experimental conditions. A high energy scattering experiment is performed and the reactions are measured as a function of the energy of the scattered particles, the scattering energy. In the corresponding theoretical model, the scattering energy is a quantity of relativistic kinematics. It is given in the center-of-mass frame of the scattered particles. One varies the scattering energy by varying the energy of the beam of a fixed-target experiment (or the beams of a collider experiment). One determines the total cross-section of all reactions or scattering events, summed up over all measured final states, as a function of the scattering energy (this simply means counting the numbers of scattering events at a given scattering energy). A resonance is measured if at a certain energy value the effective cross-section shows a *peak*, that is, if it suddenly increases by several orders of magnitude in relation to the effective cross-sections at lower and higher energies. Such a steep increase and decrease in the measured effective cross-section indicates a significant increase in the number of particle reactions at a certain scattering energy. It equals the typical resonance behavior of a forced oscillation in classical mechanics or electrodynamics; hence the name 'resonance'. Analogously, the resonances of high energy scattering experiments have been explained as the result of tuning an external force exactly to the energy of an internal state of a physical system. Such resonance behavior is also well-known from quantum mechanics. It gives rise to the excitation of electrons of the atomic shell and to induced emission of light. An unstable quantum state is generated which decays subsequently. In particle physics, the explanation of a resonance is analogous. At a certain scattering energy an unstable quantum state or a subatomic particle with a short lifetime is generated. Its decays give rise to the peak in the energy-dependent effective cross-section.

In the era of particle accelerators, the experimental examination of resonances and the measurement theory on which it is based attained a great importance, similar to the importance of the analysis of particle tracks from cosmic radiation and the corresponding measurement theory of the 1930s and 1940s.

The first particle resonance was discovered in 1952, when the first big accelerator began to operate. It was assigned to a particle of short lifetime with the spin and parity of the nucleons but a much bigger mass than the proton. Since then, the particle 'zoo' has grown immensely. It is based on a variety of further resonances which were assigned to unstable heavy particles (hadrons). Unstable particles cannot be identified one by one from particle tracks and scattering events. They are traceless like neutrons or neutrinos, but for another dynamic reason, namely their short lifetime. These resonances were classified according to conservation laws for isospin, strangeness and other quantized quantities which were first assumed ad hoc. The classification was closely connected to the early attempts to construct a quantum field theory of the strong interaction. Finally, it led to the quark model of 1964.[63]

The decays of these hadron resonances are examined in hadron spectroscopy using experimental methods analogous to those used to examine decays of the excited states of atomic shell electrons in atomic spectroscopy. Hadron spectroscopy has provided an important part of the experimental basis for particle physics over the last few decades. It finally helped the quark model to achieve a breakthrough in 1974, when the J/Ψ resonance was discovered. As it was extremely narrow, a very long lifetime was assigned to it.[64]

At first, resonances were stated merely comparatively from the functional form of a measured effective cross-section. If an observation is based on the comparison of phenomena (in the sense of Carnap's distinction of comparative observational quantities and quantitative measured quantities[65]), it is justified to talk about the *observation* of a resonance. But is it also justified to talk about the observation of the corresponding unstable quantum state or *particle*?

Resonances occur in the observational material of particle physics which is classified according to scattering events and the corresponding particle reactions. They are observed as contingent phenomena. However, their theoretical interpretation is much more theory-laden than the empirical evidence for the existence of subatomic particles discussed before. The measurement theory for the quantitative analysis of resonances is based on quantum theory. The quantum theoretical model of a resonance contains assumptions from the quantum theory of scattering, relativistic kinematics, and a theory of the excitation and decay of compound quantum mechanical systems into states of higher and lower energy which derives from quantum field theory.

---

[63]See Pickering 1984, 46–108.

[64]Ibid., 253–260. See, however, my criticism of Pickering's constructivist view of theory formation in particle physics in Sect. 2.3, as well as an experimenter's view, Riordan 1987.

[65]See Carnap 1966, Part II.

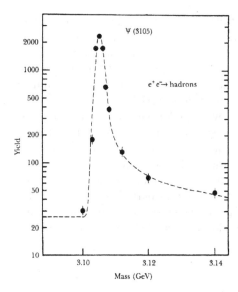

Fig. 3.5. The J/Ψ resonance
(Michael Riordan 1987, 284)

A resonance is characterized by two physical quantities: the energy E_0 of its maximum and its width ΔE which corresponds to the energy range covered at half the height of the resonance maximum. Both quantities can be read approximately from a diagram of the effective cross-section, as a function of the center-of-mass energy of the scattered particles. However, its precise determination is based on the Breit–Wigner resonance curve which derives from quantum field theory.[66]

A resonance is all the more distinctive the smaller its width ΔE is and the larger the relative frequency of reactions at the energy E_0 is in proportion to the number of reactions at much smaller or larger energies. The more distinctive it is, the more sharply its peak stands out (against a background of scattering events which do not belong to it) and the more exactly its maximum energy E_0 and its width ΔE can be determined. According to the quantum theoretical model mentioned above, a resonance is interpreted as empirical evidence for the decays of the excited state of a particle with mass $m = E_0/c^2$ and lifetime $\tau = h/\Delta E$.

According to this quantum theoretical model, a resonance is just *not* empirical evidence for a single particle which is spatio-temporally individuated. A resonance indicates a *spatio-temporal correlation of single quantum processes*, namely the decays of the excited states of quantum systems of a given kind. This correlation cannot be explained in the context of a classical particle model. If one interprets the appearance of a resonance as an empiri-

---

[66] See Perkins 1987, 128. A look at Chap. 8 of Goldberger and Watson 1975 shows the immense quantum theoretical apparatus that underlies the theory of resonances, summarized there for the needs of experimental physicists.

cal indication of the existence of a certain particle type, then apparently the *particle concept* gets a *new meaning*. By the term 'particle', Thomson meant something completely different in the theoretical interpretation of his e/m measurement. The change in the particle concept performed here will be examined in detail in Chap. 6. It indicates that the observation of a resonance is *not* like the observation of the local effect of a particle.

To interpret resonances as empirical evidence for particles already presupposes that the measurement theories of position, particle tracks and scattering events discussed above are interwoven with quantum theory. This interweaving stands only at the end of the process of theory formation which led to *quantum electrodynamics*. With quantum electrodynamics, the theorizing of the observational basis has reached a new stage in particle physics. It is no longer a pre-theoretical, metaphysical particle concept but a quantum theory which delivers the criteria for the decision of what belongs to the observational basis of particle physics.

3.4 The Track of the Positron

For a long time, the classical measurement theory for the analysis of particle tracks stood unconnected next to the knowledge that particle tracks are due to subatomic ionization processes which are subject to the laws of quantum theory. Until the early 1930s, no satisfactory quantum theoretical description of ionization processes was available at all. The exploration of cosmic rays with the cloud chamber started around 1930. At that time, only Bohr's classical ionization theory from 1913 and some quantum theoretical modifications of it were available. The position measurement theory for charged particles was based on a mixture of quantitative experimental knowledge and rather qualitative theoretical hypotheses on ionization.[67] The limitations of the classical measurement theory, however, gradually became visible over the years following the emergence of quantum mechanics. Bohr's ionization theory from 1913 was not in quantitative agreement with the half-empirical and phenomenological knowledge about particle tracks accumulated through use of the classical measurement theory. It was particularly in disagreement with the semi-empirical energy–range relation. Around 1930, the classical theory of the energy loss of charged particles in matter was known to give the correct results on average, but to be in need of quantum theoretical corrections and to have no validity for individual particle tracks.[68] Nevertheless, the experimental methods for identifying massive charged particles have long remained untouched by the theoretical revolutions of physics which gave rise to Einstein's light quantum hypothesis, Bohr's atomic model, the development

---

[67]Bohr 1913a, 1915; Rutherford et al. 1930, 434–439. See also Galison 1987, 97–100.

[68]Rutherford et al. 1930, 439.

of quantum mechanics, and finally, quantum electrodynamics. But before quantum electrodynamics could enter into the analysis of particle tracks, the positron had to appear and the muons and pions had to be identified. The path was long and difficult.

The quantum theoretical formalism for describing the interactions of charged particles with matter had existed since Born's seminal papers on the probabilistic interpretation of quantum mechanics had laid the grounds for the quantum mechanics of scattering. Using this, Bethe developed in 1930 his theory of the passage of a particle beam through matter. In it he described the interactions of massive charged particles with matter in Born's approximation.[69] Like Born's theory, however, Bethe's was a non-relativistic quantum mechanics of scattering. It did not therefore apply to the tracks of the fast, high energy particles from cosmic rays. The interactions of high energy charged particles with matter can only be described in the context of relativistic quantum mechanics and quantum electrodynamics (the relativistic quantum field theory of electromagnetic interactions). The basic equations of both theories were given by Dirac in 1927. A satisfactory approximate calculation of the ionization processes occurring in the interaction of high energy charged particles with matter was only possible after 1933. It was based on the *Bethe–Bloch formula* which derives from the relativistic Dirac equation and Bethe's quantum theory of the interaction of charged particles with matter.[70] Confidence in such calculations stood or fell with confidence in the Dirac equation. In 1930, this equation provided the only attempt at a relativistic theory of the interaction of charged particles with matter.

However, the Dirac equation did not have much credit with physicists before the discovery of the positron. Its solutions corresponding to negative energy values had no empirical interpretation. In this situation, the work of the experimenters and the theoreticians proceeded completely separately from each other. Starting out from Dirac's equation and field quantization, quantum electrodynamics was developed independently of all experimental data as an at first speculative candidate for a relativistic quantum theory. At the same time the experimenters had to do without a quantitative analysis of the tracks from the cosmic rays that was well-founded by quantum theory. Good experimental physicists only make use of safe background knowledge in their experiments. Safe background knowledge has to be theoretically convincing, well-established and empirically adequate, but the quantum electrodynamic description of the interactions of charged particles with matter fulfilled neither of these requirements in the 1930s. As long as the situation was like this, the particle tracks from the cosmic rays were only analyzed qualitatively and comparatively on the grounds of the classical measurement theory of a charged particle, making use of as few theoretical assumptions as possible. The discovery of the positron through this cautious way of analyz-

---

[69] Born 1926b; Bethe 1930.
[70] Bethe 1932; Bloch 1933.

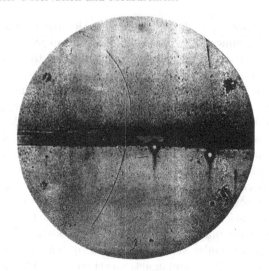

Fig. 3.6. The track of the positron (Anderson 1933, 492)

ing the tracks then decisively strengthened the empirical basis of quantum electrodynamics.

The positron was the first new particle found in cosmic radiation. It was identified with regard to charge and mass using empirical features of the particle tracks. C.D. Anderson had been working with a cloud chamber since 1931 to examine the cosmic rays at Caltech (California Institute of Technology).[71] In order to identify the sign of the charge of the particles creating tracks, a strong magnetic field was applied to the cloud chamber. On his photographs Anderson found quite a number of tracks which indicated positive particles. At first he regarded them as tracks of the proton, the only positively charged particle known at that time. There was, however, one piece of evidence that spoke against protons: the low ionization degree of the tracks indicated a mass of the order of the electron rather than the proton mass (which is almost two thousand times larger). Under the assumption that it was electrons from cosmic rays, however, the observed flight direction was not compatible with the track curvature in the magnetic field. To determine the flight direction of the particles unambiguously, Anderson finally put a lead plate of width 6 mm into the center of the cloud chamber. When passing the lead plate, the particles lost energy and momentum. This energy loss was manifested by an increase in the track curvature. Anderson discovered a track particularly suitable for particle identification in August 1932. The track on the photo showed the following phenomenological features:[72]

---

[71]Pais 1986, 351–352.
[72]See Anderson 1932, 1933.

1. Two partial tracks in the two halves of the cloud chamber met at the lead plate.
2. They were considerably different not in the magnitude but in the sign of the track curvature.
3. Both partial tracks were longer than 5 cm.
4. When both partial tracks were attributed to a single particle of negative charge, i.e., an electron, the track curvature in the magnetic field defined the flight direction exactly opposite to the direction of the energy loss due to the lead plate.

In his analysis of the track, Anderson discussed all degrees of freedom for the theoretical interpretation of the track:[73] the mass, the quantity of charge, the sign of the charge and the number of particles to which the track may be imputed. As he inferred from the ionization degree of the track, the charge could not differ in magnitude around a factor of two from the electron. The magnitude of the mass was estimated indirectly using the track length, the track curvature and the known values of the electron and the proton mass. Two theoretical assumptions were used: (i) the classical Lorentz force, according to which the given track curvature indicated, for a particle of the proton charge and mass, an energy of 300 MeV, and (ii) an extrapolation of the semi-empirical energy–range relation for protons towards low energies, according to which a proton of energy 300 MeV could only have a range of around 5 mm.[74] Due to feature (3) of the track, the mass had to be of substantially smaller order of magnitude than the proton mass. However, because of (4), the assumption that it was an electron would have implied a drastic violation of the energy conservation law: due to the curvatures of the partial tracks, an electron causing the complete track would not have lost energy in the 6 mm thick lead plate; rather, it would have been accelerated by 40 MeV, as Anderson emphasized. However, it could not consist of two electron tracks which were independent of each other and met by chance at the lead disk because the probability of such a coincidence was extremely low. Therefore, only two possibilities remained. Either the track was due to a single particle of positive charge which had lost energy at the lead plate and which had a mass and charge comparable to the electron. Or it was due to a pair of particles with opposite charges and equal mass which had arisen from one and the same reaction in the lead plate, and of which one was an electron. The existence of a positive electron, the positron, resulted from both possibilities.

---

[73] A detailed discussion of the epistemological aspects of the positron discovery is also found in Brown 1987, 186–189. Brown interprets Anderson's identification of the positron track according to a naturalistic epistemology. Similarly to Shapere's concept of direct observation from 1982 discussed in Sect. 2.5, it allows for an influence of theoretical background knowledge on observations.

[74] Rutherford et al. 1930, 294.

Anderson's identification of the atypical track of a positively charged particle of low mass shows paradigmatically that experimental physicists handle very cautiously all theoretical assumptions which do not yet belong to safe background knowledge. Anderson only made use of empirically approved classical measurement laws and phenomenological statements in order to interpret the positron track; and he used them only to compare the positron track with the well-known characteristics of proton and electron tracks. However, he did without an exact determination of the positron mass because this would have required a reliable theory of the energy loss of charged particles in matter which was not available at that time. Anderson's analysis of the positron track teaches that experimenters prefer to do *without* quantitative precision if the required measurement theory does not seem sufficiently sound.

Another fact can be learned from Anderson's way of proceeding. Just when decisive experimental discoveries are made, the analysis of experimental phenomena often proceeds quite independently of theory formation. The carrying out and evaluation of experiments is largely autonomous with regard to the simultaneous development of new theoretical approaches, as Galison's case studies in the history of particle physics also demonstrate.[75] Anderson knew the Dirac equation put forward in 1927, like many other particle physicists. If one interpreted its complete set of solutions literally, it implied the existence of particle states of negative energy. However, this knowledge in no way influenced the analysis of the positron tracks which had been available in observational material since 1931. Around 1930, experimenters were in no way interested in searching for particles with the electron mass and positive charge in order to provide solutions of the Dirac equation with an empirical basis. Quite the contrary, Anderson tried as long as possible to interpret the atypical tracks in terms of the proton charge and mass, as did Millikan in whose group he worked. Dirac himself had also tried to give a conservative interpretation to the solutions of his equation, namely to assign the solutions of the Dirac equation of negative energy to the proton.[76]

3.5 Particle Identification and Quantum Electrodynamics

Physicists are conservative. As far as possible, they try to rely only on their approved background knowledge without putting it into question. To the reliable background knowledge of 1930 there belonged the unrefuted assumption that there are only two charged massive elementary particles, namely the electron and the proton, but not the quantum theoretical approaches and

---

[75]See Galison 1987. In particular, his case study of the discovery of the neutral currents invalidates the constructivist view of recent particle physics which is held in Pickering 1984.

[76]See Pais 1986, 346–348.

methods which had been developed in the 1920s and which were swaying the whole edifice of physics as it had been known previously. In the 1930s physicists only brought new particles into play to explain experimental phenomena if such a hypothesis could no longer be avoided without endangering more fundamental principles such as energy conservation. The energy conservation law had been connected by relativity theory with that for momentum and it had survived the quantum revolution unscathed. Before the discovery of the positive electron by Anderson nobody took the whole set of solutions of the Dirac equation seriously (including Dirac).[77] The positron was only afterwards identified with particle states of negative energy belonging to the solutions of the Dirac equation. This interpretation is neither due to Anderson nor to Dirac, but to Blackett and Occhialini. They confirmed Anderson's discovery of the positron in 1933 by experimental examinations of their own and gave it a theoretical interpretation in terms of solutions of the Dirac equation.[78] This interpretation identified the solutions of the Dirac equation of negative energy values with a description of free particles with positive charge and electron mass, that is, with the idealized description of a positron that leaves an ionization track in the cloud chamber at its multiple scattering processes with liquid hydrogen.

Hence the discovery of the positron was carried out independently of the Dirac equation. It provided the uninterpreted (or *unphysical*) solutions of negative energy with an *unexpected* empirical basis. Still in 1933, Anderson was able to show that the positrons of the cosmic rays were created by *pair creation* processes in which an electron and a positron arose together. On the basis of the Dirac equation this process also found a natural interpretation.[79] On the one hand this encouraged theoreticians to develop quantum electrodynamics based on the Dirac equation, Maxwell's equations, and field quantization and to apply it to the interactions of charged particles of high energy with matter. On the other hand, in a step-by-step manner, it encouraged experimenters to use the results of theorists to analyze the observational material from cosmic rays. The further development of quantum field theory by theorists was at first completely independent of experimental particle physics. Conversely, the analysis of the observational material from cosmic rays by experimenters was not yet influenced by the first (quantum mechanical) calculations of quantum electrodynamics. However, as soon as the Dirac equation was confirmed by the discovery of the positron, both started to interweave.

Essential steps towards the theoretical description of the interactions of relativistic charged particles with matter were taken long before the theoretical and experimental consolidation of quantum electrodynamics. In 1931

---

[77] A detailed history of the physical interpretation of the Dirac equation is found in Stöckler 1984, 94–106.

[78] See Pais 1986, 362–363.

[79] See Anderson and Anderson 1983.

Møller managed to derive the scattering of two electrons from the Dirac equation in the Born approximation of quantum mechanics, in a relativistically invariant description.[80] In 1932, Bethe used Møller's scattering formula to derive a relativistic formula for the energy loss of electrons due to ionization processes in the matter of a particle detector. However, this formula could be exactly calculated only for the hydrogen atom, that is, for the energy loss of electrons in the cloud chamber.[81] In 1933, Bloch completed Bethe's formula for more complex atoms in the context of a theoretical model which describes the electrons of an atom as a Fermi gas.[82] In this way the Bethe–Bloch formula (already mentioned above) was derived for high energy charged particles, as a law describing the stopping power of atoms. In 1934, the formulae for bremsstrahlung and pair production were introduced. The former is the inverse process of Compton scattering, and the latter is the production of an electron–positron pair from a photon. Pair production was discovered in 1933 by Blackett and Occhialini, by analyzing the tracks of cosmic rays.[83] With the Bethe–Bloch formula and the formulae for bremsstrahlung and pair creation the energy loss of charged particles due to passing through matter could be completely calculated. In this way, by the middle of the 1930s the quantum theoretical knowledge which could be used for the analysis of particle tracks had increased immensely.

But at that time quantum electrodynamics was still far from consolidation. When quantum field theory was developed and refined in parallel with relativistic quantum mechanics, severe internal theoretical problems remained (which were resolved only in the 1950s by the renormalization approach). In addition, there was an experimental confusion around the analysis of the tracks from cosmic rays. There were tracks due to particles of medium mass (then called 'mesotrons', now called mesons), but they were only identified in 1936. Prior to this, they gave rise to a vicious circle in the analysis of particle tracks. In order to precisely determine the particle mass corresponding to a track, quantum electrodynamics was needed. The calculation of the energy loss of charged particles in matter stood or fell with quantum electrodynamics. Conversely, in order to trust in quantum electrodynamics, the energy loss of well-known particles in matter had to be measured. However, the predictions of quantum electrodynamics for the hitherto known particles and the unidentified meson tracks disagreed. Therefore even after the striking confirmation of the Dirac equation by the positron the validity of quantum electrodynamics remained doubtful. In particular, before the mesons were identified, it had been assumed that quantum electrodynamics fails at high energies, where the energy loss of charged particles in matter is dominated by

---

[80]Møller 1931.

[81]Bethe 1932.

[82]Bloch 1933.

[83]Bethe and Heitler 1934; Pais 1986, 375–376; see also the survey in Rossi 1952, 151, and the historical presentation in Galison 1987, 103–110.

processes of bremsstrahlung and pair production.[84] The laws which describe
the energy loss by ionization, bremsstrahlung and pair production derive
more or less in the same way from the Dirac equation in the Born approximation. However, due to the vicious circle sketched above they did not enter
the measurement theories. They were accepted as safe background knowledge
only after the end of the 1930s, to the extent that there were then additional
experimental confirmations of quantum electrodynamics.

The next confirmation was indeed the experimental identification of the
'mesotrons' or mesons. Anderson proved only in 1936 from particle tracks of
cosmic rays that there was a charged particle with a mass between the electron and proton masses. The particle he identified is now called the muon,
a heavier relative of the electron. The lack of confidence in quantum electrodynamics and in the calculation of the energy loss by scattering processes
at high particle energies delayed its identification for two years. The delay
was due to the vicious circle described above. Due to his insufficiently selective experimentation methods of 1933, Anderson found some particle tracks
which he regarded as tracks of electrons in his photographs from the cosmic
rays. However, these were the as yet unidentified muon tracks. This confirmed
the suspicion that quantum electrodynamics fails at high particle energies.[85]
Anderson further refined his measuring methods by installing a platinum
absorber of width 1 cm into the cloud chamber. This absorber stopped the
electrons and let the muons pass through. The obvious conclusion was that
the platinum absorber was passed by a charged particle which was heavier
than the electron. On the other hand, quantum electrodynamics was much
better validated for the energy loss by ionization processes; and it predicted
that the unknown particle had to be lighter than the proton (if its predictions
for particles of low to medium energy were correct).[86] In this way the confusion found an end, and confidence in quantum electrodynamics increased.

---

[84]The identification of the origin and the particle content of particle showers
from cosmic radiation caused great problems; see ibid., 110–124 as well as Cassidy
1981.

[85]Ibid., 2 and 12–15.; Anderson and Anderson 1983, 143–146. Galison 1987, 137,
emphasizes that Anderson used the same cloud chamber as for the discovery of the
positron but worked with a better developed experimenting technique.

[86]Pais 1986, 432; Cassidy 1981, 14. The theoretical context and the different
experiments belonged to two separate schools of particle physics. The historical
details of the process which led step by step to the identification of the muons cannot be explained here; see Galison 1987, Chap. 3; in particular 126–133. Galison
emphasizes that the exact instant of the muon discovery cannot be defined because
the discovery was due to a collective learning process amongst particle physicists
(as for the neutrino). In this collective learning process, more and more explanatory
options were carefully eliminated: "The move towards acceptance of the muon was
not the revelation of a moment. But by tracing an extended chain of experimental
reasoning like this one, we have seen a dynamic process that, while sometimes com-
pressed in time, has occurred over and over in particle physics. With the discovery

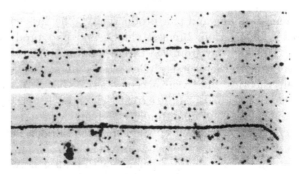

Fig. 3.7. Proton tracks measured with nuclear emulsions (Powell et al. 1959, 31)

In the 1930s, the development of quantum electrodynamics and its empirical successes took several detours which cannot be outlined in more detail here.[87] However, one conclusion is obvious. Even though the degree of acceptance of this theory increased considerably after the discovery of the positron, its empirical content was investigated carefully before any of its laws entered into the analysis of particle tracks.

When quantum electrodynamics was finally accepted, its predictions for the average energy loss of charged particles in matter gave rise to better measurement laws. Now one took advantage of the Bethe–Bloch formula, etc., for the analysis of particle tracks, or for quantitative improvements of the classical measuring theory. The quantum electrodynamic formulae for ionization, bremsstrahlung and pair production give rise to a theoretical prediction for the energy–range relation. The quantum electrodynamic prediction can be directly compared to the corresponding semi-empirical relation which is based on a classical particle model and on track lengths measured in scattering experiments. In this way it becomes possible to identify charged particles by measuring their mass. The corresponding measurement theory combines the quantum theoretical energy–range relation with the classical calculation of the track curvature in a magnetic field based on the Lorentz force. The classical measurement law is refined by a quantum theoretical correction. It is typical that such improvements in a measurement theory are based on piecemeal physics.[88] Now mass measurements became much more accurate, as compared to Anderson's earlier rough estimates of the mass of the positron and the muon which were much less theory-laden.

In addition, the semi-empirical methods for mass measurement which are completely independent of quantum electrodynamics were also further refined in the 1940s. The new nuclear emulsions had a high spatial resolution.

---

of the neutrino, for instance, one sees such a gradual elimination of alternatives." Ibid., 133.

[87] See Schweber 1994.

[88] See Chap. 4.

They allowed one to determine the mass of a particle with high precision, independently of its range, using the density of the measurement points.[89] In 1947, this method made it possible to distinguish the muon discovered in 1936 and the pion which had already been predicted in 1935 in Yukawa's theory of the strong interaction. It had been mistaken for the muon from 1936 to 1947. The muon and the pion have masses of 106 MeV/c^2 and 140 MeV/c^2, respectively.[90] Anderson could not have dreamed of resolving such a small mass difference experimentally in 1932. Further improvements of this kind in the phenomenological analysis of particle tracks became possible when Glaser invented the bubble chamber in 1954. It could register a multitude of particle tracks simultaneously and immediately make them visible on photographs. The semi-empirical energy–range relation could also be investigated with high precision for many particle types and over a large energy range once the bubble chamber was used in scattering experiments at particle accelerators. Thus on the one hand, by calculating the energy loss of charged particles in matter, quantum electrodynamics had become interwoven with the originally classical measurement theory of particle tracks. Nevertheless, on the other hand particle physics still had precise independent procedures of particle identification at its disposal. At least this is valid as far as the particle types can be identified using single particle tracks and scattering events in the observation material.

This changes only for the resonances. Their measurement is based on a quantum theoretical model of correlated decays of unstable particles. The interweaving of quantum electrodynamics with its measurement methods hereby reached a stage where it was no longer possible to identify a kind of particle independently of quantum theory. This new stage of theory-laden measurement procedures was called *T-theoreticity* by Sneed.[91] It was reached, however, only after the experimental and theoretical consolidation of quantum electrodynamics. In addition, the interpretation of resonances in terms of unstable particles also remained in many ways related to the analysis of the tracks belonging to the individual scattering events which contribute to a resonance.

3.6 Are There Subatomic Particles?

The observational basis of particle physics consists of position measurements, particle tracks, scattering events, and resonances. We have seen in this chapter how it is associated with two informal particle concepts, namely a causal and a mereological concept. According to the former, particles cause local effects in particle detectors. They give rise to the observation of a recoil as in the

---

[89]Rossi 1952, 138–142.
[90]Ibid., 162–163, and Lattes 1983.
[91]Sneed 1971.

Compton effect, clicks in Geiger counters, particle tracks, scattering events which underlie certain conservation laws, and so on. According to the latter, particles are the microscopic parts of matter. They underlie part–whole relations which are expressed in terms of certain sum rules for the constituents of a dynamic compound system. Both informal particle concepts are inherited from traditional metaphysics and philosophy of nature. We have also examined in detail the layered structure of the observational basis of particle physics which is interpreted in terms of these particle concepts. Now we are on much better grounds regarding the debate on scientific realism and the question of whether there are subatomic particles.

According to the results of Chaps. 1 and 2, empirical reality consists of contingent events and actual measurement results. According to the position of a moderate realism which is adopted here, empirical reality is theory-laden. All measurements depend on measurement theories and all measurement results are expressed in terms of physical quantities. The quantities are embedded in scales and attributed to physical objects and systems. The scales and the notion of a physical object are the basic conceptual schemes or the *a priori* of modern physics. They come hand in hand with the analytic methods and the constructive features of Galilean science. They are predominantly classical and they give rise to the generalized concept of observation discussed in Sect. 2.6. This generalized concept of observation is indeed explained in traditional metaphysical terms of substance and causality. An observable object or system in the broadest sense is considered to be an individual carrier of physical quantities, that is, a substance. These quantities are attributed to it on the basis of an individual causal story which connects it with the observable phenomena and measured data.

The conceptual schemes of classical physics have to be weakened in view of the quantum structure of the subatomic domain. Here the conceptual schemes of substance and causality no longer fully apply. The attribution of the measured values of position, momentum, mass, charge, energy and so on to the subatomic parts of matter becomes highly problematic in view of Heisenberg's indeterminacy relations. Therefore we may say that there are only subatomic particles if it is possible to tell causal stories which unambiguously trace from the data back to the physical quantities of microscopic parts of matter. Or to put it in other words, there are only subatomic particles if it is possible to observe them in a generalized sense as parts of matter which cause local effects in particle detectors.

Indeed, all particle detection methods discussed in this chapter satisfy this criterion for the existence of subatomic particles. Obviously, the discoveries of the electron and the photon were observations in the generalized sense suggested here. Thomson measured the characteristic e/m value of cathode rays in 1897. Millikan later confirmed that it can be unambiguously attributed to individual oil droplets (see Sect. 3.2.1). Einstein set up the light quantum hypothesis in 1905 and attributed a momentum $\boldsymbol{p} = \hbar\boldsymbol{k}$ to the light quantum.

Bothe and Geiger later confirmed that a light quantum transfers momentum (or loses energy and lowers its frequency) when it gives rise to a recoil electron (see Sect. 3.2.2). The theorized ways of making position measurements and analyzing particle tracks make it possible to attribute unambiguous values of mass, charge and momentum to a particle track whenever the methods of data analysis are sufficiently discriminating (Sects. 3.3–3.5). The analysis of scattering events aims at individuating the particles involved in subatomic reactions which obey certain conservation laws. Here the causal story traces from observable particle tracks back to conserved dynamic quantities such as isospin or strangeness which can be attributed unambiguously to an individual scattering event. Only resonances are different. A resonance *per se* does not indicate an individual part of matter but only a certain kind of unstable particle state. To single out the respective particle in the sense of a generalized account of observation, more detective work has to be done.

The criterion given above is obviously a necessary, but not a sufficient condition for the existence of subatomic particles. To give sufficient conditions, the ways in which the causal and mereological particle concepts apply to the phenomena of particle physics have to be investigated in more detail. The main obstacle for finding such sufficient conditions is the *quantum structure* of the subatomic domain.

Modern philosophical anti-realism, however, falls short of understanding this kind of obstacle. Recent empiricist and constructivist approaches to the philosophy of physics focus on philosophical problems which are more or less irrelevant for understanding why subatomic reality is at stake. The layered structure of the observational basis of particle physics supports the general arguments against empiricism and constructivism which have been presented in the preceding chapter. Let us recall the central empiricist and constructivist objections against physicists' inferences from experimental data to unobservable entities such as subatomic particles:

1. **The experimental data are technological products which do not exist in nature.** This criticism misses the *contingency* of the data taken by a particle detector. Even though the local effects of electrons or protons can only be observed by means of technological devices such as the bubble chamber, they cannot be arbitrarily generated. A particle track is either on the photo or it is not. A track may have small or large curvature. Its measurement points may have low or high density. And so on. The phenomenological shape of a track is not at the experimenter's disposal. In addition, the phenomena observed in a scattering experiment with the beam from the particle accelerator overlap the phenomena observed without the particle beam. There is a continuum of observed empirical structures ranging from nature outside to physics inside the laboratory. Due to the existence of cosmic rays, a particle detector still detects particle tracks when the beam is switched off. (Particle physics began with the investigation of cosmic rays and came back to it in modern astroparticle

physics.) And under well-known conditions the muon tracks on a cosmic ray photo have the same shape as the muon tracks detected by a particle detector.

2. **Unobservable causes and non-empirical structures do not belong to empirical reality but to metaphysics.** Again the *contingency* of subatomic phenomena is missed. The data of an experiment are recorded and not arbitrarily generated. The empirical structures in which they are embedded are theory-laden. But they nevertheless give rise to contingent phenomena. Any particle *track* is already fitted through the data. Nevertheless in a magnetic field, an electron track has the opposite curvature to a positron track, and a muon track shows the same sign of curvature but the curvature is much smaller. Newtonian science infers that different effects are due to different law-like causes. Causal inferences *per se* do not give rise to bad metaphysics. Only the inference that the cause of a particle track is a *classical* particle turned out to be bad metaphysics. Indeed, particle tracks show that the empirical content of a theory cannot be fixed as easily as van Fraassen believes. In the subatomic realm, the empirical content of a particle trajectory consists only of the single measurement points. The corresponding empirical structure is discontinuous, in contradistinction to the empirical structure of classical mechanics to which the planetary motions belong. The very concept of an empirical structure depends on theoretical knowledge. Continuous data models do not belong to the empirical structures of a quantum theory but to the empirical structures of classical mechanics. Therefore, van Fraassen's restrictive criterion of data models which admit only spatiotemporal structures falls short of quantum structures. This has already been noted.[92] Here an additional point can be made. Experimental discoveries may substantially change the empirical content of a theory. The discovery of the positron gave empirical content to the negative energy solutions of the Dirac equation which had previously been called *unphysical*. The non-empirical part of a theory is not bad metaphysics. It is indispensable for our description of physical reality. The description may be wrong or right. In any case, it may become testable in an unexpected way and give rise to unexpected empirical predictions.

3. **The theory-ladenness of the data and phenomena is based on a symbiosis of theory and measurement.** This objection counters the two arguments given above. If theory and measurement are symbiotically interwoven, there are no contingent data. According to Pickering's book *Constructing Quarks*,[93] the data of high energy physics are socio-theoretical construals that depend on school formation within the scientific community. However, it was shown here that the observational basis of particle physics has a layered structure which is built up ex-

---

[92] See my discussion of van Fraassen 1980, 59–60, in Sect. 2.6.
[93] Pickering 1984.

tremely carefully. In each layer there is a new interplay of observation, measurement and theory. But in each layer only independent measurement methods are considered to be reliable, as long as a theory has not yet been confirmed. The symbiosis of theory and experiment is the result and not the starting point of this development. In particle physics, the dependence of measurement on quantum field theory actually comes at a very late stage of theory formation.

If there was ever a symbiosis of theory and experiment at the beginning of theory formation, then these were the classical foundations of subatomic physics. This symbiosis, if it was one, did not prevent physicists from replacing classical mechanics and electrodynamics by the corresponding quantum theories. After all, nothing is left of the constructivist and empiricist objections against interpreting position measurements, particle tracks, scattering events and resonances as empirical evidence of subatomic particles. However, we have not yet seen *how it is possible* to interpret subatomic reality in terms of particles. The question of whether and to what extent the causal and mereological particle concepts also work in the quantum domain has remained open up to now.

4 Probing Subatomic Structure

For a century, subatomic structure was investigated by means of scattering experiments. The first scattering experiments of subatomic physics were performed in Rutherford's laboratory around 1910. In order to explain the unexpected backward scattering of α-rays at a thin gold foil, Rutherford postulated the atomic nucleus as a pointlike scattering center inside the atom. Later scattering experiments demonstrated that the atomic nucleus is not really pointlike. Like the atom, the nucleus shows internal structure in scattering experiments of sufficiently high energy. Now the effects of protons and neutrons were measured. Much later, Rutherford's backward scattering story recurred and the quark constituents of the nucleon came into play.

The scattering experiments of particle physics introduced a new particle concept. In terms of scattering theory and the related measurement methods, a particle is a *scattering center* inside the atom. Here, 'the atom' means the atoms within a macroscopic bulk of matter which may be a solid-state fixed target like Rutherford's gold foil, a tank filled with liquid hydrogen like the bubble chamber, or one of the big fixed-target detectors of the 1970s and later. This new particle concept comes in terms of spatial analogies. A scattering center may be pointlike or non-pointlike. Pointlike particles act in a scattering experiment like a point charge, while non-pointlike structures act like an extended charge distribution.

The new particle concept is theory-laden. It is based on classical and quantum mechanical *scattering theory*. It results from Rutherford's classical description of the Coulomb scattering at the atomic nucleus as well as its quantum theoretical generalizations. Indeed, it can be formulated in a model-independent way. Under well-defined experimental conditions it corresponds to the classical account of a scattering center. Therefore it may serve to interpret subatomic structures in quasi-classical terms. In the following it will be shown how the concept of a pointlike or non-pointlike scattering center is related to the informal particle concepts of the preceding chapter. How do the causal and the mereological particle concepts match in the data analysis of a scattering experiment? Is a scattering center, respectively its pointlike or non-pointlike structure, *observed* in the sense of our generalized concept of Sect. 2.5? Indeed, this question is crucial for understanding the usual popular talk of 'observing' the subatomic structure of matter by means of an electron microscope or a particle accelerator. In what sense and to what extent do the scattering experiments of particle physics compare to macroscopic observations?

Classical Rutherford scattering is in two respects paradigmatic for the investigation of the subatomic structure of matter:

(i)   The *scattering experiments* with charged particles from radioactive sources in Rutherford's laboratory proved to be the guiding experimental method for investigating subatomic structure.

(ii) The *classical assumptions about a pointlike scattering center* under which Rutherford derived his scattering formula proved to be surprisingly resistant to the transition from classical physics to quantum mechanics. They can be embedded and generalized in quantum mechanical scattering theory. Together with this theory, they deliver efficient theoretical tools for describing the scattering of charged particles at atoms and for interpreting the structure of subatomic scattering centers.

According to the model of Rutherford scattering, the structure of a scattering center is interpreted in terms of length. Depending on the results of a scattering experiment, namely the measured effective cross-section,[1] one describes the scattering center as extended or pointlike. In this way, one extends the length scale into a domain of unobservable structures in order to make them experimentally accessible. As long as no physical dynamics is available for this domain, only heuristic methods give rise to information about the spatial and dynamical structure of matter on this scale. They proceed partly by familiar models and partly by meta-theoretical considerations based on dimensional analysis.[2]

In this way, bridges are built between incommensurable theories like classical point mechanics and quantum mechanics. It is typical of current physics that bridge principles are crucially involved in theory formation and model building. They are examined here for the measurement theory based on the model of Rutherford scattering, which gives rise to the identification of particles and microscopic scattering centers. The following questions are crucial. To what extent is this measurement theory model-dependent? To what extent can the model dependence be removed by generalizations? It is obvious that any model dependence of the measurement theory also affects the theoretical structure of the phenomena. Here, a *model* is the specific description of a physical system within a specific theory. In this sense, physicists talk of *model dependence*, while they refer to more general descriptions based on meta-theoretical considerations as *model-independent*.

In terms of the realism debate, we have to state that the measurement theory of scattering experiments probing the structure of the atom gives rise to a classical construal of subatomic reality. But quantum theory tells us that

---

[1]See Appendix C.

[2]See Appendix B.

the classical construals of matter no longer work. In the framework of quantum theory, the construal of subatomic reality turns out to be *quasi-classical*. Closer examination shows that it is modelled *as if* the scattering center were a classical charge distribution. So where is the borderline between the classical and the quantum domain? How much does the classical construal of subatomic structure depend on classical assumptions? Where does it definitely break down? And to what extent do scattering experiments exhibit *any* model-independent, contingent structures of subatomic reality?

In order to answer these questions, we proceed as follows. To begin with, some typical scattering experiments of particle physics and their measurement theories are explained in Sect. 4.1. Then an important meta-theoretical concept is introduced, the concept of *scale invariance*. Scale invariance is a symmetry. It is the invariance of a measured cross-section under scale transformations. As a symmetry, it is a good candidate for a model-independent concept. However, the subject is tricky. Scale invariance enters the data analysis of scattering experiments via dimensional analysis; but dimensional analysis is a heuristic procedure which in turn depends on a quasi-classical construal of the scales of physical quantities (Sect. 4.2). Dimensional analysis and scale invariance give rise to the concept of *pointlikeness*. This is a spatial concept which cannot be arbitrarily generalized to the domain of relativistic quantum theory (Sect. 4.3). The concept of pointlikeness finally gives rise to the definition of *form factors* which describe *non-pointlike structures*. Their definitions in the non-relativistic and relativistic domains have to be carefully analyzed (Sects. 4.3.1–4.3.2). The quasi-classical construal of subatomic structure is based on a chain of models which correspond to classical Rutherford scattering and to the description of the scattering in terms of a classical charge distribution. This correspondence (and together with it any quasi-classical construal of subatomic reality) breaks down stepwise, the larger the energy of a scattering experiment, the smaller the probed structures (Sect. 4.4). It has to be concluded that scattering experiments give rise to a quasi-classical account of subatomic reality, as far as the specific effects of relativistic quantum theory are negligible. In the non-relativistic domain, form factors are the refuge of an unproblematic realism. This claim is supported by an analogy between scattering experiments and the optical microscope (Sect. 4.5). The analogy supports the metaphor of looking deeper and deeper into the atom by means of particle accelerators. This metaphor is finally discussed in terms of the generalized concept of observation developed in Chap. 2 (Sect. 4.6).

4.1 Scattering Experiments

Scattering experiments are simple. In a low-energy scattering experiment, radiation from a radioactive source is collimated and sent to some target. In the experiments performed in Rutherford's laboratory after 1908, the radiation consisted of α-particles. It took Rutherford ten years to determine the mass and charge of α-particles. Once it was established as a matter of fact that they are doubly ionized helium atoms, Rutherford's assistant Geiger started to do scattering experiments with them.[3] The source was a radium pencil, the scatterer a thin aluminium or gold foil. Single α-particles were detected using Crooke's scintillation method described in Sect. 3.3. To their complete surprise, Geiger and Marsden found backward scattering at an angle $> 90°$. Many years later, Rutherford described this unexpected discovery in a lecture as follows:

> It was quite the most incredible event that has ever happened to me in my life. It was almost as incredible as if you fired a 15-inch shell at a piece of tissue paper and it came back and hit you.[4]

This discovery was obviously *contingent*. Not only could it have happened otherwise, but it was *expected* to happen otherwise. Indeed, the measurement method was model-independent with regard to the subatomic structure of the gold foil. It only depended on background knowledge concerning the nature of the α-particles and on Crookes' scintillation method. The former had been independently established in the previous decade, whereas the latter gave rise to empirical observations in a strict sense.

The calculations of scattering theory only came afterwards. Rutherford spent two years calculating the probability of multiple backward scattering in several atomic models. (Now he proceeded according to a hypothetico-deductive model of theory formation. He did not direct Geiger and Marsden to measure the differential cross-section of the α-particles as precisely and completely as possible, in order to derive in an inductive way a theory of the positive charge distribution inside the atom. First he did his calculations. Then he instructed Geiger and Marsden to test his scattering formula.) The homogeneous charge distribution of Thomson's atomic model could not explain Geiger and Marsden's discovery. Finally he derived his famous scattering formula from an atomic model with a pointlike charge described by a Coulomb potential. This time he was able to explain the backward scattering. According to the model, the differential cross-section $d\sigma/d\theta$ goes as $\sin^{-4}(\theta/2)$ with the scattering angle θ. The formula predicts a non-negligible probability of large-angle scattering. The predictions of the formula were confirmed by subsequent scattering experiments which measured the angular distribution of the scattered α-particles, using the same measurement method as before.[5]

---

[3]See Pais 1986, 60–62 and 188–190.

[4]Andrade 1964, 111 (quoted from Pais 1986, 189).

[5]See Pais 1986, 190–192, and Trigg 1971, Chap. 2. Rutherford's model included an additional term for the shielding by the electrons which turned out to be negligible. For the Coulomb potential the calculation is found in every textbook of nuclear physics. Indeed, Rutherford had good luck with his prediction. Geiger and Marsden performed their scattering experiments with low energy α-particles of around

Scattering experiments at particle accelerators started in the 1950s. In a particle accelerator, charged particles (electrons, protons, or heavy ions) are accelerated by means of electric and/or magnetic fields to a well-defined momentum or energy. The first particle accelerator was the cyclotron. It was designed by Lawrence in 1929. The first cyclotrons were running in 1930. The very first machine had magnetic poles of diameter 10 cm. The next milestone was a nine-inch model. It accelerated protons beyond 1 MeV.[6] After the second world war, the era of the big machines began. The size of the machines was rapidly increased in order to increase the beam energy. In the 1950s, the first proton synchrotons were built. They generated beams of 1–10 GeV. In the 1970s, 500 GeV was reached.[7] Current machines generate beams of the order of 1–10 TeV.

For a scattering experiment, a particle beam with a well-defined energy is extracted from the accelerator. In a *fixed target experiment*, the beam is sent to a bulk of matter at which it is scattered. The scattered particles are detected. Either the target itself functions as particle detector, as in the case of the bubble chamber which was developed by Glaser in 1952, or the target is equipped with sophisticated particle detectors such as scintillators and photomultipliers, spark chambers, drift chambers, Čerenkov counters and so on. These new kinds of particle detector made *collider experiments* possible. In a collider experiment, two beams are crossed in an interaction zone and scattered off each other. The scattered particles are recorded by detectors which surround the interaction zone as completely as possible, in order to include all particle reactions in a solid angle as big as possible. The idea behind the design of collider experiments is again to increase the energy of the scattering. A fixed target is at rest in the laboratory, whereas a collider experiment brings two beams of high energy and opposite momentum into collision. In the center-of-mass frame of the scattering process, in a collider experiment the scattering energy is much higher.

Why were the machines made bigger and bigger, in order to generate beams of higher and higher energies? The underlying heuristics is simple. It says that the higher the beam energy is, the better the spatial resolution of a scattering experiment will be, and hence the smaller the spatial structures which may be observed. The basic idea behind this heuristic principle stems from quantum mechanics. In the quantum mechanics of scattering, the beam is described by a quantum mechanical wave function Ψ. The particles in the beam, the so-called *probe particles*, have a momentum p. According

---

5 MeV and targets with high nuclear charges. The experiments were sensitive neither to deviations from Rutherford's formula due to strong interactions between the α-particles and the gold nucleus, nor to any quantum mechanical effects. For the Coulomb potential (and *only* for the Coulomb potential) the quantum mechanics of scattering results in the same formula; see Sect. 4.3.1.

[6]Pais 1986, 408.

[7]Pais 1986, 475–478.

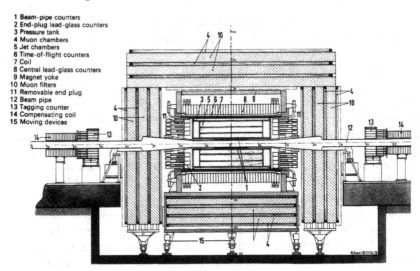

1 Beam-pipe counters
2 End-plug lead-glass counters
3 Pressure tank
4 Muon chambers
5 Jet chambers
6 Time-of-flight counters
7 Coil
8 Central lead-glass counters
9 Magnet yoke
10 Muon filters
11 Removable end plug
12 Beam pipe
13 Tagging counter
14 Compensating coil
15 Moving devices

Fig. 4.1. Collider experiment. JADE detector for the measurement of e^+e^- collisions, DESY (Perkins 1987, 63)

to quantum mechanics, this momentum corresponds to a de Broglie wavelength $\lambda = 2\pi\hbar/p$. With increasing beam energy E or beam momentum p, the de Broglie wavelength λ of the probe particles decreases. But the smaller the wavelength of the rays of an observation instrument, the smaller the structures that can be observed. The theoretical grounds for this heuristic principle are the analogies between the wave equations of quantum mechanics and optics, or a particle accelerator and a microscope.[8] However, particle accelerators are *big*. They accelerate charged particles by making them pass through macroscopic magnetic and electric fields over large macroscopic distances. Therefore the acceleration works according to the laws of classical mechanics and electrodynamics. In contradistinction to the electrons inside the atom (which have fixed quantum states), a circulating macroscopic beam of charged particles emits synchrotron radiation.

Even though the beam is generated and extracted from the accelerator according to the laws of classical physics, the scattering itself is described in terms of quantum mechanics. According to quantum mechanics, the beam is prepared in a *momentum state*. This momentum state corresponds to a *plane wave*. According to the quantum mechanics of scattering, the wave is *diffracted* by the scattering. (Therefore in the heuristic principle mentioned above the de Broglie wavelength is brought into play.) However, after the scattering *particles* or particle tracks are detected. According to the usual probabilistic interpretation of quantum mechanics, the probability of scattered particles in a given state corresponds to the squared amplitude $|\psi|^2$ of

---

[8]See Sect. 4.4.

the diffracted wave. Empirically, this probability corresponds to the relative frequency of particle detections of a given type, for a very large number of particle reactions or scattering events.[9]

The data analysis of a scattering experiment proceeds in the steps described in Sect. 3.3. Position measurements are made. They give rise to particle tracks. The particle tracks are interpreted in terms of scattering events. Today the position measurements are recorded by means of electronic devices and the particle tracks and scattering events are reconstructed out of them by means of computer programs. The particle tracks and scattering events are analyzed in terms of mass, charge, momentum, and various other kinematic and dynamic quantities. The analysis of particle tracks is theory-laden. However, only well-confirmed background knowledge and meta-theoretic assumptions concerning the unity of physics are allowed.[10] The numbers of scattering events of a certain dynamic type are counted. In the data analysis of a high energy scattering experiment, relativistic kinematics is used. The scattering events are characterized in terms of relativistically invariant quantities such as the square of the 4-momentum transfer q^2. Finally, from the relative numbers of scattering events for a given momentum transfer, the differential or total cross-section of a certain particle reaction is determined. Standard statistical methods are used to determine the measurement errors. In the end, the measurement results are given in a table of numbers for the differential and/or total cross-section of a certain kind of particle reaction, as functions of the kinematical quantities of the scattering events.

The cross-section of a scattering process is the quantity in which an experiment meets quantum theory. Conversely, it is in the differential and total cross-sections of certain kinds of scattering processes that today's quantum field theories come down to earth. The cross-section of a particle reaction is calculated from the S-matrix of the interaction term of a quantum field theory. The S-matrix gives the transition probabilities of initial quantum states to final quantum states. The initial quantum states correspond to the incoming particles of a scattering experiment, i.e., the beam particles and the target nuclei. The final quantum states correspond to the outgoing diffracted wave or the scattered particles which are detected. The S-matrix describes the scattering in the operational spirit of Heisenberg's matrix mechanics. It gives transition probabilities which correspond to measurable relative frequencies. But it treats the scattering itself as a black box. (What is going on inside the black box, i.e., during the scattering, is illustrated in terms of Feynman diagrams. However, they have no literal meaning. They are mere iconic representations of the perturbation expansion of a quantum field theory. They

---

[9]According to the usual definitions of probability, the probability of an event corresponds empirically to the relative frequency of that kind of event in the limit of infinitely many cases. See also the definition of the differential cross-section in Appendix C.

[10]The measurement method is examined in Sect. 5.3.

make the calculations easier, but they do not represent individual physical processes.[11])

Here, *today's quantum field theories* embrace quantum electrodynamics, the Salam–Weinberg theory of electroweak interactions and the current theory of strong interactions, i.e., quantum chromodynamics. They make up the current *standard model of particle physics*. According to this, there are three generations of leptons (electron, muon, tau meson, and the corresponding neutrinos) and quarks (up and down, strange and charm, bottom and top) which differ in mass; the corresponding 12 antiparticles; the exchange quanta of the electroweak and strong interactions (γ, $W^{\pm}$, W^0, 8 gluons); and the Higgs boson.[12] The standard model predicts a lot of details regarding the measured cross-sections of scattering experiments. In particular, it predicts the existence of resonances, CP violations in the decays of neutral kaons and other unstable bound quark states. However, its breakthrough only came when the scattering experiments with electrons and muons revealed the scaling behavior to be explained now.

4.2 Rutherford Scattering and Scale Invariance

Let us now have a closer look at Rutherford's classical model of the scattering process. A pointlike charged particle of mass m, constant velocity v, momentum $p = mv$, and kinetic energy $E = mv^2/2$, which is considered to be pointlike, interacts with an electrostatic central potential $V(r)$. The potential may be generated either by an extended charge distribution $\rho(r)$ or by a point charge. The former corresponds to Thomson's atomic model which could not explain the backward scattering of the α-particles. The latter is described by a Coulomb potential $V(r) = \pm Ze/r$, where Ze is the charge of the atomic nucleus given in Z units of the electron charge e, and r is the distance between the scattered particle and the atomic nucleus. The scattered particle has a classical trajectory. The scattering is the *deflection* of the trajectory according to Newton's force law. The resulting trajectory is hyperbolic. It corresponds to an asymptotically free solution of the equation of motion. (The model corresponds to the well-known Kepler problem of planetary motions which is applied to unbound celestial bodies like comets.

---

[11] *S*-matrix means *scattering matrix*. The calculation is made within the framework of the quantum theory of scattering. It is usually calculated in the Born approximation, that is, in the lowest order of perturbation theory. For high-precision measurements, higher-order corrections also come into play. The Feynman diagrams are in exact correspondence to the various mathematical terms of the calculations which describe superpositions of quantum theoretical contributions to the interaction. See Brown and Harré 1988, Falkenburg 2002a, Fox 2006, and Sect. 6.4.3.

[12] The latter has not yet been found, but in all other details the standard model is well-confirmed. Any attempts to falsify it have failed. The only new result of the past 25 years is the recent discovery of neutrino oscillations.

Comets with a hyperbolic trajectory only show up once in the solar system.)
After the scattering, the particle is deflected by the scattering angle θ. The
minimal distance between the particle and the scattering center is the classical impact parameter b. In this classical model, Rutherford calculated the
following angular distribution of the scattered particles which is the famous
Rutherford cross-section:

$$\frac{\mathrm{d}\sigma}{\mathrm{d}\Omega} = \frac{C^2}{16E^2 \sin^4 \dfrac{\theta}{2}} . \tag{4.1}$$

The differential cross-section $\mathrm{d}\sigma/\mathrm{d}\Omega$ has the dimension of an *area*. It is a
measure of the relative number of particles which are scattered into a solid
angle Ω in the direction θ.[13] It is characteristic of this expression that it
does not vanish for large scattering angles ($\theta > 90°$). That is, it reproduces
the *backward scattering* observed in the scattering of α-particles on gold foil.
The constant C is obtained from the charge numbers Z and Z' of the probe
particles and the atomic nucleus, the universal constants $\hbar$ and c, and the
coupling constant α of the electromagnetic interaction which is obtained from
the elementary electric charge e according to the relation $\alpha = e^2/\hbar c$.

The Rutherford formula has two formal features. The first is the description of the *backward scattering* which Geiger and Marsden had found
in Rutherford's laboratory. The second is the *scale invariance* of a dimensionless quantity derived from the Rutherford formula. The first characteristic
is famous. It indicated the existence of the atomic nucleus. The second characteristic is less well known. It gave rise to an important theoretical analogy
to the Rutherford formula sixty years later, when the scattering experiments
probing the structure of the nucleons, the proton and neutron, had to be
interpreted. It is a formal invariance property of the Rutherford scattering
model called *scale invariance*.

Scale invariance is a symmetry. It is the invariance of a dimensionless
quantity under scale transformations. With regard to a differential or total
cross-section,[14] it means that the dimensionless quantity obtained by multiplying it with appropriate physical quantities is invariant under transformations of the length and energy scales. A differential or total cross-section has
the physical dimension of an area. If the dimensionless quantity obtained from
a measured cross-section does *not* depend on any length, one concludes that
the scattering center and the probe particles are structureless or *pointlike*.
Pointlikeness means that structureless probe particles interact with a scattering center of size 0. Let us now see how the concepts of pointlikeness and
scale invariance are related in Rutherford's classical model of the scattering
of a pointlike charged particle at a pointlike scattering center.

---

[13]See Appendix C.

[14]The total cross section is integrated over all spatial directions, i.e., over the
solid angle 4π.

Since $C = ZZ'\alpha\hbar c$, Rutherford's scattering cross-section (4.1) can also be written in the form

$$\frac{d\sigma}{d\Omega} = \left(\frac{\hbar c}{E}\right)^2 \frac{(ZZ'\alpha)^2}{16 \sin^4 \frac{\theta}{2}}. \tag{4.2}$$

Like the cross-section, the expression $(\hbar c/E)^2$ has the dimension of an area. Therefore multiplying $d\sigma/d\Omega$ by $(E/\hbar c)^2$, the following dimensionless quantity is obtained:

$$\frac{d\sigma}{d\Omega} \left(\frac{E}{\hbar c}\right)^2 = \frac{(ZZ'\alpha)^2}{16 \sin^4 \frac{\theta}{2}}. \tag{4.3}$$

Here, the differential cross-section is made dimensionless by multiplying it by $(E/\hbar c)^2$. This dimensionless quantity no longer depends on length or energy or other dimensional quantities. It only depends on the numbers Z and Z' and the scattering angle θ. The fact that it does not depend on any magnitude with the dimension of a length is its *scale invariance*. This means that its magnitude is invariant under transformations of the length scale. At the same time it is invariant under the energy of a scattering experiment. It remains the same, no matter whether the measurement is made at high or low energies.

The concept of scale invariance should not be confused with the usual invariance of the scales of physical quantities and the laws of physics under an arbitrary choice of the *unit* in which a magnitude is expressed. Physicists believe that the laws of nature remain the same no matter whether they are expressed in the cgs system or in units of the Planck scale. Scale invariance has to be distinguished from this dimensional invariance. It is physical. In contradistinction to it, dimensional invariance is conventional. The choice of the units of length, time, mass, and temperature in which the laws of physics are expressed is arbitrary. The scale invariance of a measured cross-section, however, is *not* arbitrary. It *could* be otherwise, indicating some internal structure of the probed scattering center. Scale invariance is only a property of *some* scattering phenomena of particle physics, or of the laws of physics describing them. Dimensional invariance is expected to be valid for *all* laws of physics. It expresses the unity of physics with regard to physical quantities. The quantities which measure the properties of physical systems are expected to be of the same kind at large scales and at small scales.

Dimensional invariance gives rise to the dimensional analysis of specific physical systems and models. Dimensional analysis argues with the algebraic properties of the laws and models of physics.[15] These algebraic properties are expressed in terms of the dimensions length $[L]$, time $[T]$ and mass $[M]$,

---

[15]It is based on the so-called Π-theorem. See Appendix B. The algebraic structure underlying dimensional analysis is investigated in Krantz et al. 1971, Chap. 10.

or combinations of them. It is an inference from the dimensions in which a physical phenomenon is characterized to the dimensions needed for the description of its cause or vice versa. In the case of Rutherford scattering, the phenomenon is the Rutherford formula and its cause is a pointlike charge or the Coulomb potential. The phenomenon has dimensions of area, L^2. A dimensionless quantity is obtained from it by multiplying by $(E/\hbar c)^2$ which turns out to have the dimensions L^{-2}. The dimensionless quantity obtained in this way depends neither on any length nor on the scattering energy of the probe particles.[16]

Scale invariance is exactly this formal property of being independent of length and scattering energy. It is a formal property of the Rutherford cross-section. The dimensionless magnitude obtained from the Rutherford cross-section only depends on the scattering angle θ. The differential cross-section calculated from a non-pointlike classical charge does *not* have this formal property. If it is made dimensionless by multiplying with $(E/\hbar c)^2$ the quantity one obtains does not only depend on θ. Dimensional analysis shows that in addition it contains a dimensionless factor which varies with the kinetic energy E, and which depends on a hidden length R that must be attributed to the scattering center or the probe particles.[17]

According to this reasoning, the dimensionless quantity obtained from a measured cross-section indicates whether the scattered particles are pointlike or not. In the case of pointlike probe particles, scale invariance indicates a pointlike scattering center, whereas scaling violations indicate a non-pointlike scattering center. Scale invariance of the dimensionless cross-section and pointlikeness of the scattered particles are equivalent. In Rutherford's classical scattering model, scale invariance is a necessary and sufficient condition for pointlikeness of the probe particles and the scattering center. The scattered particles interact in a pointlike manner if and only if the dimensionless quantity obtained from a measured differential cross-section according to (4.3) does not depend on the scattering energy.

In a heuristic account of scale invariance, physicists are inclined to say that the probe particles do not 'see' any spatial structure at a given scattering energy. In a more precise approach, one has to state that in the case of scale invariance a measured cross-section does not exhibit any effects that may be caused by a non-pointlike scattering center. However, the approach is based on many idealizations. Of course the equivalence of scale invariance of the cross-section and pointlikeness of the scattered particles only holds *ceteris paribus*. It only holds in a model which does not take into account other factors that are relevant for the scattering and might cause or compensate scaling violations. In addition, it only holds in the *classical* model of Rutherford scattering. The whole reasoning sketched above is based on the classical scales of physical quantities. One must ask how the method of dimensional

---

[16]For details see Appendices B and D.
[17]See Appendix D.

analysis and the concepts of scale invariance or pointlikeness can be extended to the *quantum mechanics of scattering*.

4.3 Pointlikeness in the Quantum Domain

In the late 1960s, the concept of scale invariance was generalized to the high-energy scattering processes of particle physics. In the classical model, it is indeed allowed to infer from a measured cross-section to the spatial structure of the particles involved in the scattering. In the heuristics of higher and higher energies mentioned above and in a problematic counterfactual argument, this kind of inference takes on the following meaning. A beam of particles of arbitrarily high energy or vanishing de Broglie wavelength *would* interact with the scattering center in such a way that there is nothing which causes an effect of the dimension of a length in the measured cross-section.

The argument meshes a classical account of the spatial structure of a scattering center and the quantum mechanics of scattering, according to which the charged 'particles' of a beam are *diffracted as if they were waves* in a scattering process. This is nothing but the usual quantum schizophrenia which is at work in physical practice. In the heads of the experimenters and theoreticians, the incompatible particle and wave pictures of classical physics have a peaceful coexistence. Modern physicists *presuppose* wave–particle duality. According to their physical intuitions, the wave and particle pictures of classical physics exclude as well as complement each other.[18] However, the peaceful coexistence of the classical pictures (or models) of particle and wave in particle physics would not be possible if there was not *some* correspondence *between* the two pictures. For the concept of scale invariance, this correspondence is crucial.

One crucial difference between classical point mechanics and the quantum mechanics of scattering is the existence or non-existence of a particle trajectory. In a quantum mechanical model of scattering the dimensional analysis must do without the classical impact parameter b of a particle trajectory. Only then can the results of dimensional analysis be carried over to the quantum mechanics of scattering. In addition, the criteria for a pointlike scattering center must remain the same in the quantum case. That is, there must be some *correspondence* between classical and quantum scattering at a Coulomb potential. This is indeed the case. By chance for a Coulomb potential the classical calculation and the calculation in the quantum mechanics

---

[18]This kind of quantum schizophrenia follows the lines of Bohr's principle of complementarity. Most often it is tacitly used, in a heterogeneous view of the phenomena of particle physics; see the next chapter. In the last chapter of this book it is made explicit.

of scattering give the same results. Both calculations lead to Rutherford's formula.[19]

Given that correspondence, the dimensional analysis of a measured cross-section also remains valid for the quantum mechanics of scattering. It therefore works as a meta-theoretical method for analyzing the data of subatomic scattering experiments. Dimensional analysis applies not only to classical Rutherford scattering but to all theoretical models of a scattering process which are characterized by the physical quantities of this model (except the impact parameter). The same is true of the concept of scale invariance. The scale invariance of a dimensionless cross-section is equivalent to pointlikeness of the scattered particles, in a whole class of models characterized by the scattering energy E, the multiples Z and Z' of some generalized charges of the scattered particles, and a dimensionless coupling constant α which is characteristic of the kind of interaction or dynamics involved in the scattering. Hence the concept of scale invariance is restricted neither to classical models of a scattering process nor to models of the electromagnetic interaction.

In an appropriate frame the theoretical interpretation of a measured cross-section in such terms was also generalized to the relativistic quantum mechanics of scattering.[20] Therefore, the argument according to which a scale invariant dimensionless cross-section suggests that pointlike particles are involved in the scattering process can be carried over to the relativistic case. It was also applied to the results of high-energy scattering experiments in 1968, when scale invariance was considered to indicate pointlike structures inside the proton and neutron.[21]

However, in the domain of quantum theory one has to be more careful. According to dimensional analysis, scale invariance indicates the absence of a structure described in terms of length. Without additional assumptions which specify the model of a scattering process, this means nothing but what it says, namely *structurelessness*. The concept of pointlikeness is more specific. It implies locality. It is based on an assumption which is closely related to the informal causal particle concept discussed in the last chapter, namely that the interaction between the scattered particles is local. One might be inclined to look for additional, model-independent criteria for interpreting scale invariance in terms of pointlike scattering, or the effect of a scattering

---

[19]Indeed, in the quantum mechanics of scattering the Rutherford cross-section is obtained from the exact solution of the Schrödinger equation for the Coulomb potential, as well as from the Born approximation of a screened potential. See Born 1926c (in Herrmann 1962, 78–92), and Mott and Massey 1965, 55–58 and 111–112. The reason is presumably that the Coulomb potential fulfils the condition mentioned in Mott 1965, 4 for the quasi-classical behavior of a particle beam in an external field. Indeed, it gives rise to a quasi-classical description of the atomic nucleus; see the definition of classical form factors explained in Sect. 4.3.1.

[20]One chooses the frame in which the energy component of the relativistic 4-momentum transfer disappears (Breit system; see Breit and Wheeler 1934.)

[21]See Riordan 1987, 156–188.

center of negligible size. The unexpected feature of the backward scattering observed at first in Rutherford's laboratory was such a criterion, but in a classical particle model. Within the quantum domain, a spatial interpretation of pointlike as well as non-pointlike structures is affected by several ambiguities. According to the quantum mechanics of scattering, a scattered particle is diffracted like a wave. In scattering processes, particles are associated with two characteristic wavelengths, namely the de Broglie wavelength and the Compton wavelength. In relativistic quantum theory, length is replaced by spatio-temporal distances. In the domain of a quantum field theory beyond the Born approximation, the physical meaning of the locality assumption hidden in the concept of pointlikeness also becomes ambiguous.[22]

The concept of pointlikeness stems from Rutherford's classical model of scattering. It was associated with the concept of scale invariance which is based on dimensional analysis. Dimensional analysis is a meta-theoretical method which trusts in the unity of physics. Quantum mechanics, however, demonstrates that this unity of physics is no longer granted. We are *not* committed to believe in a physical reality with a thoroughly classical, local structure; indeed quite the contrary. Therefore one should be very careful with a literal interpretation of the concept of pointlikeness, as it is used by particle physicists. From the critical point of view adopted here, *pointlikeness* is a particle property which belongs to a *quasi-classical construal* of subatomic reality. In other words, particle physics attempts to construct the results of high energy scattering experiments in terms of pointlikeness. The higher the scattering energy, the more this construal becomes problematic, as will be shown in the following. Deviations from pointlike scattering behavior, i.e., scaling violations, count as *non-pointlike structures*. These structures are described in terms of non-relativistic or relativistic *form factors*. The higher the scattering energy, however, the more difficult it is to interpret the form factors literally.

4.3.1 Classical Form Factors

In the Rutherford model of the atom, the atomic nucleus is a classical point charge which generates a Coulomb potential. This model of the atomic nucleus is also the basis for describing the nucleus as a non-pointlike structure in terms of *form factors*. In the classical model of scattering, structurelessness means pointlikeness. Conversely, scaling violations of a dimensionless cross-section indicate an internal structure of the scattering center. This structure is described by an extended charge distribution $\rho(r)$ rather than a point charge. Rutherford first found deviations from his scattering formula in 1919, when he made scattering experiments with α-particles and hydrogen.[23] He interpreted them in his classical model as indications of nuclear force effects.

---

[22]See the definition of generalized form factors discussed in Sect. 4.3.2.
[23]Pais 1986, 237–240.

Since the classical Rutherford formula can also be derived within the non-relativistic quantum mechanics of scattering,[24] the classical interpretation of an extended charge distribution in terms of form factors was carried over to quantum mechanics. The classical model of the atomic nucleus as a charge distribution turned out to be surprisingly stable under the transition to quantum mechanics. Even though quantum mechanics does not admit of a classical spatial interpretation of an effective cross-section, it admits of a precise definition of what is a pointlike or non-pointlike scattering center. In the non-relativistic case, a scattering center is pointlike if and only if it gives rise to the Rutherford differential cross-section in the scattering of charged particles.

In the classical model of scattering, the phenomenon of backward scattering is due to the impenetrability of the scattering center for scattered particles with small impact parameter. In quantum mechanics, the scattered particles are described in terms of waves and the scattering behavior (including backward scattering) is a diffraction phenomenon. By chance, the equation of motion gives the same angular distributions in the particle and wave models of Coulomb scattering. For a Coulomb potential, the two cross-sections are equal.

In the classical case as well as in the quantum case, the scattering center is described in terms of a fixed electrostatic potential. This is a *classical* picture of the scattering center. It is described in terms of classical electrostatics. In quantum mechanics, this means that the scattering center is assumed to be *localized*. Strictly speaking, such a description gives rise to a *semi-classical* model of the scattering. Only the probe particles are described in terms of a quantum mechanical wave function Ψ, while the scattering center is not. It is described either as a point charge or as an extended quasi-classical charge distribution. In the latter case, the differential cross-section is *not* scale invariant. Its deviations from the Rutherford formula are expressed in terms of a form factor.

The classical definition of a form factor is as follows. In classical electrostatics, an extended charge distribution is described by means of a *charge density* $\rho(r)$.[25] In the classical model, the charge distribution $\rho(r)$ may be understood either as a continuous charge density or as a probability density of the positions of discrete point charges. $\rho(r)$ generates the following electrostatic potential $V(r)$:

$$V(r) = C \int \frac{\rho(r')\mathrm{d}r'}{|r - r'|} . \tag{4.4}$$

---

[24]See note 19 above.

[25]Here and in the following, non-radial contributions to the charge density $\rho(r)$ and the potential $V(r)$ are in general suppressed. Whenever they come into play, it is indicated by the vector representations $\rho(\boldsymbol{r})$ and $V(\boldsymbol{r})$, respectively.

Here, $C = ZZ'\alpha\hbar c$ is given in terms of the charge numbers Z and Z', the coupling constant α and the constants $\hbar$, c. In the classical model, $\alpha\hbar c$ can be expressed in terms of classical electrodynamic units. In the quantum mechanical model, one assumes that a charged scattering center is described by a wave function $\psi(r)$ which gives rise to a quasi-classical charge distribution $\rho(r)$ in accordance with the usual interpretation of quantum mechanics:

$$\rho(r) = |\psi(r)|^2 . \tag{4.5}$$

Together with (4.4), this interpretation gives rise to a potential $V(r)$ which is expressed partially in terms of classical concepts and partially in terms of a quantum mechanical wave function. Equation (4.5) goes beyond the usual probabilistic interpretation of quantum mechanics. It is based on the counterfactual claim that the quantum mechanical wave function of the scattering center *would* give rise to a probability density $\rho(r)$ in the case of *many* position measurements *within* the scattering center. (Indeed, the description of the scattering center is based on a generalized version of Bohr's correspondence principle which will be discussed in Chap. 5.) In this way it becomes possible to give a consistent quantum mechanical description of the scattering of charged particles at a compound system which is described in terms of a quantum mechanical many-particle wave function. For the Coulomb potential $V(r) = C/r$, the charge density reduces to Dirac's delta function, $\rho(r) \equiv \delta(r)$. For an extended charge distribution, an exact solution of the Schrödinger equation is no longer possible in the quantum mechanics of scattering. In the Born approximation, the potential (4.4) gives rise to the following differential cross-section:[26]

$$\frac{d\sigma}{d\Omega} = \frac{\pi^2 C^2}{E^2 \sin^4 \frac{\theta}{2}} \int \rho(r) \frac{\sin(qr)}{q} r \, dr . \tag{4.6}$$

Here, q is the momentum transfer from the probe particles to the scattering center. It is related as follows to the momentum p and the kinetic energy E of the probe particles and the scattering angle θ:

$$q = 2|p| \sin \frac{\theta}{2} = 2\sqrt{2Em} \sin \frac{\theta}{2} . \tag{4.7}$$

Comparing (4.6) with Rutherford's cross-section given in (4.1) [which is called $(d\sigma/d\Omega)_R$ in the following] shows that $d\sigma/d\Omega$ and $(d\sigma/d\Omega)_R$ are identical

---

[26]Mott and Massey 1965, 86–88. Two warnings should be added here. First, the Born approximation fails for the Coulomb potential since it only holds for potentials that decrease faster than $1/r$. However, the exact solution differs only in an imaginary phase that does not change the cross-section. Second, the scattering at a delta potential gives rise to a zero cross-section. Both points show that a literal interpretation of calculations is problematic. Note that they do not affect my general statements about pointlikeness and locality.

up to the *form factor* $F(q)$ which depends only on the momentum transfer q:

$$F(q) = \frac{4\pi}{q} \int \rho(r)\sin(qr)r dr , \qquad (4.8)$$

$$\frac{d\sigma}{d\Omega} = \left(\frac{d\sigma}{d\Omega}\right)_R |F(q)|^2 . \qquad (4.9)$$

According to (4.8), the form factor $F(q)$ is the *Fourier transform of the charge distribution* $\rho(r)$. The way in which it depends on the momentum transfer q is a measure of the spatial structure of a non-pointlike classical central charge, as far as this structure is due to electromagnetic interactions. According to (4.7) and (4.9), the form factor $F(q)$ depends on the kinetic energy E and the scattering angle θ, giving rise to scaling violations of the Rutherford cross-section $(d\sigma/d\Omega)_R$ which was made dimensionless by multiplying by $(E/\hbar c)^2$:

$$\frac{d\sigma}{d\Omega}\left(\frac{E}{\hbar c}\right)^2 = \frac{(ZZ'\alpha)^2}{\sin^4\frac{\theta}{2}}|F(q)|^2 . \qquad (4.10)$$

For the Coulomb potential with the pointlike charge distribution $\rho(r) \equiv \delta(r)$, the form factor reduces to 1, giving rise once again to the scale invariant Rutherford cross-section:

$$F(q)_R \equiv 1 , \qquad (4.11)$$

$$\left(\frac{d\sigma}{d\Omega}\right)_R\left(\frac{E}{\hbar c}\right)^2 = \frac{(ZZ'\alpha)^2}{16\sin^4\frac{\theta}{2}} . \qquad (4.12)$$

In this way, the equivalence of structurelessness and scale invariance which is predicted by dimensional analysis is reproduced. In the classical model, where the structure of a scattering center is described in terms of a form factor, a structure and a spatial charge distribution $\rho(r)$ are equivalent, whereas structurelessness means pointlikeness. According to this classical model of the scattering center, the charge distribution $\rho(r)$ can be obtained from a measurement of the differential cross-section and from the Rutherford formula, according to (4.8) and (4.9). According to (4.8), the form factor $F(q)$ is an *experimental magnitude* which is obtained by normalizing a measured cross-section $(d\sigma/d\Omega)_{Exp}$ to the theoretical cross-section $(d\sigma/d\Omega)_R$ of a pointlike scattering center:

$$|F(q)_{Exp}|^2 = \frac{(d\sigma/d\Omega)_{Exp}}{(d\sigma/d\Omega)_R} . \qquad (4.13)$$

According to (4.7), one obtains the radial charge distribution $\rho(r)$ from the measured form factor $F(q)_{\mathrm{Exp}}$ by means of a Fourier transformation:

$$\rho(r) = \frac{1}{2\pi^2 r} \int F(q)_{\mathrm{Exp}} q \sin(rq) \mathrm{d}q \ . \tag{4.14}$$

According to (4.13) and (4.14), the radial charge distribution of the electronic shell of atoms can be measured in atomic physics. This has been done by diffraction of low energy electrons and X rays. The structure of the atomic nucleus has been investigated at much higher scattering energies.[27] The form factors involved in these experiments have to be defined within the *relativistic* quantum theory of scattering.

4.3.2 Relativistic Generalizations

The theoretical relation between scale invariance and structurelessness is based on general dimensional arguments. In contradistinction to this, the definition of pointlikeness and non-pointlikeness given above is *model dependent*. It depends on a classical description of the scattering center in terms of a localized charge distribution, the fixed potential, and the form factor to which it gives rise.

However, the concepts of scale invariance and pointlikeness were generalized to the relativistic domain as well as the concepts of scaling violations and form factors. These generalizations were crucial for the development of particle physics in the 1950s and in the late 1960s. In the 1950s, Hofstadter and his group in Stanford measured the form factors of the atomic nucleus beyond Rutherford's scattering energies. They determined the form factors of many kinds of atomic nuclei. However, their experiments which probed the structure of the proton and the neutron for the first time became most famous.[28] These experiments showed that the electromagnetic structure of the nucleons can be expressed in terms of two electromagnetic form factors which describe deviations from pointlike scattering. The first is related to the charges 1 and 0 of the proton and neutron, respectively. The second is related to the magnetic moment of the nucleons, thus explaining their anomalous magnetic moment.[29]

In 1968, a discovery like the one in Rutherford's laboratory recurred. In a high-energy scattering experiment at the SLAC (Stanford Linear Accelerator), large angle scattering was observed and the measured total cross-section turned out to be scale invariant. Bjorken and Feynman interpreted this scale invariance analogous to the Rutherford scattering as an empirical indication for the existence of pointlike nucleon constituents. After carrying out other

---

[27] See Sect. 4.5, where the heuristic principle of increasing the scattering energy in order to get better spatial resolution is finally explained.

[28] Hofstadter 1956, Hofstadter et al. 1958, Hofstadter 1989.

[29] Rosenbluth 1950; Hofstadter 1958.

scattering experiments of a similar type and after accumulating much more additional experimental evidence, these so-called partons were identified with the quarks, that is, the fractional charges of the quark model established in 1963.[30]

In order to interpret these experiments of the 1950s and 1960s, Rutherford's scattering model and the concept of form factors were generalized to relativistic quantum theory as follows. In relativistic quantum mechanics, the scattered probe particles are described by the *Dirac equation* instead of the Schrödinger equation. The connection with the non-relativistic description of the scattering is established by describing the scattering in an appropriate frame. One chooses the Breit frame in which the energy component of the relativistic 4-momentum transfer disappears.[31] The scattering center is represented in this frame by a potential $V(r)$ in which a spatial charge distribution $\rho(r)$ may be inserted. In a first approach it is described by a fixed Coulomb potential, that is, as a very heavy but pointlike atomic nucleus. Two idealizations are involved in this description. (i) The so-called infinite mass approximation according to which the mass m of the probe particles is negligibly small in comparison to the mass M of the scattering center ($m \ll M$ or $M \to \infty$); and (ii) the assumption that the scattering center is pointlike. It is obvious that these two assumptions do not actually go together. They give rise to a simple but unrealistic model of the scattering. In the Born approximation, one obtains a simple expression for the differential cross-section, the Mott scattering formula.[32] The Mott cross-section corresponds to the classical Rutherford scattering in the following sense. For low energies of the probe particles ($v \ll c$ or $2E \ll c^2$), it reduces to the Rutherford cross-section. For high particle energies, however, the dimensionless quantity obtained from it is again scale invariant.

The next generalization proceeds from heavy nuclei to the lightest nucleus, the proton. The scattering is again described in terms of unrealistic idealizations. The model is based on the so-called *Dirac proton*, that is, a pointlike proton without anomalous magnetic moment which is subject to the Dirac equation. (The anomalous magnetic moment indicates some internal structure of the proton. It is neglected in this model. The proton is assumed to behave like a heavy positron.) In the Born approximation, the calculation of the differential cross-section now results in a more complicated expression. However, it is scale invariant like the Mott and Rutherford scat-

---

[30] For the historical details of this thrilling story of an unexpected discovery and its interpretation see Riordan 1987, Chap. 6–8. The report of the experimental results is Panofsky 1968. The original articles of the theoretical interpretation are Bjorken 1969, Bjorken and Paschos 1969, Bjorken 1970, and Feynman 1969.

[31] See Breit and Wheeler 1934; Nachtmann 1990, 254.

[32] See Itzykson and Zuber 1985, 98; Nachtmann 1990, 86; Perkins 2000, 150. For the following, see also the excellent overview in Povh et al. 1999, 43–111.

tering formulas. It reduces to the Mott scattering formula for low mass m of the probe particles ($m \ll m_{\text{proton}}$).[33]

Both scattering formulas calculated in relativistic quantum mechanics represent *relativistic generalizations of the Rutherford formula*. They share with it the formal property of scale invariance. Under well-defined approximation conditions, they reduce to it as a non-relativistic classical limit. Correspondingly, both formulas can be inserted instead $(d\sigma/d\Omega)_R$ in (4.13). On this basis form factors are introduced into the scattering models. The introduction of a form factor de-idealizes the description of the scattering center. However, this de-idealization pays the price of introducing a *quasi-classical charge distribution* into the *relativistic quantum mechanics of scattering*. The resulting model of the scattering is obviously *semi-classical*. (This is the price which has to be paid for keeping the correspondence with Rutherford's classical model.) This semi-classical approach makes it possible to measure form factors in high energy scattering experiments performed at relativistic particle energies. The dependence of the form factors on the relativistic 4-momentum transfer q can then be interpreted as evidence of the non-pointlikeness of the scattering objects:

$$|F(q)_{\text{Exp}}|^2 = \frac{(d\sigma/d\Omega)_{\text{Exp}}}{(d\sigma/d\Omega)_{\text{RQM}}} \ . \tag{4.15}$$

Here, $(d\sigma/d\Omega)_{\text{RQM}}$ is a differential cross-section calculated in relativistic quantum mechanics from the Dirac equation. It describes the scattering of charged probe particles with spin 1/2 at a scattering center which is represented by *any* electromagnetic potential. The relations (4.14) and (4.15) provide the theoretical foundations for the scattering experiments in which Hofstadter and his group measured the form factors and related charge distributions of many types of atomic nuclei at Stanford in the 1950s.[34] The scattering experiments probing the electromagnetic structure of the proton and the neutron became famous. They demonstrated that the proton and the neutron can be described in terms of two form factors which express deviations from pointlike structures, in correspondence to their anomalous magnetic moments. The first form factor is associated with the charge and the second with the magnetic moment.[35] The electromagnetic form factors of the nucleons were at first phenomenologically determined according to a formula of the type (4.15), from the relation of the measured cross-section to a cross-section calculated for pointlike charged particles. The form factors measured in this way are *quasi-classical quantities*. They express the extent to which a measured structure deviates from a *classical magnetic moment*. In a semi-classical model of the scattering, they are related to the electric

---

[33] Bjorken and Drell 1964, Chap. 7.4.

[34] Hofstadter 1956, Hofstadter et al. 1958, Hofstadter 1989.

[35] Rosenbluth 1950; Hofstadter et al. 1958.

charge distribution and the density of the magnetic moment of the proton or neutron.[36]

The next step in the generalization, however, dispenses with the semi-classical models. It is the step from relativistic quantum mechanics to quantum electrodynamics, the quantum field theory of electromagnetic interactions. In relativistic quantum field theory, it is indeed possible to give a precise general interpretation which is *no longer model-dependent* to the form factors of the proton and neutron measured according to (4.15).

In order to derive model-independent expressions for the form factors, one starts from the *general principles of relativistic quantum field theory*, namely Lorentz invariance, gauge invariance, and the validity of the Dirac equation for particles of spin 1/2. In this way one obtains a very general description of the scattering process which is no longer tailored to semi-classical situations. The description consists of two Lorentz invariant terms. They only depend on the (Lorentz invariant) square q^2 of the relativistic 4-momentum transfer q. They can be identified with the terms of the form factors of the nucleons obtained according to (4.15). In this way one gets rid of the model dependence which had first been put into the definition of the form factors. Thus, in the context of relativistic quantum field theory, it becomes possible to define what pointlike particles are in a scattering process *without* quasi-classical presuppositions.[37] The electromagnetic structure of an arbitrary scattering object is described in terms of two form factors $F_1(q^2)$ and $F_2(q^2)$, which correspond to the charge and magnetic moment of the scattering object for the 4-momentum transfer $q = 0$. The scattering object turns out to be *pointlike* if and only if its form factors are constant. It has *electromagnetic structure* if and only if the form factors $F_1(q^2)$ and $F_2(q^2)$ are functions varying as q^2.

In 1969, Bjorken and Feynman suggested explaining the unexpected scale invariance found at much higher scattering energies in terms of pointlike nucleon constituents. The explanation was associated with a further generalization of the relativistic definition of the electromagnetic form factors. This time the model was extended to the description of *inelastic scattering processes* in which the probe particles transfer energy to the scattering center, giving rise to excited particle states and their subsequent decays accompanied by secondary processes like gamma radiation or the production of a hadronic shower. In contradistinction to elastic scattering (which is characterized by a 4-momentum transfer q alone), inelastic scattering is also characterized in terms of the relative excitation energy ν which the probe particles transfer to the scattering center, relative to the beam energy E. For proton–electron

---

[36] The phenomenological access to the form factors preceded the theoretical understanding of these quantities historically, too. The physical interpretation of the electromagnetic form factors of the nucleons in the context of relativistic quantum mechanics was given only some years after the measurement, in Ernst et al. 1960.

[37] See Drell and Zachariasen 1961.

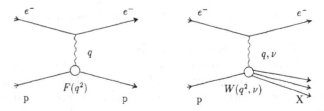

Fig. 4.2. Elastic and inelastic electron–proton scattering (q = 4-momentum transfer, ν = relative energy transfer)

scattering, the corresponding Feynman diagrams of the perturbation expansion for the scattering process in the Born approximation are as shown in Fig. 4.2.[38]

According to the quantum mechanical model of a scattering process, inelastic scattering also gives rise to resonances. In the scattering processes discussed here, the energy transfer to the scattering center is very high. In contradistinction to the production of resonances, they are called *deep inelastic* scattering.

For inelastic electron–nucleon scattering, the most general invariant description of electromagnetic interactions gives rise to two form factors $W_1(q^2, \nu)$ and $W_2(q^2, \nu)$. They depend on the two relativistically invariant quantities q^2 and ν. These were measured in the 1960s. After Hofstadter's experimental results for the electromagnetic form factors of the proton and neutron, nobody expected to find scale invariance. Indeed, none was found – until the now famous experiment of Taylor and his group at SLAC. The surprising results were announced at the Vienna conference in 1968.[39] This experiment was in several respects a repetition of the experiment of 1909 whose surprising results finally made Rutherford assume the existence of an atomic nucleus and calculate his scattering formula. After Hofstadter's form factor measurements of the proton and neutron, the structure of the nucleons was thought to be as homogeneous as the positive charge distribution of the atom according to Thomson's atomic model in 1909. Taylor and his group assumed that the nucleons contain a more or less homogeneous extended charge distribution. To their complete surprise, this assumption was falsified in a domain which had previously remained unexplored. The discovery was again indicated by large scattering angles for which no scattered particles were expected, including unexpected backward scattering. However, the theoretical interpretation differed from 1909. This time the assumption of pointlike scattering centers was not justified by means of a detailed theoretical model of the scattering centers. It was rather based on the inference from the scale invariance of dimensionless quantities to the existence of structureless

---

[38] For elastic and inelastic lepton-nucleon scattering, see, e.g., Perkins 2000, 154–160.

[39] Panofsky 1968; see also Riordan 1987, 154–155.

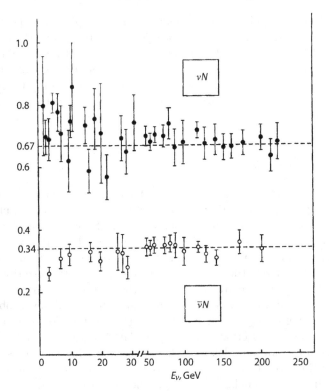

Fig. 4.3. Scale invariance of deep-inelastic electron–nucleon scattering (Perkins 2000, 163)

or pointlike scattering centers. The respective dimensionless quantities calculated from the measured cross-section corresponded to the two form factors $W_1(q^2, \nu)$ and $W_2(q^2, \nu)$ of inelastic electron–nucleon scattering. (These form factors only depend on the Lorentz invariant quantities q^2 and ν. Therefore they are also Lorentz invariant quantities.) The theoretical interpretation of their scale invariance was given by Bjorken and Feynman. They concluded that the proton and the neutron must have pointlike constituents.[40] In this way the most important step was taken towards the experimental proof of the existence of fractional charges, namely the quarks, inside the nucleons.

Later further kinds of lepton–nucleon scattering experiments were performed. In particular, deep inelastic scattering was repeated with muons and neutrinos. (Neutrino beams were generated from proton beams scattered at a target in such a way that many secondary particles arise and only neutrinos remain after a certain decay length.) For these experiments, the model was further generalized from quantum electrodynamics to the quantum field

---

[40]Bjorken 1969, Bjorken and Paschos 1969, and Feynman 1969; cf. also Riordan 1987, 156–188.

theories of the current standard model of particle physics. The resulting model contains three so-called *structure functions* $W_1(q^2, \nu)$, $W_2(q^2, \nu)$ and $W_3(q^2, \nu)$. They are interpreted in terms of the momentum distributions of pointlike quark constituents of the nucleon. However, scaling violations were found as well, and they were interpreted in terms of energy-dependent effects of quantum chromodynamics which contribute to the momentum and spin of the nucleons.[41]

4.4 A Chain of Models

We have seen how the concept of form factors was generalized stepwise. The sequence of generalizations demonstrates the way in which physicists made new domains of scattering theory accessible. Starting from the familiar grounds of classical scattering theory, they constructed a sequence of models. Each model stands in a relation of approximate reduction to its neighbour of lower scattering energy, the last-but-one model, e.g., Mott scattering, stands in a relation of approximate reduction to the Rutherford formula. In this way they obtained a chain of models for scattering (pointlike) probe particles of increasing energy at a pointlike charge or in the Coulomb potential. Relative to the cross-section formulas calculated in these models, different kinds of form factors were defined. They describe scaling violations in terms of non-pointlike structures, giving rise to semi-classical models of subatomic structure. In these semi-classical models, the measured scaling violations or non-pointlike structures were interpreted in terms of electric charge distributions and magnetic moment distributions. The higher the scattering energy, the smaller the measured subatomic structures, as the de Broglie wavelength $\lambda = 2\pi\hbar/p$ would suggest.

The chain of these semi-classical scattering models builds a bridge between the classical and the non-classical descriptions of scattering centers. Each of the semi-classical models in the chain employs a generalized version of Bohr's correspondence principle,[42] given that they describe the electromagnetic structure of the atom in terms of charge and magnetic moment distributions and form factors rather than in terms of a quantum mechanical many-particle wave function. However, the chain is made up of more and more general models of the scattering process, and the most general model is no longer semi-classical. It no longer depends on a specific description of the scattering process and a semi-classical model of subatomic structure. It is only based on the general features of a quantum field theory, namely Lorentz invariance, gauge invariance and a relativistic field equation.

---

[41] Perkins 2000, 154; Povh et al. 1999, 107–111. According to the predictions of QCD (quantum chromodynamics, the quantum field theory of strong interactions), they are due to interactions of the quarks in which virtual gluons are emitted and absorbed, giving rise also to virtual quark–antiquark pairs.

[42] See Sect. 5.4.2.

Indeed, this chain of models builds semantic bridges between the domains of incommensurable theories in Thomas Kuhn's sense.[43] These bridge the conceptual gaps between classical point mechanics, classical electrodynamics, non-relativistic quantum mechanics, S-matrix theory and relativistic quantum field theory. Classical point mechanics describes the individual trajectories of classical charged particles. Classical electrodynamics describes the electrodynamic structure of the atom, nucleus, or nucleon in terms of a Coulomb potential or in terms of form factors. Non-relativistic quantum mechanics dispenses with particle trajectories and describes the scattered beam in terms of a quantum-mechanical wave function. S-matrix theory is the general tool of the relativistic quantum mechanics of scattering. Quantum field theory dispenses with quasi-classical potentials and semi-classical descriptions of a scattering center.

Along the chain of models, more and more classical assumptions are abandoned. Indeed, Rutherford's scattering model is theory-laden in a substantial way and the semi-classical models of subatomic structure are as well. In the stepwise generalization of the model, however, one tries as far as possible to get rid of the specific classical assumptions employed in the models. In particular, a non-relativistic model of *spatial* structure is first replaced by a relativistic model of *spatio-temporal* structure. The relativistic model is then reinterpreted in terms of *electrodynamic* structure and finally generalized to a general model of *dynamic* structure. The most general Lorentz invariant definition of the form factors, however, is still based on the following four crucial assumptions:

(i)   Before and after scattering, the scattered particles are asymptotically free (i.e., in unbound states), according to the Born approximation.

(ii)   The scattered particles underlie relativistic kinematics. Their scattering conserves the relativistic 4-momentum.

(iii) The interactions of structureless unbound particles are described in relativistic quantum field theory.

(iv)   The interaction of two structureless particles can be calculated in a perturbation expansion. The contributions of the scattered particles to the S-matrix factorize.

Assumptions (i) and (iv) grant *separability*. They make it possible to ascribe separate dynamic structures to the scattered particles. (i) says that the scattered particles are separate long before and after the scattering, i.e., their quantum states are not entangled. (This corresponds to experimental evidence, i.e., to the measurement of individual position measurements and particle tracks.) (iv) says that their contributions to the scattering can also be separated, i.e., the interaction term of the Lagrangian of the corresponding quantum field theory factorizes (in the Born approximation) in two inde-

---

[43]Kuhn 1962, 1970.

pendent subspaces of Hilbert space. (This is an idealization. However, the
neglected effects can be added in terms of higher-order corrections.)

Now, what about the possibilities of interpreting the form factors in a
literal way? What kind of subatomic structure does a measured form factor
describe? Does the semi-classical construal of a spatial charge distribution
correspond to subatomic reality? If the theory based on the assumptions (i)–
(iii) is empirically adequate, it supports the attribution of contingent measured form factors to subatomic particles. Then a literal interpretation of the
form factors is tenable. In this case, the electromagnetic structures described
by relativistic form factors may be considered to be true at least in a certain relativistic frame. However, such an interpretation only holds in terms
of *critical* realism. A measured form factor expresses an actual structure of
subatomic reality. This subatomic structure is regarded as the cause of measured scaling violations. These scaling violations, however, are interpreted in
relation to what counts as scattering at a *pointlike* charge. The realistic interpretation of the measured subatomic structures is *relational* in four respects,
i.e., subatomic structures are defined in relation to

1. the domain accessible at a given scattering energy,
2. the frame which admits of a spatial interpretation,
3. the model of pointlike scattering in this domain,
4. the perturbation order employed in the calculations.

1. The higher the scattering energy, the smaller the measured subatomic
   structures, according to the de Broglie wavelength $\lambda = 2\pi\hbar/p$ of the beam
   of a scattering experiment. The theoretical foundations of this heuristic
   consideration will be examined in the next section. For the moment another point is crucial. Subatomic structure does not really exist *per se*.
   It is only exhibited in a scattering experiment of a given energy, that
   is, due to an interaction. The higher the energy transfer during the interaction, the smaller the measured structures. In addition, according to
   the laws of quantum field theory at very high scattering energies, new
   structures arise. Quantum chromodynamics (i.e., the quantum field theory of strong interactions) tells us that the higher the scattering energy,
   the more quark–antiquark pairs and gluons are created inside the nucleon. According to the model of scattering in this domain, this gives rise
   once again to scaling violations which have indeed been observed.[44] This
   sheds new light on Eddington's old question of whether the experimental method gives rise to *discovery or manufacture*. Does the interaction
   at a certain scattering energy *reveal* the measured structures or does it
   generate them? Without giving any prompt answer, it has to be noted
   that the issue is grist for Kantian mills. According to Kant's version of
   critical realism, empirical reality (or phenomenal substance) is a "sum

---

[44]Perkins 2000, 154; Povh et al. 1999, 107–111.

total of relations" rather than a substance in the traditional sense of a thing-in-itself, i.e., a Cartesian *ens per se*.[45]

2. In the non-relativistic model, a form factor is simply the Fourier transform of a spatial charge distribution. But in the relativistic models the form factors do not in general admit of a spatial interpretation. Their spatial interpretation is only possible relative to a very special frame: the Breit frame, where the energy component of the relativistic 4-momentum transfer q disappears.[46] This frame is in general not identical with the rest frame of the scattering center. Therefore it is unclear how a relativistic form factor (and the corresponding spatial charge distribution in the Breit system) relates to the actual shape of a particle or scattering center *per se*, in its rest frame. The two frames only agree if the mass of the probe particle is negligibly small compared with the mass of the scattering center, that is, if the latter may be considered to be fixed in space (infinite mass approximation.)[47] However, this is exactly the relativistic model of scattering at a fixed potential (Mott scattering) which establishes the connection with non-relativistic Rutherford scattering. Beyond Mott scattering, there is no obvious spatial interpretation of the scattering center. In particular, the electromagnetic form factors of the proton or neutron do not express frame-dependent spatial structures but Lorentz invariant electromagnetic structures.

3. In addition, subatomic structures are only defined relative to a certain model of the scattering. What a non-pointlike structure is depends on the model of scattering at a pointlike charge. It has already been shown that the original concept of a form factor is semi-classical and non-relativistic. That is, it is defined relative to the classical or quantum mechanical Rutherford scattering formula. Its generalizations are defined relative to the Mott scattering formula and to the scattering of electrons at a Dirac proton. Pointlikeness and non-pointlikeness are then defined in terms of scale invariance or scaling violations of the calculated and measured cross-sections. The most general definitions of Lorentz invariant form factors and structure functions are model-independent, but they only hold relative to the four assumptions (i)–(iv) discussed above. These assumptions are indeed crucial for calculating particle interactions according to a quantum field theory in the framework of S-matrix theory in the Born approximation. However, they are obviously not the most general approach one may imagine to a fundamental quantum theory of the four known interactions.

---

[45] See Kant 1787, B 323.

[46] See Breit and Wheeler 1934; Nachtmann 1990, 254.

[47] In an intuitive, classical particle picture of the scattering, this means that one can neglect the recoil of the scattering center in comparison to that of the scattered particle.

4. In particular, the Lorentz invariant definition of the nucleon form factors and structure functions rests on the Born approximation of the quantum theory of scattering. However, through this assumption, the model dependence (shaken off at first sight) comes back in through the back door. The definition presupposes the perturbation expansion of the interaction. As a theoretical description of non-pointlike structures, the form factors are calculated from the lowest order of perturbation theory. In the models beyond Rutherford scattering, they are also defined relative to the Born approximation of the scattering at pointlike particles. As long as the Born approximation is good enough, their definition is unambiguous. But in a high precision scattering experiment, higher-order terms of the perturbation expansion also come into play. For a scattering formula which includes higher-order terms, *the definition of form factors becomes ambiguous*. Now the question arises: should the form factors be understood as *perturbative quantities* (i.e., related to the Born approximation of a scattering formula for pointlike particles)? Or should they be understood as *non-perturbative quantities* (i.e., related to a scattering formula which also includes higher-order corrections)? In the case of a scattering process described by the Dirac equation, these options are as follows. In the first case, the higher-order contributions to the scattering have to be expressed in terms of form factors. In the second case, the cross-section of two pointlike or structureless particles also contains all relevant higher-order contributions to the scattering.

This ambiguity obviously brings a conventional element into the definition of form factors and the structure of subatomic particles. According to the perturbative definition of the form factors, an electron shows electromagnetic structure at sufficiently high scattering energy.[48] In textbooks this formal issue is often interpreted in terms of *virtual particles* and it is said that the higher-order structure of an electron corresponds to a cloud of virtual particles surrounding the naked electron. Although theoreticians have no problem with the formal description of the scattering processes behind this informal talk, the meaning of this way of talking is unclear. To attribute a cloud of virtual particles to an electron, confuses the concepts of an electron *per se* and an electron as a relational entity, namely as an agent in a scattering process.

Hence, the definition of subatomic structure is not only relative to the scattering energy, to a specific relativistic frame, and to a chain of models of pointlike scattering. In addition, it becomes ambiguous beyond the validity of the Born approximation. However, the problems of interpreting the form factors of subatomic particles only arise in the domain of *relativistic quantum field theory*. In the domain of the non-relativistic quantum mechanics

---

[48] Perturbative form factors are defined for quantum electrodynamics, e.g., in Itzykson and Zuber 1980, 340. The definition is related to the renormalization program.

of scattering as well as in the domain of the infinite mass approximation, it is still possible to maintain a semi-classical construal of subatomic structure in terms of electromagnetic form factors. In both cases, it is still possible to consider the scattered particles as well-separated entities and to give a quasi-classical description of the scattering center. Only in the domain of a relativistic quantum field theory does one get seriously involved with the semantic problems neglected in this idealization.

4.5 Analogy with the Optical Microscope

In the non-relativistic case, a scattering center may be interpreted without any problems as a spatial structure. The reason is the *classical analogue* of describing scattering centers in terms of form factors in the quantum mechanics of scattering. In a quantum mechanical scattering process, the electromagnetic structure of the scattering center is scanned by a beam of charged particles as *by light waves in a microscope*. The total effect of a scattering center on a particle beam is measured. Here, the particle beam is described by a quantum mechanical wave function. The measured effective cross-section results from a diffraction process. The scattering center acts on the diffracted wave (respectively, the beam) as a whole, i.e., as an integral spatial structure. Hence, the quantum scattering of a particle beam at a non-pointlike structure is analogous to the optical sampling in a microscope.

This is more than just a heuristic analogy which does not work upon closer consideration. With high energy electromagnetic waves (i.e., with an X-ray microscope), one has independent experimental access to the atomic structure. In addition, at the level of the field equations there is a formal analogy between the quantum mechanics of scattering in the Born approximation and the diffraction of electromagnetic waves at a pointlike or extended charge distribution. If there are charges with charge density ρ, the wave equation of classical electrodynamics for the scalar potential ϕ is:[49]

$$\Delta\phi - \frac{1}{c^2}\frac{\partial^2\phi}{\partial t^2} = 4\pi\rho . \tag{4.16}$$

The wave equation (4.16) is solved by a retarded potential. The scalar potential and the charge density in (4.16) can be expressed as follows:

$$\phi(\boldsymbol{r},t) = \phi_0(\boldsymbol{r})e^{-i\omega t} , \tag{4.17}$$

$$\rho(\boldsymbol{r},t) = \rho_0(\boldsymbol{r})e^{-i\omega t} . \tag{4.18}$$

Inserting ϕ and ρ from (4.17) and (4.18) for an arbitrary instant t into (4.16), one obtains the following equation for ϕ_0 and ρ_0:

---

[49]The formal analogy is spelt out in full detail in Pinsker 1953, 143–151.

$$\Delta\phi_0 + \left(\frac{\omega}{c}\right)^2 \phi_0 = \rho_0 \ . \tag{4.19}$$

This equation has the same form as the Schrödinger equation, if the conditions for Born's approximation are satisfied, i.e., if the scattered spherical wave ψ^{sc} is small compared with the incoming plane wave ψ^{in}, if the potential $V(r)$ decreases sufficiently fast, and if terms in $\psi^{\mathrm{in}}V(r)$ are negligible:[50]

$$\Delta\psi^{\mathrm{sc}} + k^2\psi^{\mathrm{sc}} = \left(\frac{2m}{\hbar}\right)^2 V(r)\psi^{\mathrm{in}} \ . \tag{4.20}$$

Due to the formal correspondence of (4.19) and (4.20), the solution $\psi = \psi^{\mathrm{in}} + \psi^{\mathrm{sc}}$ of the Schrödinger equation in the Born approximation is obtained, if k is substituted for ω/c and $(2m/\hbar)^2 V(r)\psi^{\mathrm{in}}$ for $-4\pi\rho_0$ in the solution of (4.19). With the potential (4.4) of a radial charge distribution, one obtains the term (4.8) for the form factor which expresses the deviation of the Rutherford scattering. The scattering of electromagnetic waves at a charge distribution $\rho(r)$ on the one hand, and the scattering of charged particles at a potential $V(r)$ on the other hand, are therefore described by the same formal field equations. The formal solutions for the scalar potential ψ_0 and the scattered wave ψ^{sc} in the Born approximation are also the same. Hence there is a precise formal analogy between the two kinds of scattering processes.

This analogy is far-reaching. In particular, a form factor $F(k)$ can also be defined for the intensity $I(\theta)$ of electromagnetic waves scattered (or diffracted) at a charge distribution $\rho(r)$. $F(k)$ expresses the deviation of the intensity $I(\theta)_{\mathrm{T}}$ in the Thomson scattering of electromagnetic waves at a point charge (which is analogous to the Rutherford scattering of an electron beam):[51]

$$I(\theta) = I(\theta)_{\mathrm{T}}|F(k)|^2 \ , \tag{4.21}$$

with

$$I(\theta)_{\mathrm{T}} = I_0 \left(\frac{e^2}{mc^2}\right)^2 \frac{1}{2}(1 + \cos^2\theta) \ , \tag{4.22}$$

and

$$F(k) = \frac{4\pi}{k} \int \rho(r)\sin(kr)r\,\mathrm{d}r \ . \tag{4.23}$$

Here, the quantity

$$k = \frac{4\pi}{\lambda}\sin\frac{\theta}{2}$$

---

[50]Ibid. 144.
[51]Compton and Allison 1935, 116–118 and 134–140.

is, according to the de Broglie relation $|\boldsymbol{p}| = 2\pi\hbar/\lambda$, analogous to the momentum transfer

$$q = 2|\boldsymbol{p}| \sin\frac{\theta}{2}$$

of charged probe particles. I_0 is the intensity of the incoming electromagnetic wave. e and m are the charge and mass of the point charge. Equation (4.21) only describes the *coherent* contribution to the scattering, i.e., the contribution *without energy transfer*. Analogously, (4.9) only describes the *elastic* scattering of charged particles. The expressions (4.21) and (4.22) also result from the quantum mechanics of scattering. For Thomson scattering and its generalization to a non-pointlike charge distribution, the classical and the quantum mechanical calculations again give exactly the same results (as in the case of Rutherford scattering). In both cases, there is an exact correspondence between the classical and quantum mechanical scattering formulae.

The form factors (4.8) and (4.23) are formally identical. A Fourier transformation of $F(q)$ or $F(k)$ results in the same charge distribution $\rho(r)$. Thus, the formal comparison between the quantum mechanics of scattering and the scattering of electromagnetic waves at a charge distribution reveals more than a mere formal analogy. Classical electromagnetic waves, light quanta or photons, or an electron beam scan the structure of subatomic scattering centers not only in an analogous way but in exactly the same way (under ideal circumstances and as far as the Born approximation is valid). Classical electrodynamics and quantum mechanics predict that electromagnetic waves and charged particles give rise to the same pictures of subatomic structures. This prediction has been empirically well-confirmed for a long time. In particular, measurements by X rays or electron beams exhibit the same shell structure of atoms. These experimental results also confirm the predictions from the many-particle wave function Ψ of quantum mechanics (in the usual probabilistic interpretation), namely the prediction that $|\Psi|^2$ corresponds to the spatial charge distribution inside the atom. This exact correspondence between classical electrodynamics and quantum mechanics underlies the usual popular representations of subatomic structure, e.g., in school books or chemistry text books, as well as the use of an *electron microscope*.

The analogy between the electrodynamics of diffraction and the quantum mechanics of scattering has important heuristic consequences. The analogy with the optical microscope makes it possible to define a *spatial resolution* for a scattering experiment with charged particles. According to quantum mechanics, the spatial resolution of a scattering experiment is limited by the *de Broglie wavelength of the probe particles*. Below the de Broglie wavelength, spatial structures are no longer measurable. Here, a spatial structure which cannot be measured corresponds to a form factor $F(q) \equiv 1$, that is, to Rutherford's scattering formula. The spatial resolution ΔR of the optical microscope is proportional to the light wavelength λ:

$$\Delta R = A\lambda \, . \tag{4.24}$$

Here, the constant A contains quantities specific to the measuring device such as the aperture of the microscope. The spatial resolution is indirectly proportional to the wave number $2\pi/\lambda$ of light. Due to the exact correspondence between the diffraction of electromagnetic waves and the quantum scattering of charged particles at a spatial charge distribution, the spatial resolution of an electron microscope or a scattering experiment is, by analogy with (4.24), proportional to the de Broglie wavelength λ as well as to the momentum $p = \hbar k$ of the charged particles and, in the relativistic domain, to their kinetic energy $E = pc$:

$$\Delta R \sim \frac{1}{p} \, , \tag{4.25}$$

$$\Delta R \sim \frac{1}{E} \, . \tag{4.26}$$

Equations (4.25) and (4.26) express the heuristic relation between increasing beam energy and improved spatial resolution mentioned before. However, the relativistic formula (4.26) is only valid for the frame in which the energy transfer between the probe particles and the scattering center disappears, i.e., in the Breit system.[52] Only in this frame can the spatial interpretation of the form factors be generalized. The larger the momentum and related energy of the beam of a high energy scattering experiment, the smaller the spatial structures that can be measured. This generalization gives rise to the metaphor of 'looking' deeper and deeper into the atom at higher and higher energies. The metaphor fits in well with the fact that the scale invariance which indicates pointlike structures or quark constituents inside the proton and neutron were only measured at much higher scattering energy than the energy available to Hofstadter in his form factor measurements of the 1950s.

However, the non-relativistic relation (4.25) between spatial resolution and beam momentum has nothing to do with Heisenberg's uncertainty relation. Similarly, the relativistic relation (4.26) between spatial resolution and beam energy has nothing to do with the classical turning point of the Rutherford scattering.[53]

---

[52] Breit and Wheeler 1934; Sects. 4.3 and 4.4.

[53] The conditions $V = mv^2/2$ and $C/r = E$ define the *turning point* $R = C/E$ of charged particles in classical Rutherford scattering. In the classical particle picture, the turning point $R = C/E$ can be considered as the *effective size* of a pointlike scattering center which is described by the Coulomb potential $V = C/r$, since it defines the order of magnitude of a non-pointlike charge distribution which gives rise to measurable deviations from Rutherford's scattering formula. If the size of an extended charge distribution is much smaller than the effective size $R = C/E$, the trajectories of the probe particles do not come close enough to the scattering

In the end, what should we conclude from this analogy between classical diffraction and the quantum mechanics of scattering? The radial distribution $\rho(r)$ obtained from a measured form factor exactly corresponds to the charge distribution measured using a gamma microscope, and the latter in turn works exactly like an optical microscope. This correspondence supports the following conclusion. The charge distribution $\rho(\boldsymbol{r}) = |\psi^N(\boldsymbol{r})|^2$ inside an atom inferred from the form factors measured in a scattering experiment with charged particles is no less real than a spatial structure observed through a microscope. It belongs to empirical reality, in the sense of an actual spatial structure inside the atom. Is there any cut in the continuum of observation using 'optical devices' such as the naked eye, the magnifying glass, or the microscope, down to the electron microscope, and the scattering experiments with gamma rays or the particles of high energy physics? From a naturalistic point of view, one might argue as follows. Observation by the naked eye obeys the laws of geometrical optics. Thus, *if* any cut is made in the continuum of observation, then it must be made where the laws of geometrical optics end and when the image of a spatial structure is due to the complicated effects of diffraction. In this case, the cut should be made between the magnifying glass and the optical microscope. But this would mean that microbes, etc., observed in a drop of water through the microscope do not belong to empirical reality.[54] Therefore, if one is *not* willing to make such a cut, then one should admit that the subatomic structures scanned by particle beams in electron microscopes or particle accelerators are as real as the objects or persons visible in a photograph. In a like manner, the latter does not give rise to immediate perceptions, but to a true (though two-dimensional and perspective) representation at another scale.

---

centers to give rise to such deviations. In his 1911 paper, Rutherford derived an upper limit for the size of the atomic nucleus from such considerations. In this sense, the magnitude $R = C/E$ gives the spatial resolution of classical Rutherford scattering. However, the agreement with (4.26) is accidental.

The non-relativistic relation (4.25) was called the uncertainty relation of scattering energy and spatial resolution. It was 'derived' from the quantum mechanical uncertainty relation for position and momentum. This heuristic legitimation of (4.25), however, is misleading. No scattering experiment gives rise to position measurements (of the scattering center) in the usual sense. In addition, (4.25) does not stem from quantum mechanics. It derives from wave optics and its analogy to the quantum mechanics of scattering which gives rise to a surprising robustness of (4.25) against changes between very distinct models of scattering.

[54]See the instructive discussion of seeing through the microscope in Hacking 1983, 186–209.

4.6 Looking Into The Atom

Let us recall at this point the results of Chap. 2. From an empiricist point of view (like the one adopted by van Fraassen[55]), observation is restricted to sensory perception. Since such a view is at odds with the methods of modern science, a generalized view of observation should be employed in order to take into account the de-anthropomorphization of what counts as an observation in physics. The generalized concept of observation developed in Sect. 2.5, however, does *not* support the conclusions drawn above. On the lines of Shapere's account of observation, we suggested two necessary conditions for an observation of a physical object or system x: (i) the physical quantities characteristic of x can be measured and (ii) they can actually be attributed to x on the grounds of an individual causal story.

Indeed, the measurement of pointlike or non-pointlike structures in a scattering experiment does *not* count as an observation of scattering centers or their spatial structure in this sense. The second condition is violated. It is *not* possible to trace a measured cross-section back to its individual cause. No causal story relates a measured form factor or structure function to its cause. The attempt to tell an individual causal story about the scattering center fails. The reason is not that the scattering obeys the laws of quantum mechanics and that quantum mechanics only has probabilistic meaning. There is a perfect analogy between the diffraction of a particle beam and the diffraction of light. The light diffracted in a microscope makes an individual object visible. In exactly the same way, the beam in an electron microscope scans an individual object. In this way, the charge structure of single atoms at the surface of a solid was made visible at the nanoscale. In a scattering experiment, however, the beam is not diffracted at an individual scattering center but in a macroscopic bulk of matter. Rutherford's cross-section was not measured by scattering α-rays at an individual gold atom but on a gold foil. In Hofstadter's form factor measurements, the form factors of 'the' proton and 'the' neutron were measured from the scattering of an electron beam at a big target. The causal story which can be told about a measured cross-section is only about the *total* effect of a macroscopic target on a particle beam.

About the action of individual atoms or subatomic structures on the beam, no individual causal stories can be told. They can only be told about individual particle tracks and scattering events which contribute to a measured cross-section. These individual causal stories are indeed told in terms of conservation laws and sum rules for dynamical quantities such as mass, charge, spin, etc. They support not a causal but a *mereological* account of subatomic scattering centers.[56] Hence the analogy between the diffraction of light and the diffraction of an electron beam does *not* work at the level

---

[55]Van Fraassen 1980, 13–19.

[56]See Chap. 7.

of representing individual subatomic structures. In a high energy scattering experiment, subatomic structures are not observed in a generalized sense. They are inferred from measured cross-sections on the theoretical grounds explained in this chapter. With the beams generated in particle accelerators, one can neither look into the atom, nor see subatomic structures, nor observe pointlike structures inside the nucleon. Such talk is metaphorical.[57] The only thing a particle beam makes visible is the macroscopic structure of the target. (Indeed, twenty years ago, some members of the CDHS collaboration used the neutrino beam of the CDHS scattering experiment to take a picture of a hydrogen target. As expected, the neutrino beam made the walls of the tank much better visible than the gas inside. The picture was shown just for fun in a meeting. It served as a striking demonstration that the experiment was efficient enough to measure *any* cross-section of neutrinos on hydrogen.)

After all, the quasi-classical picture of subatomic reality obtained from form factors and structure functions is (and remains) a *construal*. This construal of subatomic reality rests on the following four theoretical pillars:

(i)   **Dimensional Invariance.** The meta-theoretical assumption that the laws of nature are dimensionally invariant is a general invariance principle of physics. Like many symmetry principles, it serves as a tool for unification. Dimensional invariance is a necessary condition for attributing quantities like mass, momentum and charge to entities at a microscopic scale.[58]

(ii) **Correspondence.** The correspondence between classical scattering and quantum mechanical models of scattering is based on formal intertheoretical relations. Indeed, this correspondence is twofold. The quantum mechanics of scattering at a classical potential is in exact formal analogy with the diffraction of a classical light wave. For the Coulomb potential, it is in exact correspondence with the Rutherford formula for classical charged particles.

(iii) **Pointlikeness.** The concept of pointlikeness is based on the inference from a scale invariant effective cross-section to a pointlike or structureless scattering center. This inference is based on the first two assumptions. Correspondence gives rise to the analogy with the optical microscope. According to this, a higher beam energy yields better spatial resolution. Dimensional invariance gives rise to dimensional analysis. According to this, the energy independence of a dimensionless cross-section indicates a pointlike scattering center.

(iv) **Form Factors.** The definition of the form factor is semi-classical. Form factors describe the deviations of measured cross-sections from Rutherford scattering or its relativistic generalizations. They are interpreted in

---

[57]My attention was drawn to this point by Fox 2006. In contradistinction to Falkenburg 1995, 176–177, I no longer claim that scattering experiments are observations in Shapere's sense; see below.

[58]See Krantz et al. 1971, Chap. 10, and my remarks in Sects. 5.4 and 5.5.

terms of the charge distributions of non-pointlike scattering centers. In the relativistic case, however, the interpretation only holds for the Breit frame in which the energy transfer of the scattering disappears.

The construal of subatomic reality is expressed in terms of form factors and structure functions. (For a pointlike structure, the form factor is 1.) As we have seen in the preceding sections, the definition of subatomic structures in such terms is *relative* in four respects. Subatomic structures are defined relative (1) to the *scattering energy*, (2) to a *relativistic frame*, (3) to a *model of pointlike scattering* and (4) to the *perturbation order*.

The spatial interpretation of form factors is restricted to relativistic quantum mechanics in the infinite-mass approximation and in the Born approximation. For obvious reasons, the probe particles of a form factor measurement should also be chosen in such a way that no typical quantum effects occur (no exchange effects or scattering of identical particles; no spin effects). Thus, the quasi-classical construal of *spatial* subatomic structures holds *if and only if the scattering has classical correspondence*. Beyond correspondence with the classical case, the concept of spatial structure has to be replaced by an abstract concept of electromagnetic structures. This concept can be generalized to other interactions, that is, to dynamic structures in a general sense. However, any interpretation of such structures must take into account that they are *relational* entities. Only in certain energy domains do they act as invariant structures. With increasing scattering energy of the probe particles, the probed structures change. They are not like a Cartesian *ens per se* which exists independently of any interactions. They are rather like a thin glass target which is probed by billiard balls. At low scattering energy, the glass holds. At high scattering energy, it breaks, showing a completely different scattering behavior.

It was argued above that the popular talk of 'looking into the atom' with the help of high energy particle beams is just a metaphor. Metaphorically, we may state: by means of a particle accelerator, one 'sees' subatomic dynamic structures inside the targets of scattering experiments. Expressed in terms of analogy, they are shown as by a microscope. The higher the scattering energy, the smaller the 'observable' structures. The 'observation' is all the more faithful the better the quasi-classical conditions of Rutherford scattering are satisfied, i.e., the better the correspondence with the classical case. Beyond that correspondence, the faithfulness of the 'observation' gets worse and worse. In the domain of relativistic quantum field theory, our classical construal of physical reality necessarily gives a distorted picture of subatomic structure. It simply makes us look into the atom through the wrong glasses. Unfortunately, we do not have any better tools.

5 Measurement and the Unity of Physics

The preceding chapters showed that the measurement theory of current particle physics has two predominant features. First, it is heterogeneous. To put it in Nancy Cartwright's terms, it has a piecemeal structure.[1] It combines non-relativistic and relativistic laws from classical physics and several quantum theories. Second, it is based on powerful unifying principles. The construction of the length, time, and mass scales establishes at least a semantic unity of physics. The use of such a heterogeneous measurement theory is no grist for Nancy Cartwright's mills. It is based on trust in a hidden unity of physics. In this chapter I will investigate the heuristic assumptions and the unifying principles behind this trust.

The disunified structure of the current quantum theories and their measurement theories is due to the quantum measurement problem. Up to the present day no quantum theory of measurement bridges the explanatory gap between the quantum mechanical wave function Ψ and the individual measurement results.[2] At the present stage of quantum physics this explanatory gap seems irreducible. Any attempts at closing it face serious obstacles, the most serious being the relation of the reduction of the wave function to Lorentz invariance. Indeed, the current philosophical interpretations of quantum theory which aim at closing the gap cannot cope with relativistic quantum field theory. In addition, they violate the principle of ontological parsimony. In order to re-establish *some* kind of *un*critical realism, they must invent bewildering non-empirical structures.[3]

---

[1]See Cartwright 1999.

[2]In the quantum theory of measurement, this explanatory gap is called the 'objectification problem'; see Busch et al. 1991, 75–83, 99–137; Mittelstaedt 1998, 65–102.

[3]In this regard, Putnam 1990, 10–14, criticizes Bohm's theory of hidden variables and Everett's *many worlds* interpretation as attempts to save classical metaphysical realism. The latter invents a multitude of parallel worlds which are generated by measurements in a branching quantum universe. The former re-establishes particle trajectories and assumes a non-local potential which gives rise to action-at-a-distance. This is at odds with the principle of Lorentz invariance. This principle, however, is also violated by collapse theories lacking such metaphysical surplus structure.

As long as there is no way out of this disunity, it is helpful to adopt some crucial ideas of Niels Bohr's quantum philosophy, in particular his view of the language of physics. Bohr's claim was that the *classical language is indispensable*. This has remained valid up to the present day. At the *individual* level of clicks in particle detectors and particle tracks on photographs, all measurement results have to be expressed in classical terms. Indeed, the use of the familiar physical quantities of length, time, mass, and momentum–energy at a subatomic scale is due to an extrapolation of the language of classical physics to the non-classical domain. But quantum theory is also indispensable. Without it no high precision measurements of particle physics would be possible. Quantum theory applies at the *ensemble* level of measuring cross-sections, resonances, form factors, and so on.

In face of this theoretical disunity it is problematic to assume a semantic unity of physics at a large scale and at a small scale. However, the measurement theory of current particle physics does just this. Thus, it conceals problems of incommensurability in Thomas S. Kuhn's sense.[4] It combines incommensurable classical and quantum laws. How does this work? Obviously, incommensurability is a philosophical problem. It does not arise with the use of measurement laws in physical practice. In many areas of current physics, physicists combine classical and quantum descriptions and call the resulting models *semi-classical*. So is incommensurability *only* a philosophical problem? Apparently not. It raises the question of whether and why the usual practice of extending the familiar classical quantities to the subatomic domain is legitimate. And any answer must be given in terms of physical concepts.

In the quantum domain, the key to the incommensurability problem is to understand that the operational, axiomatic, and referential aspects of physical concepts fall apart. However, from an axiomatic point of view they belong together. After pointing this out as a matter of principle in Sect. 5.1, I will analyze the heterogeneous measurement theory of particle physics. It reflects the layered structure of the phenomena sketched above in Chap. 3. The layers are connected in terms of the length, time, and mass scales. The connections are legitimated by a generalized correspondence principle which brings coherence and redundancy into the heterogeneous measurement theory of particle physics (Sect. 5.2). However, it is far from being obvious whether the correspondence between classical and quantum descriptions holds at the level of individual particle tracks. The data analysis of high energy scattering experiments combines classical and quantum laws in a very peculiar way. Probabilistic quantum corrections are added to classical laws which are in turn applied to individual tracks. How does this work? We shall see that, similarly to the case of form factors, a chain of models establishes a correspondence between the quantum calculations and a classical picture (Sect. 5.3).

---

[4]Kuhn 1962, 1970.

Indeed, several unifying principles support the measurement theories of particle physics and the application of quantum theories to individual particle tracks and scattering events. Amongst them are the generalized correspondence principle suggested by Heisenberg in 1930, symmetries and the corresponding conservation laws, and the principle that the laws of physics are dimensionally invariant. The latter principle is powerful. It underlies the construction of the scales of physical quantities and the heuristic method of dimensional analysis. All these unifying principles work top–down, in contradistinction to the decoherence approach which gives a bottom–up explanation, although only at the probabilistic level (Sect. 5.4). Taken together, these principles make it possible to construct the scales of physical quantities. At this point we once again face the challenge of constructivism. The unity of the length, time, and mass scales is a theoretical construal. But is it *only* a construal (Sect. 5.5)? Here, the crucial questions concern semantic consistency. At the probabilistic level quantum theory is semantically consistent in a strict sense. At the level of individual particle tracks and scattering events, only weaker semantic conditions are satisfied. Indeed, correspondence is needed as a bridge principle which makes it possible to apply quantum theory to individual systems.

5.1 Incommensurability and Measurement

How is it possible to construct a unified physical language which overcomes the semantic disunity of incompatible theories? According to Thomas S. Kuhn's view of scientific revolutions, the theories involved in the measurement of subatomic particles are *incommensurable*. The measurement of subatomic particles started with Thomson's classical e/m measurement, that is, before the quantum revolution was in sight. Classical measurement methods and their results gave rise to the quantum revolution. But many classical measurement laws survived the quantum revolution and the new quantum laws were *added* to them instead of replacing them. In physical practice, Kuhn's incommensurable rival theories are not *really* competing. Quite on the contrary, they have been living in peaceful coexistence for decades.

Kuhn's incommensurability thesis is based on an analogy between mathematics and empirical science. It claims that competing theories which differ in structure do not have a common measure. They are as incommensurable as the cathete and the hypotenuse in the right-angled triangle. If we take the term 'measure' on the physics side of this analogy literally, we may relate it to the meaning of physical quantities which are implicitly defined by the laws of competing axiomatic theories. From an axiomatic point of view, the experimental practice of combining laws which derive from incommensurable theories is far from being obvious.

According to Kuhn, incommensurability has three aspects: change of the problems which the scientists attempt to solve, change in the meaning of crucial theoretical concepts, and change of the world within which science is practised. Here we are concerned with the change of meaning and the associated problems of translatability on which the 1969 postscript to Kuhn's book focuses.[5] Competing theories are associated with different classification systems regarding the phenomena. Indeed, we have seen in Chap. 2 that all phenomena of modern physics are already structured by some theory. Recall the discussion of particle trajectories. In classical mechanics they count as empirical. (They are best exemplified in the trajectories of celestial bodies.) In quantum mechanics they do not: a particle track is nothing but a sequence of position measurements. Kuhn is absolutely right in claiming that it is only partially possible to communicate in a neutral language about observational or experimental evidence. Following Quine's work on ontological relativity, he argues that as a consequence the usual (Tarskian) concept of truth as correspondence to what is 'really there' is no longer tenable.[6] Indeed, any substantial change in the conceptual foundations of a discipline raises serious problems for scientific realism. And the most serious problems are the ones with which we are dealing here, the ontological problems raised by the quantum revolution. Incommensurability challenges scientific realism by generally denying the possibility of a unique scientific language. If there is no language to bridge the conceptual gap between competing theories, Kuhn is right in claiming that it is neither possible to explain whether theories refer nor why one of them should come closer to truth than the other.

The laws of classical physics and of quantum theory have a very different physical semantics. The semantics of classical physics is associated with the familiar scales of length, time, and mass. The semantics of quantum theory is associated with the uncontroversial *probabilistic minimal interpretation* according to which the only physical meaning of the quantum mechanical wave function (or the empirical content of the theory) lies in the *expectation values of quantum mechanical observables*. This difference in semantics has serious ontological consequences. To make them clear, semantics and ontology should be distinguished as follows.

The semantics of a physical theory is the uncontroversial physical meaning of its quantities. Here, the term 'uncontroversial' is related to the scientific community of physics, to physical practice, and in addition to technological practice. (The uncontroversial meaning of physical quantities is pinned down in the international system of physical units.) Semantics involves weaker metaphysical claims than ontology.

---

[5]See Kuhn 1970, 198–204, compared to 148–150. See also Hoyningen-Huene 1993, 206–222.

[6]See Kuhn 1970, 206: "There is, I think, no theory-independent way to reconstruct phrases like 'really there'; the notion of a match between the ontology of a theory and its 'real' counterpart in nature now seems to me illusive in principle."

The ontology of a physical theory is the domain suggested by it and the kind of entities to which it refers.[7] The ontology of classical physics consists of particles and waves in spacetime. The ontology of quantum physics is unclear. According to the probabilistic minimal interpretation, the ontology of quantum theory is nothing but probabilistic wave functions and quantum theoretical expectation values which express conditional probabilities. In comparison to classical particles or waves this is quite a meager ontology. Therefore, *beyond* the uncontroversial minimal interpretation the ontology of quantum theory is subject to the never-ending debates on the interpretation of the quantum mechanical wave function and to a bewildering variety of invented ontological structure.[8]

There is a crucial formal difference which corresponds to the different semantics of classical and quantum theories. The quantities of classical physics are represented by *real-valued functions*, the quantities of a quantum theory, however, by *operator-valued measures*.[9] Empirically, the classical quantities are attributed to *individual phenomena* but the observables of a quantum theory to *probabilistic ensembles*, that is, to *many* measurement results obtained under the same experimental conditions. Of course this semantic difference between the theories transfers to the measurement laws of the two theories. And of course this semantic difference is a strong case for incommensurability in Kuhn's sense.[10] Indeed we will see in the following that it is the main obstacle to the semantic unity of physics.

Does incommensurability really have such dramatic consequences? Most physicists would agree that a change in the conceptual foundations of science changes our world view dramatically. But most of them would never agree that theoretical revolutions preclude the existence of terms in which the change of physical concepts can be expressed and in which the empirical consequences of theories may be compared. Richard P. Feynman (whom no one could suspect of having been affected too much by professional philosophy of science) wrote about the change in the meaning of mass in the transition to relativistic mechanics:

> [...] *philosophically we are completely wrong* with the approximate law. Our entire picture of the world has to be altered even though the mass changes only a little bit. [...] Even a small effect sometimes requires profound changes in our ideas.[11]

These words confirm Kuhn's incommensurability thesis only partially. Feynman makes a distinction between the *philosophical import* of a law and the

---

[7] This corresponds to Quine's account of ontology. According to Quine, to be is to be the value of a bound variable; see Quine 1953, 15.

[8] See above note 1.

[9] See, e.g., Busch et al. 1991, 10–11.

[10] See Scheibe 1999, 174.

[11] Feynman et al. 1963–1965, Vol. I, 1-2.

language of physics in which it is expressed. Like Kuhn, he emphasizes that the world views associated with non-relativistic mechanics and special relativity are completely different, even though approximately equal laws derive from both theories. On the other hand, he speaks about *the mass* which is subject to a small quantitative change when we pass from the non-relativistic to the relativistic description of a motion. Kuhn's account of scientific revolutions tells us that these descriptions rely on incommensurable concepts. A non-relativistic mass is constant but a relativistic mass depends on its velocity in a given inertial frame. One might therefore suspect that in speaking about *the mass* Feynman uses the concept ambiguously.

In scientific practice, however, *both* concepts seem to be at work in pre-established harmony. Physicists have been using the non-relativistic and relativistic concepts of mass for decades, without running into ambiguity or confusion. Quite often they use both concepts simultaneously. Even in view of the quantum revolution, scientific practice somehow overcomes the semantic problems posed by incommensurability. The measurement theory of current particle physics combines non-relativistic and relativistic, classical and quantum laws. The mass scale covers all mass values assigned to any object in the subatomic, macroscopic or cosmological domain. Physicists suppose that the scale represents a class of physical properties of material things and their parts. They assign mass values to electrons, billiard balls, and black holes. In doing so, they suppose that these numbers represent *commensurable* physical quantities.

It is not necessary, however, to interpret such scientific practice as a case against Kuhn's incommensurability thesis. One might cite his 1961 paper *The Function of Measurement in Modern Physical Science*. Once the phenomena can be measured, competing theories *must* have almost identical empirical content. Their quantitative predictions must agree approximately for many observable phenomena. They compete only with regard to the anomalies of the rival theories. Thus according to Kuhn's views, measurements put very strong constraints on theory development. Measurement results force the quantitative predictions of incommensurable theories to come as close to each other as possible. Surely he would never say that the theories have to come close to *truth*.

Kuhn is right on incommensurability and most probably on the illusiveness of the truth of full-fledged theories. However, he underestimates the unifying power of the language in which measurements are expressed. There is more behind measurement than axiomatic systems, their logical incompatibility and quantitatively compatible numbers. Measurement has operational, axiomatic, and referential aspects which cannot be separated, as shown by *empiricist measurement theory*:[12]

---

[12]See Krantz et al. 1971. Empiricist measurement theory as such is metaphysically neutral; see Falkenburg 1997 and Appendix A. It is an axiomatic theory which deals with the formal concatenation structure of empirical operations. Correspond-

1. It has an *operational basis* which gives rise to a well-defined empirical ordering of the phenomena.
2. It depends on *axioms* which are strong enough to determine a numerical representation of this ordering.
3. It implies the *referential claim* that the axioms and their numerical representation express the operational basis in an empirically adequate way.

The traditional views about the meaning of physical quantities tended to overemphasize or underestimate some of these aspects of measurement. In this way they made decisive steps towards one-sided metaphysical views about physical quantities. This is the reason why they diverged substantially.

Newton's definition of mass in the *Principia* was *referential*. He defined mass as volume times density. This makes sense if and only if the density is explained in terms of the number of massive particles per volume. Obviously, this explanation exceeds the modest referential claims of modern empiricist measurement theory. It refers to unobservable entities such as atoms. In particular, Newton's mass definition makes implicit use of a non-empiricist principle of inference to the best explanation, namely his third *rule of reasoning*.[13] Rule III says that extensive quantities which are found in all experimentally accessible bodies can be attributed to *all* bodies and also to the smallest parts of bodies, that is, to the atoms. Conversely, according to Newton's comment on rule III, the size and mass of macroscopic bodies are the sum of the size and mass of the atoms which constitute this body. Thus, in the last analysis Newton was suggesting to define mass in terms of *atomic units*. A modern correlate is to define the mass of bodies in terms of the proton mass. The mass definition in the *Principia* is circular if and only if the concept of density on which it is based is cut off from reference to the number of atoms in a given volume.

Mach was aware of this when he criticized Newton's concept of mass.[14] He criticized Newton's atomism from an empiricist point of view. (In 1883, atoms were still unobservable in a strict sense as well as in a generalized sense.) He asked for *operational* definitions and suggested defining mass in terms of the mutual acceleration which bodies give rise to in a gravitational system. Whereas Newton's mass definition is absolute (if understood in terms of atomic units), Mach's mass definition is relational. It only gives rise to the definition of mass ratios. Mach's operational definition of mass was a model for Einstein's operational definition of simultaneity in the famous 1905 paper on special relativity. Mach's and Einstein's operational definitions in turn influenced Pauli, Heisenberg, and Bohr in their views about quantum mechanics. Heisenberg's matrix mechanics was only expressed in operational terms such as the energy and polarization of the light emitted by atoms.

---

ingly, it gave rise to abstract measurement theory which is now a branch of applied mathematics; see Narens 1985.

[13]Newton 1687, 795–796.
[14]See Mach 1883, 210–211.

Bohr's famous Como lecture which founded the Copenhagen interpretation of quantum mechanics explicitly draws the parallel with Einstein's special relativity.[15] However, any operational definition is based on a measurement law which has certain axiomatic foundations. Otherwise, one would end up in Bridgman's radical operationalism according to which every different measurement law defines a different quantity.[16] This would give rise to a complete disunity of physics. The definition of scales of physical quantities would not be possible.

Einstein definitely did not defend strict operationalism. Indeed, he was unhappy with the operational foundations of quantum mechanics and the Copenhagen interpretation. Against Heisenberg's matrix mechanics he claimed that it should be the theory which tells us what can be measured and not the other way round.[17] He thought that the *axiomatic* view of physical concepts is primary and that operational definitions should be based on them. In his 1905 paper on special relativity theory, the operational definition of simultaneity was based on a heuristic principle which asked for a more general measurement method in order to unify classical mechanics and electrodynamics. The requirement of unification worked as a theoretical constraint which told one what could be measured from a more general point of view, namely the propagation of light rather than the mechanical motions of clocks.

However, the axiomatic method developed by Hilbert suggested that theoretical concepts cannot be explicitly defined in the way the definitions in Newton's *Principia* had suggested. They are only implicitly defined in terms of axioms.[18] According to this, the dynamical concepts of mass and force as well as the kinematical concepts of velocity and acceleration are implicitly defined by the system of axioms of classical mechanics (Newton's second and third law and their successors, the Lagrange equations or Hamilton equations of analytical mechanics[19]).

Taken all together, these views again demonstrate that any one of the referential, operational, or axiomatic views taken *on its own* is *one-sided*. Physical practice *combines* them. The problem is that up to the present day *the referential, operational, and axiomatic features of quantum concepts fall apart*. This first became obvious with the operational foundations of Heisenberg's matrix mechanics and Einstein's discomfort with it. Since then, the problem has not been resolved. In particular, it remained for the concepts of

---

[15]Bohr 1928. At this point it should be emphasized that the Copenhagen interpretation consists of the informal philosophical views which Bohr, Heisenberg, Pauli, and others associated with quantum mechanics in the usual probabilistic Born–von Neumann interpretation. The latter is better called the *orthodox view*. In addition, it should be noted that the so-called Copenhagen interpretation is more due to Heisenberg than to Bohr himself; see Howard 2002.

[16]Bridgman 1928.

[17]See Heisenberg 1969, 91–92.

[18]Hilbert 1918, Bernays 1922.

[19]See Schmidt 1993.

particle physics. The attribution of physical properties to subatomic particles is not based on a unified theory but on *several incommensurable theories*. The current concepts of subatomic particles have axiomatic aspects which stem from quantum field theory, operational aspects which rest on classical measurement laws, and referential aspects which are unclear.

5.2 A Heterogeneous Measurement Theory

While the observation of particle tracks on a photo plate is also possible for the layman, the data analysis of these tracks makes use of a very complex measurement theory. This measurement theory today is a heterogeneous structure of laws which belong to structurally different theories. Its core is *classical*. It consists of laws of *classical point mechanics* and *relativistic kinematics* which serve to describe particle trajectories. At the level of individual particle tracks, *quantum electrodynamic corrections* obtained from transition probabilities for subatomic scattering processes are added to this classical core of assumptions about the trajectories of single particles. The measurement theory is further refined by *conservation laws* for quantized quantities such as spin, parity, and isospin. At the ensemble level, all these laws are embedded into quantum mechanical scattering theory and completed by quasi-classical expressions for *form factors* together with the quantum laws for the decay of *resonances*. In this way, the classical measurement theory of a point particle is attached to several laws which derive from quantum mechanical scattering theory and the latter is in turn combined with quasi-classical assumptions about the charge distribution which describes the electrodynamic structure of a scattering center. Up to the present day, the measurement theory of particle physics is *semi-classical*. It has the following basic ingredients:

1. The law of the *classical Lorentz force* $\boldsymbol{F} = (q/c)(\boldsymbol{E} + \boldsymbol{v} \times \boldsymbol{B})$. This connects the mass m and charge Q of a classical particle to its velocity $\boldsymbol{v}$ and acceleration $\boldsymbol{F}/m$ in electric and magnetic fields $\boldsymbol{E}, \boldsymbol{B}$.
2. *Relativistic kinematics.* This describes the spatio-temporal propagation of a high-energy particle and demands that its interactions obey relativistic energy–momentum conservation.
3. A large number of *phenomenological laws*. Several of these (such as the empirical energy–range relation[20]) are obtained from experiments on the basis of a classical particle model. Physicists call such laws *semi-empirical*.
4. Quantum mechanical *conservation laws* for quantized quantities like spin, parity or isospin. According to the Noether theorem these are connected to the symmetries of a quantum dynamics of subatomic particles.
5. Laws for the *dissipation of energy* and the *deflection* of charged particles in matter. These are based on quantum mechanical scattering

---

[20]See Rutherford et al. 1930, 294, and Sect. 3.4.

theory and on the quantum electrodynamic description of ionization, bremsstrahlung, pair creation, and multiple scattering.[21]

6. *Statistical laws* for calculating the relative frequencies of scattering events of a given dynamic type and the effective cross-sections of the respective particle reactions.[22]

7. *Quantum mechanical scattering cross-sections* for pointlike or structureless particles. The formulae belong to the chain of models described in Sect. 4.4: the cross-sections of Rutherford scattering, Mott scattering, Dirac scattering on the proton, etc.

8. The Breit–Wigner formula for the mean energy and width of a resonance. This stems from the theory of quantum decays and it relates the energy and width of a resonance to the mass and lifetime of an unstable quantum mechanical state.[23]

The statistical assumptions (6) stem from probability theory. The rest of the measurement theory is patched together from a multitude of specific laws which are based on classical point mechanics, relativistic kinematics, non-relativistic quantum mechanics, quantum electrodynamics, and more recent quantum field theories. This heterogeneous measurement theory obviously reflects the layered structure of the phenomena described in Sect. 3.3. In addition, it reflects the growth of safe background knowledge during the historical development of particle physics. First, there were the classical laws (1)–(2) describing individual particle tracks and scattering events. They belong to the first stage of theorizing the observational basis for particle physics, that is, they were the only measurement laws available in the early days before quantum electrodynamics was established. Then, there were the phenomenological or semi-empirical laws (3). In the days of confusion about cosmic ray tracks and particle identification during the 1930s and 1940s, they gave an independent empirical basis to quantum electrodynamics (see Sects. 4.4 and 4.5). Later, more and more quantum mechanical conservation laws (4) were introduced. Their application to scattering events helped to bring order into the *particle zoo* of the 1950s and 1960s. After the 1960s, high energy scattering experiments gave rise to large numbers of particle tracks which were analyzed using statistical methods. In order to obtain high precision measurements of effective cross-sections, the energy loss and multiple scattering of charged particles in matter (5) were taken into account and sophisticated methods of statistical analysis (6) came into play. This was only possible after quantum electrodynamics had been accepted as a well-established theory. Finally, the formulae (7)–(8) were added. They are based on quantum me-

---

[21] See Rossi 1952, 22–90; Perkins 2000, 349–355; Particle Data Group 2004, 238–255.

[22] See Cowan 1998; Particle Data Group 2004, 283–299.

[23] See Perkins 2000, 55–58, in comparison to Goldberger and Watson 1975, Chap. 8. See Sect. 3.3.4.

chanical scattering theory and quantum field theory. Like the laws (5)–(6), they only apply at the probabilistic level.

It is instructive to see which of these concepts and laws belong to the individual and which to the probabilistic level. All quantities attributed to particle tracks and scattering events belong to the level of *individual* events. This level is essentially *classical*. The measurable quantities of a particle track are *mass, relativistic four-momentum, and charges*. In addition, classical considerations suggest that the track length corresponds to the absorption length of a stable particle or to the lifetime of an unstable particle. The measurable quantities of scattering events are the quantities assigned to individual particle tracks *plus* several quantities derived from them using the conservation laws (2) and (4), namely isospin and hypercharge, and other dynamical quantities (strangeness, charm, etc.), which were associated with internal particle symmetries and gauge invariant quantum field theories, respectively. These dynamical quantities are generalized charges. (In the current standard model of particle physics, they are associated with the coupling strength of the electroweak and strong interactions.)

In turn, all quantities which genuinely belong to a *quantum theory* are *probabilistic*. They can only be attributed at the ensemble level, that is, to the measured relative frequencies of the particle tracks and scattering events resulting from a particle beam which was prepared in a well-defined quantum state. The most important measurable quantity of a quantum dynamics is the *effective cross-section*. As an empirical quantity, it is obtained by counting the relative frequency of scattering events of a given dynamic type. As a theoretical quantity, it is calculated within quantum field theory from the S-matrix element of the corresponding particle reaction. Thus, in predicting the effective cross-section of a particle reaction or the corresponding relative frequency of scattering events, quantum field theory and the quantum theory of scattering meet their empirical basis.[24] A cross-section measurement underlies the measurement of a form factor as well as the measurement of a resonance. In the classical picture, a form factor is a probabilistic measure of the size of a scattering center. Resonances do not have an obvious classical correlate (even though the term has a classical origin, too). The mean energy and width of a resonance are probabilistic measures of the mass and lifetime of an unstable quantum state.

How do the classical and the quantum laws fit in with the individual and probabilistic levels, respectively? Quantum theory is incompatible with classical particle trajectories. Classical particle propagation is assumed according to (1) and (2), whereas the quantum theory of scattering underlies the assumptions (5), (7) and (8). In the laws (3)–(5), however, the individual and probabilistic levels are interwoven. There are links between the classical and the quantum laws. They make it possible to apply the quantum theory of scattering under certain conditions to individual particle tracks. On the one

---

[24]See Appendix C.

hand, there are quantum laws which *correspond perfectly* to classical laws. On the other hand, there are quantum mechanical *superselection rules* for conserved quantities such as spin, parity, and so on. Correspondence and superselection rules hold for particle tracks in the non-relativistic domain, i.e., at low scattering energy.

The exact agreement of the classical and quantum mechanical Rutherford formulae smoothly connects quantum mechanical scattering theory to the familiar classical measurement theory of particles. This smooth connection is obviously established at the probabilistic level. Exact correspondence holds for the relative frequencies of particle deflections. At the probabilistic level, the classical expression of the Lorentz force is approximately obtained from quantum mechanics too, as long as the external electric and magnetic fields are weak in comparison to the subatomic fields.[25] In the early days of quantum mechanics, it seemed justified to extrapolate this correspondence of classical and quantum mechanics from the probabilistic level to the level of individual particle tracks. In his 1930 book, Heisenberg formulated a generalized version of Bohr's correspondence principle which is closely related to this belief:

> The correspondence principle, which is due to Bohr, postulates a detailed analogy between the quantum theory and the classical theory appropriate to the mental picture employed. This analogy does not merely serve as a guide to the discovery of formal laws; its special value is that it furnishes the interpretation of the laws that are found in terms of the mental picture used.[26]

The principle and its implications will later be analyzed in detail.[27] Here, it has only to be noted that it supports a *trust in the unity of physics* which is based on *belief in correspondence* between the applications of quantum theory and the classical theories. This trust made physicists believe that the transition probabilities obtained from quantum mechanical scattering theory also apply to the scattering processes along an individual particle track.

In the non-relativistic domain this trust is unproblematic and legitimate. (In the relativistic domain, however, this is *not really* the case. The data analysis of particle tracks again indicates what we have already noticed. The classical picture of reality does not break down at once. It breaks down stepwise.[28]) The experimental practice of analyzing particle tracks relies on this belief. Indeed it is empirically supported by the energy–range relation which belongs to the semi-empirical laws (3). The range of particles of a given energy within matter can be measured in a particle accelerator with the help

---

[25]Mott and Massey 1965, 4.

[26]Heisenberg 1930b, 105. In the German version, the principle is expressed in a different way; see my own translation of this passage in Sect. 5.4.2.

[27]See Sect. 5.5.

[28]See Sects. 4.4 and 5.3.

of a detector which functions as a fixed target. The measurement investigates how the energy of a given particle beam is connected to the average particle range and track length in matter. Such a measurement makes it possible to test the quantum electrodynamic predictions for energy loss directly. That is, the measured average particle track length can be compared to the quantum mechanical expectation value for the range predicted by quantum electrodynamics.[29]

Hence, the measurement theory of particle physics has remarkable features. It is not only *heterogeneous*, but also *redundant*, that is, the quantum electrodynamic laws which enter the measurements are supported by semi-empirical laws. These semi-empirical laws make it possible to test the quantum electrodynamic formulae independently. As shown in Sects. 3.4 and 3.5, during the phase of consolidation of quantum electrodynamics experimenters like Anderson made substantial efforts to determine the mass and charge of particles by improvements of such independent semi-empirical measurement procedures. After the consolidation of quantum electrodynamics, the semi-empirical methods remained in the measurement theory. Up to the present day, they make it possible to perform several consistency checks on the measurements. And all the tests hitherto made confirmed that the measurement theory sketched above is *coherent*. The combination of classical laws and quantum laws gives consistent quantitative measurement results, even though in the relativistic domain the quantum laws do no longer hold at the level of individual tracks but only at the probabilistic level of *many* particle tracks.

The measurement theory of particle physics is therefore a *heterogeneous* structure of measurement laws which contains *redundancies* and which gives *coherent* quantitative results at the probabilistic level. It is worth comparing with Ian Hacking's fantasy of a physics in which every law was as specific as possible, no law was redundant, and any two laws were inconsistent with each other:

> God did not write a Book of Nature of the sort that the old Europeans imagined. He wrote a Borgesian library, each book of which is as brief as possible, yet each book of which is inconsistent with every other. No book is redundant. For every book, there is some humanly accessible bit of Nature such that that book, and no other, makes possible the comprehension, prediction and influencing of what is going on.[30]

The measurement laws of particle physics are thoroughly distinct from such a *totally* disunified structure. They make up a theory. This measurement theory is highly redundant. Its redundant parts are quantitatively compatible even if they are incommensurable. Unlike Hacking's Borgesian fantasy, it points out that the contents of the book of nature *could* be formulated in a coherent language, if only we were able to speak it. It suggests that a Galilean book

---

[29]See Sect. 5.3.3 and notes 20–21.
[30]Hacking 1983, 219.

of nature exists, even though up to the present day we have only deciphered it incompletely and in incompatible mathematical fragments.

5.3 Particle Tracks

The foundations of this heterogeneous, redundant, and coherent measurement theory were laid in the 1930s and 1940s. In the course of the analysis of tracks from cosmic rays, the laws (3) and (5) were added to the classical laws (1)–(2). In the early 1930s, Bethe, Bloch, and Heitler calculated the laws (5) for the dissipation of energy along a particle track on the basis of Born's quantum mechanical scattering theory and Møller's paper on scattering and quantum electrodynamics.[31] Since for various reasons nobody had confidence in quantum electrodynamics around 1930, the formulae of (5) were not used for the analysis of cosmic ray tracks at that time. Only after their validity was confirmed by additional independent measurement methods based on the semi-empirical laws (3) in the 1940s did they finally enter the measurement theory of particle physics.[32]

The formulae for energy dissipation (or, as the particle physicists say, *energy loss*) are genuinely probabilistic. Nevertheless, it became experimental practice in particle physics to apply them to individual particle tracks. Indeed all the calculations mentioned above were based on trust in Bohr's correspondence principle, or Heisenberg's generalized version of it. Indeed Mott and Heisenberg showed in 1930 that Born's quantum mechanics of scattering predicts particle tracks with a classical shape. Thus, the energy loss of charged particles in matter was calculated on the basis of a naive, quasi-classical, realistic picture of subatomic reality. The underlying assumptions give insight into the physicist's trust in a hidden unity of nature, the reasons for which it is justified, and the limitations of such trust. As in the case of form factors,[33] there is a chain of models which establishes a correspondence between the classical case and the quantum calculations. Again, this chain of models demonstrates that in the subatomic domain the classical picture of reality does not break down at once, but only step by step. However, the subsequent case study is more delicate. It deals with *individual* particle tracks whereas the probabilistic interpretation of quantum theory operates at the *ensemble* level. From a philosophical point of view, the crucial question concerning energy loss is: What is the quantum ensemble to which the quantum theory of scattering applies?[34]

---

[31] See Born 1926a,b; Bethe 1930; Møller 1931; Bethe 1932; Bloch 1933; Heitler and Sauter 1933; Bethe and Heitler 1934.

[32] See Sects. 3.4–3.5.

[33] See Sects. 4.3 and 4.4.

[34] The subsequent sections are based on Falkenburg 1996. In the following, however, I hope to give a much clearer analysis of this case study and to correct some of my previous errors.

5.3.1 Mott's Prediction of Classical Tracks

Shortly after the development of quantum mechanics it was shown that the generation of particle tracks in the Wilson chamber is perfectly compatible with quantum mechanical scattering theory. As Heisenberg stressed in his 1930 book on quantum mechanics, the probability of α-particle deflection due to repeated ionization of molecules in the vapour is non-zero

> only if the connecting line of the two molecules runs parallel to the velocity direction of the α-particles.[35]

The corresponding calculation was carried out for the first time by Mott in 1929.[36] It was based on Born's 1926 quantum mechanics of scattering which is the theoretical framework of Born's probabilistic interpretation of quantum mechanics.[37] According to quantum mechanical scattering theory, the scattering is not due to an impact but due to *diffraction*. In particular, the quantum mechanical description of scattering lacks the classical trajectory of a deflected particle and the corresponding classical impact parameter. The squared wave function predicts only the probability (and hence the relative frequency) of particle detections at a certain scattering angle, as shown in Born's seminal 1926 papers.

Mott calculated the probability for two subsequent collisions of an α-particle and a hydrogen atom with the effect of the ionization of both atoms. The ionized atoms give rise to observable measurement points. They are the core of droplets which condense in the vapour of the Wilson chamber. The observation of a droplet is a *position measurement*. But the observation of the particle deflection which is given by straight lines drawn between the adjacent droplets is a *momentum measurement*. Heisenberg showed in his 1930 book by a heuristic consideration that the uncertainty relation for position and momentum holds for any ionization process along the track. Due to the finite size of the water molecules in the Wilson chamber, the position and momentum measurement cannot both be sharp.[38] Thus, the quantum mechanical explanation of the single measurement points of a particle track is in

---

[35]Heisenberg 1930a, 53 (my translation); see Heisenberg 1930b, 70.

[36]Mott 1929.

[37]Born 1926a,b. The probabilistic interpretation was later generalized and put in more precise terms by von Neumann; see von Neumann 1932. In current discussions, Born's second 1926 paper is rarely taken into account. (Hence the importance of quantum mechanical scattering theory for the interpretation of quantum mechanics is neglected.) Its formal parts are discussed in Jammer 1966, 284–285. In addition, it contains informal speculative ideas about the interpretation of the wave function which are criticized in Jammer 1974, 42–44. See also Beller 1999.

[38]See Heisenberg 1930a, 18; 1930b, 24. This heuristic argument tacitly makes use of the correspondence principle. However, it is supported by wave mechanics employing Huygen's principle. The inaccuracy of the position measurement for the single measurement points of a particle track and the measurement error of the particle momentum obtained from a curved particle track using the expression for

perfect correspondence to the classical particle picture. It only differs from a classical trajectory with regard to what cannot be observed, namely a classical path *between* the position measurements. Thus, for all practical purposes it predicts a classical particle track. In this way, it supports the application of quantum mechanical scattering theory to individual particle tracks.

Mott's calculation is probabilistic. It is not concerned with individual measurement results, that is, with the discontinuous change of the wave function which corresponds to the observation of a droplet in the vapour. It is performed in a quantum mechanics without measurement. According to this, the wave function scatters at two atoms at a given distance R. Born's quantum mechanics of scattering predicts the diffraction of the wave function. First and second order perturbation theory give two different kinds of contribution to the scattered wave. Both are due to inelastic scattering at both atoms. (Here, 'inelastic' means 'with energy loss'. The energy loss is equal to the ionization energy. Elastic scattering does not give rise to ionization. Thus, it does not result in observable droplets in the Wilson chamber.) According to *first-order* perturbation theory, only one atom is ionized after the scattering and the twofold scattering must consist of *two incoherent single scattering processes* which give independent, i.e., non-interfering contributions to the scattered wave. Hence, the first order contribution involves adding up twice the probability of scattering at just one atom. Only via the *second order* term are both atoms ionized at once in coherent scattering. Mott's main result is that *both kinds of contribution predict a classical track*. To first order, the outgoing wave is concentrated in a cone of very small angle behind the ionized atom, in the direction of the incoming wave. To second order, the contribution is non-zero if and only if both atoms lie inside that angle in the same direction.[39] Generalized to N atoms, Mott's results predict the following: To first order, the scattering probability is N times the probability for incoherent scattering at a single atom. Coherent scattering at more than one atom contributes only to second, third, ..., Nth order. However, to *any* order the scattered wave propagates along a classical path.

Mott's results are completely in the spirit of Bohr's correspondence principle and Heisenberg's generalized version of it quoted above in Sect. 5.1. The quantum mechanics of a particle track corresponds to a classical path. To be more precise, the quantum mechanics of scattering predicts an *ensemble of possible tracks* which are *identical for all practical purposes* and which correspond to a classical trajectory. This prediction establishes correspondence

---

the Lorentz force are typically more than 12 (!) orders of magnitude larger than Heisenberg's uncertainty relation.

[39] See Mott 1929; Heisenberg 1930a, 56, and 1930b, 75–76. Heisenberg's presentation of Mott's results is clearer than modern textbook accounts. See for example Messiah 1964, Sect. 19.3.10–11. Here, the classical shape of the track is only explained from the second order contribution, separating it misleadingly from the preceding discussion of the incoherent first order scattering, which for obvious reasons is the *dominating* contribution.

between classical and quantum mechanics *at the level of individual tracks*. The probability of two subsequent particle deflections along a straight line differs from 1 by an extremely small value which may be neglected *for all practical purposes*. Therefore, any effect of the *measurement* which causes an observable droplet in the vapour may be disregarded.

It has to be noted that Mott's 1929 calculation is based on an 'unrealistic' idealization. The *energy loss* associated with ionization is not taken into account. The particle is described as if it did not *really* transfer a definite amount of energy to the hydrogen atom when ionizing it. Although the calculation deals with the amplitudes of inelastic collisions, it is performed as if the momentum state of the charged particle remained unaffected by the energy transfer to the hydrogen atom which gives rise to ionization. That is, the α-particle which gives rise to subsequent ionization processes and observable droplets in the vapour is treated *as if its collisions with the hydrogen atoms were elastic*. Admittedly, this 'unrealistic' idealization (which is nothing but the neglect of the momentum transfer) looks reasonable once we notice that the energy loss of an α-particle due to ionization of hydrogen atoms can be neglected. The ionization energy of hydrogen is *very* small compared to the kinetic energy of the α-particle. Therefore the momentum of the α-particle remains practically unchanged along its track in the Wilson chamber.

However, with regard to the inter-theoretical relation between classical and quantum mechanics the neglected quantity is crucial. Recall Feynman's dictum quoted above in Sect. 5.1. In quantum mechanical scattering theory, it reads as follows: *Our entire picture of the world has to be altered even though the momentum changes only a little bit*. An ionization process which gives rise to an observable droplet is a measurement. It obeys Heisenberg's uncertainty relation, that is, the momentum of the α-particle after ionization is unsharp. In addition, it is an irreversible process associated with energy dissipation. If this energy loss is neglected, the momentum is regarded as constant along the whole track. To put it another way, it is assumed that the particle state after any position measurement still belongs to the ensemble of the initial momentum state, or that the preparation of the α-particle does not change along the track, or that the measurements are *repeatable*. This assumption is obviously at odds with the fact that the observed track is due to a sequence of quantum transitions which are associated with *non-repeatable* measurements and energy dissipation along the track. This is a matter of principle, even though the results of Mott's calculation show that the idealization is good enough *for all practical purposes*.

However, under such idealized conditions the classical and the quantum description of a track agree for any sequence of measurement points, the classical and the quantum model of the track seem to be in pre-established harmony, and their incommensurability does not matter. This correspondence between the classical and the quantum case holds not only for the straight particle tracks calculated by Mott and Heisenberg, but also for the curved tracks in a magnetic field. For a weak external field, the Schrödinger equation for a stationary beam of particles predicts approximately the classical beam deflection which is described by the Lorentz force.[40]

5.3.2 Bethe's Calculation of Energy Loss

In the case of a substantial amount of energy loss along a particle track, the agreement of the classical and the quantum descriptions vanishes. Nevertheless, the neat quasi-classical picture of the scattering processes along an observable track was maintained in the later theoretical treatments of the energy loss of charged particles in matter. According to the intuitive classical picture which still underlies these calculations up to the present day, a charged particle loses its energy along a measured track as follows. The particle is slowed down repeatedly by inelastic collisions with detector atoms, the track curvature in an external magnetic field increases, and the track ends when the particle has transferred its total kinetic energy and momentum to the detector atoms. The discovery of the positron was based on such an effect. In order to identify the sign of the charge of the particles from cosmic rays, Anderson measured the flight direction by putting a lead plate into the Wilson chamber. The lead plate causes a substantial energy loss and gives rise to an observable increase of the curvature of particle tracks in a magnetic field.[41]

The first reliable calculation of this non-negligible energy loss was given by Bethe in 1930.[42] Prior to 1930, there was only one calculation, namely Bohr's classical theory of ionization dating from 1913–15. It was known to give wrong results for fast particles.[43] For the energy loss along a particle track, Bohr's correspondence principle in general *fails*. Bethe's 1930 calculation of energy loss due to excitation and ionization of the detector atoms results in an expression which agrees with Bohr's classical result only for *vanishing* particle velocity v and energy loss.[44] This limit of *no* velocity and *no* energy loss is exactly the idealized case of Mott's calculation.

In 1930, Bethe published a seminal paper about the energy loss of charged particles in matter. The calculation is made in a *semi-classical model*. The model adds classical assumptions about the individual scattering processes along a particle track to Born's quantum mechanics of scattering. (The formulae are given here in modern notation. The calculation is performed in time-

---

[40]See Mott and Massey 1965, 4.

[41]See Sect. 3.4.

[42]Bethe 1930.

[43]See Bohr 1913a, 1915.

[44]The two results differ in a term with logarithmic dependence on the particle velocity v. Bethe's formula contains an expression $\ln v^2$ where Bohr's formula contains $\ln v^3$; see Bethe 1930, 360, note 5. Thus, correspondence holds only for $v \to 0$. For slow heavy particles, Bohr's classical formula does indeed give approximately correct results; see Jackson 1975, Sect. 13.3.

dependent perturbation theory.) Before the scattering, the charged particle is described by a quantum state

$$\phi_k(r,t) = e^{ikr - \omega t} \, , \tag{5.1}$$

which is a plane wave with momentum p, wave number $k = p/\hbar$ and angular frequency $\omega = \hbar k^2/2m$. It is scattered at a detector atom in the quantum state

$$\varphi_0(r,t) = \Psi_0(r)e^{-(i/\hbar)E_0 t} \, , \tag{5.2}$$

where the index 0 denotes the ground state of the atom. Before the scattering (which begins at t_0), the charged particle and the detector atom are in uncoupled states. For $t \ll t_0$, the wave function of the combined system particle-plus-atom factorizes:

$$u(r,t) = a_0(k)\phi_k(r,t) \otimes \varphi_0(r,t) \, . \tag{5.3}$$

As modern textbooks tell us in an appealing expression for the *start* of an *asymptotic* process, the interaction is 'switched on' at $t = t_0$. Now, the wave functions of the detector atom and the incoming particle form a compound system. For $t \geq t_0$, the wave function of the compound system particle-plus-atom develops as predicted by the time-dependent Schrödinger equation:

$$u(r,t) = \sum_{n,k'} a_n(k',t)\phi_{k'}(r,t) \otimes \varphi_n(r,t) \, . \tag{5.4}$$

The time-dependent amplitudes $a_n(k',t)$ of the coupled system are given by time-dependent perturbation theory. (Bethe could not yet use this theory in a sound way because the δ-distribution was not yet available to him.) After the interaction, for $t \gg t_0$, it is assumed that the particle is asymptotically free. By the detection of the particle at a time t_1, the wave function collapses to a single component $\phi_q(r,t) \otimes \varphi_n(r,t)$, where $q = \hbar(k-k')$ is the momentum transfer between the particle and the atom and n labels the excited state of the atom in the case of inelastic scattering. At time $t \geq t_1$ the collapsed wave or reduced quantum state has the form:

$$u_{n,q}(r,t) = a_n(q,t_1)\phi_q(r,t) \otimes \varphi_n(r,t) \, . \tag{5.5}$$

This wave function depends only on the state n of the atom and the momentum transfer q, but no longer on the momentum p of the incoming particle. That is, the transition probability for a certain amount of energy loss does *not* depend on the momentum of the incoming charged particle. This is crucial. It makes the situation similar to Mott's calculation, justifying by hindsight the idealization on which Mott's model was based. In Mott's model, the momentum transfer was neglected. Here, it turns out that it plays no role for the calculation of energy loss. (Thus, once again it seems unproblematic to forget about the effects of measurement and to apply the results of the calculation to individual particle tracks.) The scattering amplitude of the diffracted wave is given by the amplitude $a_{n,q}(r, t_1)$ of the collapsed state $u_{n,q}(r, t)$ at the time t_1 of particle detection:

$$a_n(q, t_1) = c(q, k)|V_{0n}|^2 \delta(W - W_0) . \tag{5.6}$$

Here, V_{0n} is the matrix element of the transition of an atom from the ground state $\phi_0(t)$ into an excited state $\phi_n(t)$, and $\delta(W - W_0)$ guarantees energy conservation for the total energies W_0 and W of the particle-plus-atom system before and after the scattering, respectively.

The transition rate per incoming particle per unit time T, that is, the differential cross-section $d\sigma/dq$, is directly related to the scattering amplitude. With appropriate normalization, and expressed as a function of the momentum transfer q, it becomes

$$\frac{d\Phi_n(q)}{dq} = |a_n(q, t_1)|^2 . \tag{5.7}$$

Here today's expression σ for the total cross-section is replaced by Born's *Ausbeutefunktion* $\Phi_n(q)$, which is a function of the momentum transfer q and the nth excitation of the atom. This expression from Born's 1926 papers on the quantum mechanics of scattering is also used by Bethe.

From this result, Bethe derives the crucial formula for the energy loss of a charged particle in matter as follows.[45] The formula is probabilistic. It gives an expression for the *quantum mechanical expectation value*

$$\langle E_n - E_0 \rangle = \langle u_{n,q}(t)| E_n - E_0 | u_{n,q}(t) \rangle \tag{5.8}$$

of the energy transfer from the scattered particle to the detector atom due to excitation of the atom from the ground state ϕ_0 to the nth state ϕ_n:

$$\langle E_n - E_0 \rangle = (E_n - E_0) \int \frac{d\Phi_n(q)}{dq} dq . \tag{5.9}$$

To obtain the expectation value for the energy loss due to a transition to *any* of the excited states of the atom, we have to add the probabilities for all transitions from the ground state into an arbitrary discrete excited state with energy E_ν or into the continuum. The sum may be written as[46]

$$\langle E \rangle = \left(\sum_\nu + \int d\nu \right) (E_\nu - E_0) \int d\Phi_n(q) . \tag{5.10}$$

According to the usual probabilistic interpretation of quantum mechanics, the formula has the following physical meaning: $\langle E \rangle$ is the mean energy loss

---

[45] Bethe 1930, 351–360.
[46] See equations (55a) and (55b) in Bethe 1930, 359.

per atom and per incoming particle (in the limit of infinitely many incoming particles, $N_{in} \to \infty$). This interpretation applies at the ensemble level, that is, to particles which are prepared in the same quantum state. Since neither the expectation value of the energy loss nor the transition rates which contribute to it actually depend on the momentum of the particle, it seems that the momentum does not matter. Therefore with respect to its energy loss, the particle may be considered to belong to one and the same ensemble along the whole track.

Indeed, Bethe took it for granted that (5.10) applies to every scattering along an individual particle track, be it with or without an observable effect. Now he made two further steps in order to apply his formula (5.10) to individual particle tracks. (With them, the classical part of his semi-classical modelling enters.) First, he calculated the number of observable and unobservable inelastic collisions along a track as $N\Phi$, where Φ is the total transition rate and N is the number of atoms in a given material per volume Δx^3.[47] Then he gave the following simple expression for the *mean energy loss* ΔE *per length* Δx *of matter*:[48]

$$\frac{\Delta E}{\Delta x} = N\langle E \rangle . \tag{5.11}$$

Here, $\langle E \rangle$ is given by (5.10) and N is the number of atoms per volume Δx^3. In giving (5.11), Bethe claims that the expectation value $\langle E \rangle$ has the meaning of the *average energy loss of a charged particle by successive scattering from many detector atoms along an individual track*, normalized to the number of atoms per path Δx. That is, the individual track is considered to stem from *many* scattering processes of a particle in a quantum state which belongs to one and the same statistical ensemble.

Of course, Bethe discussed neither the physical interpretation of his results nor their philosophical justification. He simply suggested that $\langle E \rangle$ applies to the subsequent individual quantum transitions along a track, whether their results are observable or not. This was completely in the spirit of Mott's quasi-classical results. They showed that to first order the scattering probability of subsequent ionizations is simply the sum of incoherent scattering at several atoms separately and that the coherent higher order scattering terms are negligible. Bethe's 1930 calculation is also made in first order perturbation theory. As in Mott's calculation, for obvious reasons, the incoherent first-order contributions to the scattering amplitude dominate. In accordance with this, Bethe obviously maintained an intuitive quasi-classical, realistic picture of subatomic scattering processes. His derivation of (5.11) is based on the following implicit assumptions:

(i)   A particle excites and ionizes several atoms during its passage through matter.

[47] Bethe 1930, 358.
[48] Bethe 1930, 360, equation (56).

(ii)   The observed measurement points of the track are due to ionizations but there is also unobservable energy loss due to excitation.

(iii) All of the excitations and ionizations of the detector atoms are independent of each other, and they have a probability described by $\langle E \rangle$ according to (5.10).

The second and third assumptions are crucial. (ii) is simply based on a neglect of the effects of measurement, whereas (iii) is justified by the fact that the coherent higher-order contributions to the scattering are negligibly small.

Indeed, one of Bethe's quantitative results is the *ratio of observable to unobservable collisions*. For hydrogen, 28.5% of all inelastic collisions give rise to ionization,[49] causing observable droplets in the vapour of the Wilson chamber. The calculation shows that a substantial amount of the energy lost along a particle track in the Wilson chamber is indeed *measured*, in contrast to Mott's idealized model in which the energy loss along a track was neglected. However, Bethe's 1930 calculation additionally results in a hidden momentum dependence of $\langle E \rangle$ which also violates Mott's idealization and which is only negligible for fast particles. It stems from the upper limit $q_{\max}$ of the crucial integral

$$\int \mathrm{d}\Phi_n(q) = \int_0^{q_{\max}} \mathrm{d}\Phi_n(q) \tag{5.12}$$

in (5.10). A particle can lose no more energy than it has. Therefore, with increasing total energy loss along the track, the expectation value varies as $(\ln E_{\mathrm{kin}})/E_{\mathrm{kin}}$ with the remaining particle energy E_{kin}.[50] Hence, the situation is no longer comparable with Mott's idealized model in which the preparation of the particle and the expectation value of the scattering results do not change along the track. The particle states after the measurement points do indeed no longer belong to the same quantum ensemble with respect to $\langle E \rangle$ when this quantity varies along the track with decreasing particle momentum.

For this reason, Bethe's semi-classical model for the energy loss of a charged particle in matter becomes obviously incoherent. Calculating the quantum mechanical expectation value $\langle E \rangle$ according to (5.10) is at odds with applying it to the individual measurement points along a track according to (5.11), in the light of the usual probabilistic (i.e., ensemble) interpretation of quantum mechanics. In assuming that (5.11) gives the average energy loss of a particle along an individual track, Bethe relies on Mott's results in a case which is *not* compatible with the idealized conditions of Mott's model. Within a *quantum mechanics without measurement*, he calculates a formula which holds for the *energy dissipation due to the position measurements along a track*. Indeed for the fast particles which were the subject of Bethe's calculation, this is once again just a matter of principle (at least as long as the

---

[49]Bethe 1930, 360.

[50]See Bethe's result for hydrogen in Bethe 1930, 360. See also Falkenburg 1996, 349–350.

particles are not *too* fast, i.e., have relativistic velocity!). Since the momentum dependence of $\langle E \rangle$ is very weak, this incoherent method works perfectly for all practical purposes. Hence, the particle states after the measurement points of an individual particle track belong *almost* to the same quantum ensemble with respect to $\langle E \rangle$. What looks queer from a philosophical point of view looks like a good approximation from a pragmatic point of view, that is, for the demands of physical practice.

5.3.3 How the Classical Picture Breaks Down

Mott's and Bethe's calculations discussed above are only valid for the non-relativistic domain. According to Bethe's 1930 results, energy loss due to ionization is small and the shape of particle tracks is smooth. For relativistic particles, however, the quasi-classical picture breaks down. Quantum electrodynamics predicts that a particle does not lose its energy smoothly. Due to quantum fluctuations, the energy loss along a particle track may become completely irregular and extreme deviations from the classical path may occur. Several kinds of processes may give rise to large fluctuations in the energy loss. In addition to energy loss due to ionization, quantum electrodynamics predicts processes of bremsstrahlung and pair creation, that is, the emission of a photon or an electron–positron pair, respectively. These processes are associated with large fluctuations in the energy loss along a particle track. They give rise to irregular deflections which violate the classical shape of a track predicted by Mott in 1929 and presupposed also by Bethe in his 1930 energy loss calculation.

Energy loss due to ionization is in general small in comparison to the particle energy. Bethe's 1930 calculation does treat it on a par with excitation of the detector atoms, within Born's quantum mechanics of scattering. Ionization is excitation of an electron to a continuum state, i.e., an unbound state of a free electron outside the atom. According to Bethe's results, ionization does not yet affect the quasi-classical smooth shape of a particle track (even though it *should* affect the trust in a sound ensemble interpretation of quantum mechanical scattering theory). Energy loss due to bremsstrahlung or pair creation, however, is substantial in comparison to the particle energy. It gives rise to a sudden loss of a large fraction of the kinetic energy of a charged particle at a certain point of a track. The resulting tracks are unsmooth. They have an observable kink. After Møller had shown in 1931 how Dirac's theory of the electron can be combined with Born's quantum mechanics of scattering, Bethe, Bloch, and Heitler extended Bethe's 1930 calculations to such processes which destroy the classical picture of a particle track completely. The quantum electrodynamic expectation values of these scattering processes depend on the energy of the charged particles which pass through matter. At non-relativistic particle energies, ionization is predominant and the relative frequency of *bremsstrahlung* or pair creation is negligible. In the relativistic domain, the relative frequency of the latter processes increases rapidly with increasing particle energy.[51] Thus, in the transition from the non-relativistic to the relativistic domain the smooth quasi-classical shape of the particle tracks predicted by Mott's and Bethe's 1929/30 calculations gets lost for an increasing number of particle tracks.

Therefore, with increasing particle energy the correspondence between the shape of individual particle tracks and the classical case breaks down stepwise. At very low particle energies, the scattering processes along a track which give rise to observable droplets in the Wilson chamber may be regarded as quasi-elastic, that is, the momentum transfer and energy loss due to ionization may be neglected. In this case, the quantum mechanics of scattering predicts particle tracks with a perfect classical shape, as Mott's 1929 calculation demonstrates. Bethe's 1930 calculation shows that, in the limit $v \to 0$ of vanishing particle velocity v, there is even a correspondence with Bohr's classical calculation of energy loss. For fast non-relativistic particles, Bohr's classical theory of ionization breaks down but Bethe's 1930 semi-classical calculation of energy loss due to ionization demonstrates that in the non-relativistic domain the shape of a particle track still fits in with the classical picture *for all practical purposes*.

In the relativistic domain, the omnipresent quantum fluctuations may cause an unpredictable, abrupt, and substantial energy loss. Due to this, more and more particle tracks violate the classical picture. With increasing particle energy, the relative frequency of such 'deviant' tracks with a kink increases. Therefore the quantum electrodynamic formulae derived by Bethe, Bloch, and Heitler in 1932–34 for the energy loss of relativistic electrons due to ionization, bremsstrahlung, and pair creation are no longer reliable for the individual tracks. They apply only at the *probabilistic* level of *many* particle tracks. At the probabilistic level, the correspondence to the classical description of particle tracks breaks down gradually. With increasing particle energy, an increasing number of 'deviant', unsmooth tracks fall off the average behavior, violating the prediction of quasi-classical, smooth, trajectory-like sequences of observable measurement points which corresponds to the classical case in the limit of vanishing particle velocity. In turn, for decreasing particle energy, correspondence is first re-established at the probabilistic level in terms of a decreasing number of particle tracks which fall off the smooth average prediction of the energy loss.

However, the quantum electrodynamic formulae are still associated with Bethe's semi-classical understanding of the energy loss due to repeated single scattering processes along an individual track. In particular, they are simply inserted into Bethe's expression (5.11) for the *mean energy loss ΔE per length Δx of matter*. Again, this procedure is justified by the incoherent first-order contributions to the quantum mechanics of scattering. The fact that they dominate makes it unproblematic to apply classical stochastic methods to the analysis of particle tracks. Bethe's and Bloch's 1932–1933 calculations

---

[51] See Rossi 1952, 29–30 and 60.

gave rise to the Bethe–Bloch formula for the mean energy loss of fast charged particles per path length in matter which consists of heavy atoms.[52] The Bethe–Bloch formula makes it possible to calculate a theoretical value for the average range of charged particles in a given kind of matter or detector material.[53]

5.3.4 Data Analysis in Scattering Experiments

After the consolidation of quantum electrodynamics, the semi-classical formulae discussed above were used for the data analysis of particle tracks. Even though in the relativistic domain the classical picture of energy loss along a particle track breaks down, it became experimental practice to apply them to individual particle tracks in the spirit of Bethe's semi-classical calculation. The resulting semi-classical formulae for $\Delta E/\Delta x$ have been used for decades. They belong to the standard background knowledge of particle physics. Hence, up to the present day they can be found in the most recent version of the *Particle Physics Booklet*.[54] At the individual level, this procedure may give wrong results because the energy loss ΔE per length Δx is a probabilistic mean value whereas the single particle tracks are due to stochastic scattering effects. In addition, at high energies the distribution of the actual energy loss per detector length around the mean value is highly non-Gaussian due to the processes of bremsstrahlung and pair creation which give rise to large fluctuations. This means that in any high energy scattering experiment there is a very small fraction of very short particle tracks which belong to very high energy particles and strongly violate the semi-classical energy–range relation derived from the Bethe–Bloch formula. In a high precision measurement of a differential cross-section, the measured relative frequencies of particles of different initial momentum are distorted due to these tracks with very large energy loss, and their effect has to be corrected at the probabilistic level by taking into account the predictions of quantum electrodynamics.

This is possible since the predictions of quantum electrodynamics for ionization, bremsstrahlung and pair creation are well-known and well-tested. In addition, the deflection of charged particles in matter due to elastic multiple scattering processes is calculated from quantum electrodynamics. Since today's particle accelerators generate particle beams with high intensity and high precision beam energy, the predictions of quantum electrodynamics have been tested not only with regard to the mean energy loss but also with regard to the variance around the mean value and the precise non-Gaussian shape of the quantum fluctuations.

In the practice of the data analysis of high energy scattering experiments over the last few decades, the predictions of quantum electrodynamics and the

[52]Bloch 1933; for a semi-classical calculation see Rossi 1952, 17.
[53]See Rossi 1952, 22–27.
[54]See the formulae for $-dE/dx$ in PDG 2004, 238–255.

semi-classical formulae for the energy loss of charged particles in matter have come together as follows. The Bethe–Bloch and the Bethe–Heitler formulae for the energy loss $\Delta E/\Delta x$ of a given type of charged particles in a given kind of matter are used to infer the initial momentum at the beginning of a particle track from the actual shape of the track. The way this is done is a paradigm of 'fudging' pieces of incompatible theories together according to an intuitive classical realism. (This is indeed a paradigm of what Nancy Cartwright calls *piecemeal physics*.) The average energy loss ΔE per finite detector length Δx is generously understood as the *differential energy loss per path length*, dE/dx. This quantity is then inserted into the classical expression $\boldsymbol{F} = (q/c)\boldsymbol{v} \times \boldsymbol{B}$ for the *Lorentz force* in the following way:

$$\frac{d\boldsymbol{p}}{dt} = \frac{q}{m}\boldsymbol{p} \times \boldsymbol{B} + \frac{|\boldsymbol{p}|}{m}\frac{d\boldsymbol{p}}{dx} . \tag{5.13}$$

Here, the velocity $\boldsymbol{v} = \boldsymbol{p}/m$ is obtained from dx/dt, where $\boldsymbol{p}$ and m are the momentum and mass, $d\boldsymbol{p}/dx$ stems from dE/dx, and dE/dx is calculated according to Bethe's formula (5.11) for the energy loss of charged particles in matter. The differential equation (5.13) describes the mean decrease in the momentum of charged particles in matter due to energy dissipation along the track. It is implemented in every computer program which calculates the initial particle momentum at the beginning of a track which is relevant for the measurement of a differential cross of a certain kind of scattering event. (Hence, the measurement method described here underlies every high precision measurement of a differential or total cross-section, including the measurements described in Sects. 4.3 and 4.4.) However, the calculation implements only the statistical mean value of an irregular stochastic process. Therefore in every computer program for the statistical analysis of the particle tracks of a high precision scattering experiment, statistical corrections are also implemented. They take into account the difference between the statistical mean value and variance, that is, the difference between the quantum mechanical expectation value $\langle E \rangle$ and the non-Gaussian distribution of energy loss around the mean value which is due to quantum fluctuations caused by processes of bremsstrahlung and pair creation. At the probabilistic level of data analysis, the 'deviating' tracks with non-average (i.e., non-classical) shape are taken into account in order to correct the measured relative frequency of particles of a certain momentum for the quantum electrodynamic effects which can only be appropriately understood at the probabilistic level.[55] This procedure of data analysis is based on the following line of reasoning:

(i)   The individual particle tracks are analyzed according to semi-classical formulas with respect to the energy dissipation along the track. Solving the differential equation (5.13) for the parameters of a given track results

---

[55] See Falkenburg 1996, 360–363, in comparison to Cowan 1998, 153–187.

in a measurement of the value of the initial particle momentum at the beginning of this track.

(ii) For high energy particles or relativistic initial particle momentum, the correspondence with the classical picture of a particle track breaks down. This means that, with increasing particle energy, for an increasing number of particle tracks the data analysis according to (i) gives a wrong result and the measured relative frequencies of particles of a certain initial momentum are biased by this measurement error. Correspondingly, the measured relative frequencies give rise to a distorted differential cross-section measurement.

(iii) However, at the probabilistic level quantum electrodynamics predicts the number of particle tracks for which the quasi-classical data analysis (i) is expected to fail, as well as the stochastic distribution of the deviations around the average. Thus the statistical variance of the deviating particle tracks from the quasi-classical average gives rise to a probabilistic prediction for the distortion of the differential cross-section measurement. Such a measurement error, however, may be corrected at the probabilistic level by means of simple statistical methods.

Indeed, this kind of reasoning and the corresponding statistical methods of the data analysis of high precision cross-section measurements give rise to *very* coherent measurement results for the differential cross-section measurements discussed in Sect. 4.3.2. Here the method of data analysis cannot be explained in detail. Suffice to say that in the face of the very complex technical devices and theoretical background knowledge involved in such a measurement, physicists always implement as many kinds of redundant measurement methods and internal consistency checks of their data as they can. In current particle physics this redundancy is built into the very design of any experiment. There are independent measurements of the detector properties (including the calibration of the energy scale covered by a detector) and there are as many internal consistency checks of the data as possible. In addition, no experimental result is taken for granted which has not been confirmed by *another* experiment based on *independent* experimental methods. At this point, the quest for redundancy in the face of complex measurement methods meets the well-known quest for the reproducibility of experimental results.

5.4 Building Bridges: Unifying Principles

The resulting method of data analysis has the heterogeneous, redundant, and coherent structure sketched above in Sect. 5.2. It has semi-classical foundations: It combines a basically classical method of analyzing individual particle tracks with probabilistic quantum predictions and corrections. It is redundant: It makes it possible to compare predictions of quantum electrodynamics with semi-empirical formulae based on detailed measurement results which are *not* based on quantum electrodynamics. It gives coherent results: The agreement of independent experiments which measure the same quantity, the comparison of measurement laws to independent measurements, and the redundant methods built into any data analysis of particle physics support each other.[56]

One point should be clear from the discussion in the preceding sections. The heterogeneous structure of the measurement theory of current particle physics is *not* primarily due to its historical development. Rather, it is due to the fact that up to now, there has been no unified theory of the classical and the quantum domains into which its fragments might be integrated.

In turn, the coherence of this measurement theory is based on the fact that its fragments are successfully connected by unifying principles. In particular, Bohr's generalized correspondence principle has the function of building semantic bridges between incommensurable theories when other unifying principles fail. Such unifying principles may be called *bridge principles*. They establish inter-theoretical relations between disunified theories. However, they should not be confused with what the empiricist philosophy of science called bridge principles, namely the rules of correspondence that link the theoretical language of physics and the observational language of experiment and measurement. In Nagel's famous book *The Structure of Science*, this empiricist understanding of bridge principles was explained using Bohr's atomic model of all things, thus provoking confusion with Bohr's correspondence principle.[57] The latter principle was a heuristic principle that made it possible to construct the models and axioms of old quantum theory prior to the quantum mechanics of 1925. In the following, the term *bridge principle* is only used in this heuristic sense. The bridge principles at work in quantum physics establish bridges between classical and quantum physics, that is, inter-theoretical relations. They should not be confused with empirical correspondence rules. Indeed, in modern physics empirical correspondence rules do not have the importance which logical empiricism ascribed to them. The reason is that the phenomena of physics are far from being empirical in a strict sense. They are phenomenological structures.[58]

5.4.1 Bohr's Correspondence Principle

Bohr's original version of the correspondence principle referred to the frequencies of radiative transitions of atoms.[59] According to Bohr's atomic model of 1913, a quantization rule introduced *ad hoc* postulates that the electrons in the atomic shell cannot lose energy by radiation like classical charges. Otherwise, they would crash into the atomic nucleus within a very short time.

---

[56]The insight that experiment (like any work of man) is not immune to bias, error, deception, and fake is no case against this claim.

[57]Nagel 1961, 94–95.

[58]See Sect. 2.4.

[59]For the following, see Darrigol 1992, Part B, and Falkenburg 1998.

According to Bohr's quantum postulates, atoms only emit radiation when the electrons jump from a quantized energy level to another level with lower energy. The quantized radiative transitions violate the classical theory of radiation. The correspondence principle now establishes an analogy between classical and quantized radiation. This analogy refers to the emitted light frequencies. It states that, for high energy levels and small light frequencies, the light frequencies in the radiative transitions of an atom correspond to the radiative frequencies of a classical charge orbiting around a charge center. In 1920, Bohr formulated the correspondence principle as follows:

> [...] there is found to exist a far-reaching correspondence between the various types of possible transitions between the stationary states on the one hand and the various harmonic components of the motion on the other hand. This correspondence is of such a nature that the present theory of spectra is in a certain sense to be regarded as a rational generalization of the ordinary theory of radiation.[60]

The correspondence ensures that the quantum theory of radiation can in a certain sense be joined seamlessly to classical electrodynamics. It permits the application of the concepts of classical radiation theory to atomic spectra through analogy. Bohr called this analogous application a "rational generalization" of the classical theory of radiation. To regard the quantum theory of atomic spectra in this sense as an extension of the classical theory of radiation has a formal and a qualitative aspect. The *formal aspect* concerns the numerical correspondence of formal expressions. Formally, the analogy consists in a quantitative relation of approximation between the quantum and the classical laws of electromagnetic radiation. It states under which formal conditions ($n \to \infty$, $\Delta\nu \to 0$) the quantized frequencies of radiation pass over approximately into classical frequencies. The *qualitative aspect* concerns the physical interpretation of these expressions in terms of the same physical quantity. Qualitatively, the analogy fills a quantum concept (the formal expression $\Delta E = h\nu$ for the radiative frequencies of an atom) with classical physical meaning. It states that the discrete frequencies of radiation from the line spectrum of an atom represent the same type of physical property as the continuum of frequencies of radiation of oscillating classical charges.

Only the two aspects taken together connect the frequencies of quantum theory and classical electrodynamics to a homogeneous class of physical properties. Conversely, taken together they allow for the non-optical spectrum of electromagnetic microwave and long-wave radiation to be regarded as an extension of the optical spectrum of light emitted by atoms. Obviously Bohr's correspondence principle is not an empirical rule of correspondence. It does not (only) assign the empirical concept of a 'line in the spectrum' to the formal law of radiation $\Delta E = h\nu$ which implicitly defines the wavelengths of the

---

[60]Bohr 1920.

lines, as Nagel put it in his 1961 book.[61] It does not primarily tie together the concepts of a theoretical language and an observational language. Rather, it establishes a numerical and physical analogy between classical radiation theory and quantum theory, that is, an inter-theoretical relation. This two-fold analogy allows for the continued use of the classical concepts of 'frequency' and 'wavelength' in the atomic model of Bohr and Sommerfeld. Bohr's original version of the correspondence principle quoted above belonged to old quantum theory. It supported the development of atomic models prior to the rise of quantum mechanics. It was a heuristic principle which made it possible to combine the classical orbits of the Bohr–Sommerfeld atom with the quantum postulates of Bohr's 1913 *Trilogy*. Thus, even though in old quantum theory the correspondence principle has the status of an internal principle, it expresses nothing but an inter-theoretical relation. It is a bridge principle which has a heuristic function in theory formation on the one hand, while it allows for the physical interpretation of the models of old quantum theory on the other.

5.4.2 Correspondence Generalized

Heisenberg's later generalization reinterpreted correspondence as a principle which deals explicitly with inter-theoretical relations. In 1930, Heisenberg generalized the correspondence principle in such a way that the formal and the qualitative aspect of the analogy between classical and quantized radiation frequencies may be extended to *all* physical quantities:

> In its most general version, Bohr's correspondence principle states a qualitative analogy (which can be carried out in detail) between the quantum theory and the classical theory belonging to the respective picture employed. This analogy does not only serve as a guide for finding formal laws, rather, its special value is that it furnishes at the same time the physical interpretation of the laws that are found.[62]

According to this, the concepts of classical physics stand in a precise correspondence with the concepts of quantum mechanics. On the one hand, this correspondence has a heuristic function in the search for the formalism of the new theory. With the help of quantization rules, one is able to construct a quantum theory from a classical theory, where the statements of the classical and the quantized theory stand in well-defined quantitative relations of approximation. On the other hand, the generalized correspondence principle has a semantic function in providing the 'physical interpretation' of the abstract

---

[61] Nagel 1961, 95.

[62] Heisenberg 1930a, 78 (my translation). The English version quoted above in Sect. 5.2 neglects three points: (i) the *generalization* of Bohr's original principle, (ii) the *qualitative* features of the analogy, and (iii) the term 'physical' concerning the interpretation. Heisenberg quotes Bohr 1923, 117.

quantum theory and its models. The principle states that it is permissible to interpret the abstract formalism of a quantum theory obtained through the quantization of a classical theory at least partially in terms of the familiar classical concepts. The 'qualitative analogy' between quantum and classical descriptions ensures that quantum theory is intuitive in a sense required by Heisenberg, i.e., that we have a qualitative understanding of its experimental consequences on an operational basis.[63] The generalized correspondence principle is thus a semantic principle of continuity which guarantees that the predicates for physical properties such as 'position', 'momentum', 'mass', 'energy', etc., can also be defined in the domain of quantum mechanics, and that one may interpret them operationally in accordance with classical measurement methods. It provides a great many inter-theoretical relations, by means of which the formal concepts and models of quantum mechanics can be filled with physical meaning. This physical meaning again has two aspects, one formal and one informal.

As in Bohr's original version, the *formal aspect* of correspondence lies in relations of approximation between the classical and the quantum laws. These laws on their own define the respective physical concepts axiomatically or implicitly, through the axioms of the respective dynamics. The formal meaning which a quantum concept obtains through formal correspondence with a classical concept, however, is different. As in Bohr's 1920 version of the correspondence principle, it lies in the formal conditions under which the underlying quantum law approximately reduces to a classical law. In a formal approach to theory reduction, classical and quantum concepts may be partially joined through a correspondence-confirming standard identification of observables.[64]

The informal aspect of the meaning of quantum concepts is operational rather than axiomatic. To be more precise, it lies in the familiar classical scales of physical quantities and their operational basis. According to Bohr and Heisenberg, the physical meaning of quantum concepts provided by the correspondence principle is operational.[65] It lies in the way in which the values of the corresponding quantities are measured with a classical (or quasi-classical) measurement method (that is, in the way, in which according to Carnap the intension of a concept is pragmatically anchored[66]). Bohr and Heisenberg both called this physical meaning 'intuitive', even though in quite a different sense.[67] Bohr's as well as Heisenberg's uses of the term 'intuitive' indicate that they mean the informal, respectively qualitative aspects of physical con-

---

[63] Heisenberg 1927.

[64] Scheibe 1999, 224–225.

[65] It should already be noted here that for Bohr this is not sufficient. The correspondence principle also underlies his complementarity philosophy which aims at giving reference to quantum concepts in terms of quantum phenomena and their classical analogues; see Sect. 7.2.2 and Pringe 2006.

[66] Carnap 1947.

[67] See Bohr 1928; Heisenberg 1927; see also Falkenburg 2005.

cepts. (Recall that their views on the physical meaning of quantum concepts are in strong opposition to Einstein's axiomatic view of the operational aspects of physical concepts.[68]) However, the informal or non-axiomatic aspects of physical concepts are not yet exhausted by their operational meaning. In addition, there is the question of whether and how these concepts refer. This is a point not focused on by Heisenberg, and to which I shall return later. Conversely, it is well known today that the correspondence principle does not completely exhaust the domain of quantum theory. There are many quantum phenomena without classical correspondence.[69]

Heisenberg, however, ascribed to the generalized correspondence principle the function of giving a physical interpretation of the abstract formalism of a quantum theory. This means in particular that the formalism is interpreted in such a way that it can be applied against the background of classical physics and on semi-classical conditions. (Indeed, all experiments of atomic, nuclear, and particle physics are subject to semi-classical conditions as far as they use macroscopic measurement devices and model the scattering of subatomic particles at the atoms inside these devices. The crucial classical assumption is the locality of the detector atoms inside the macroscopic bulk of matter.) We have already investigated how this works in two case studies. In the first case, the correspondence principle has a decisive referential import which was not noticed by Heisenberg. In the second case, the correspondence principle establishes reference of the quantum description of a particle track to a classical path only *for all practical purposes*. In both cases, correspondence breaks down in the domain of relativistic quantum theory.

1. In the definition of subatomic *form factors*, two applications of the generalized correspondence principle are involved. The *first* is the chain of models described in Sect. 4.4. These models of quantum mechanical scattering theory correspond approximately to classical Rutherford scattering under well-defined conditions. Exact correspondence between the classical and quantum mechanical differential scattering cross-sections is given in the case of the Rutherford formula, that is, for the Coulomb potential, for non-relativistic probe particles, and in the absence of quantum mechanical spin or exchange effects. The *second* is the physical practice of describing the charge distribution inside the atom by a classical form factor, as explained in Sect. 4.3. It is based on a correspondence between the quantum mechanical many-particle wave function $|\Psi(r)|^2$ of charged subatomic particles and a classical charge distribution $\rho(r)$ which is the Fourier transform of a classical form factor $F(q)$. Here, the correspondence principle serves the classical representation $\rho(r)$ of a non-classical

---

[68]See Sect. 5.1.

[69]For example, EPR correlations (Einstein et al. 1935 and Aspect et al. 1982), the Bohm–Aharanov effect, the Bose–Einstein condensate, the double slit with single photons or electrons, and the recent *which-way* experiments of quantum optics. See also Sects. 7.3–7.5.

probability density $|\Psi(r)|^2$ which has *no* operational meaning as far as *bound* charges are considered. According to the usual operational interpretation of quantum mechanics, $|\Psi(r)|^2$ means the probability of position measurements. However, non-destructive measurements of *single* charges inside the atom are not possible. Thus, to identify $|\Psi(r)|^2$ with a (quasi-)classical charge distribution inside the atom exceeds the usual probabilistic Born–von Neumann interpretation of quantum mechanics with the help of the correspondence principle.

2. The analysis of *particle tracks* is based on a similar chain of models which relate the quantum mechanics of scattering to the corresponding classical case. Here, the model which corresponds to the classical case, namely Bohr's 1913 calculation of energy loss by ionization, is highly unrealistic, as shown in Sects. 5.3.1–5.3.3. In addition, it turned out that Mott's calculation *confirms* the generalized correspondence principle whereas Bethe's calculations *trust* in it even in a domain where it no longer holds.

Both cases show that correspondence is a powerful principle. As far as it holds, the classical picture of reality may be maintained *for all practical purposes*. Indeed, the correspondence principle prevents classical realism[70] from breaking down at once in the quantum domain. In both the cases discussed above, there is a chain of models based on a model with classical correspondence. Along this chain of models, the classical picture of reality only breaks down stepwise. The correspondence principle claims that subatomic charge distributions, scattering centers, and quasi-classical particle tracks exist at least in the non-relativistic domain, *as if there were classical particles.*

Here, one remark has to be added. The operational interpretation of quantum mechanics according to Born and von Neumann is not sufficient for the physical interpretation of many experiments of atomic, nuclear and particle physics. In addition, the generalized correspondence principle and other unifying principles are needed.

In particular, the Born–von Neumann interpretation does not *really* establish reference to subatomic charge distributions. It is *counterfactual* to interpret the squared amplitude $|\Psi(r)|^2$ of the many-particle wave function Ψ of the electrons or nucleons inside the atom as a spatial charge distribution. This means assuming that the charge distribution interacts *as if* many position measurements of electrons giving rise to ionization *were* performed. The corresponding interpretation for the protons and neutrons inside the atomic nucleus is even more artificial. The nucleon correlate to the position measurement of an electron by ionization is nuclear fission. In the case of the quark distributions inside the nucleons, due to quark confinement there is no correlate of such a position measurement at all. Indeed, there is a further crucial explanatory gap. The high energy scattering experiments in which the nucleon form factors and structure functions are measured cannot be inter-

---

[70]In the sense of Sect. 1.5, that is, the traditional assumptions about physical objects as causes of the phenomena.

preted in terms of a many-particle wave function of the proton or neutron. The nucleon structure functions are expressed in terms of free quark momentum distributions, that is, in terms of *plane quantum waves*. These quark momentum distributions are extracted from the measured differential cross-sections relative to a model of pointlike scattering, which in turn is in terms of scale invariance, that is, based on the heuristic method of dimensional analysis.[71]

Physical practice obviously does not care about the philosophical problem of counterfactuals. It is based on the tacit use of bridge principles on the one hand and a certain tacit realism on the other hand. In many cases, the referential import of the generalized correspondence principle is sufficient *for all practical purposes*. Where the correspondence with the classical cases breaks down, however, the tacit realism is *no longer classical* but a realism about quantum waves that is based on wave–particle duality.[72]

Thus, the correspondence principle bridges some crucial semantic gaps between quantum theory and the classical theories. Unlike a correspondence rule of the empiricist philosophy of science, it does not assign observational concepts to theoretical concepts. Rather, it creates quantitative and qualitative analogies between two theories or their models. It establishes inter-theoretical relations between certain statements and models of a classical and a quantized theory. By doing so, it links the language of classical physics with the language of a quantum theory. Indeed, the realistic interpretations of quantum mechanics beyond the Born–von Neumann interpretation are in no way better. They do not have *any* more referential import than the generalized correspondence principle. With regard to the physical interpretation of subatomic structure, the correspondence principle is even more powerful. In addition, the correspondence of the models of quantum mechanical scattering theory to classical scattering centers and classical particle tracks only breaks down where the realistic interpretations do as well, namely in the domain of relativistic quantum field theory. (In particular, this is true of Bohm trajectories or many worlds.)

5.4.3 Other Unifying Principles

Let us now have a closer look at the relation between the generalized correspondence principle and some other unifying principles of physics. We have already seen how the correspondence of the quantum theory of scattering with the classical picture of scattering centers and particle tracks breaks down stepwise in the relativistic domain. Hence, the generalized correspondence principle is necessary but not sufficient for establishing a semantic unity of physics. Additional unifying principles are needed in order to define length, time, and mass from the cosmological event horizon down to the

---

[71] See Sects. 4.3 and 4.4, and Appendices B and D.

[72] See Chap. 7, in particular Sect. 7.3.1.

quarks (not to speak of the size of the universe and the Planck scale). The following principles also contribute to establishing a certain semantic unity of physics. However, none of them is powerful enough to support the referential claims of the measurement methods discussed above:

1. **Ehrenfest's Theorem.** This derives from the correspondence principle. It states that the quantum mechanical expectation value of an observable is subject to the same law of motion as the corresponding classical physical quantity. For the subatomic scattering processes discussed in this book, however, it is not valid. The subatomic forces are too strong. In particular, it is not valid for the mean energy loss $\langle E \rangle$ of a charged particle in matter calculated by Bethe in 1930 and used for the analysis of particle tracks up to the present day.

2. **The Ergodic Hypothesis.** This states that the time average of the dynamics of the members of a statistical ensemble is equal to the ensemble average. This claim bridges the gap between the levels of the development of an individual system and the ensemble behavior. As a principle of classical statistics, however, it too cannot be more powerful than the correspondence principle. In the case of particle tracks, the ergodic hypothesis trivially applies to Mott's idealized quantum mechanical description. As long as energy dissipation along a particle track is neglected, the particle is considered not to change its quantum state. Therefore in Mott's model the measurement points along a particle track (which correspond to the time average) stem from one and the same quantum state. Hence they may be considered to contribute to the ensemble average.

3. **Symmetries and Conservation Laws.** Symmetries and conservation laws have been very powerful formal tools of particle physics for decades. They underlie the construction of gauge invariant quantum field theories from the early days of the 1963 quark model up to the present formulation of the superstring theory. Symmetries fix classes of theories. They are obviously more powerful than the correspondence principle. In particular, they give rise to the most general axiomatic particle definition. In 1939, Wigner characterized the irreducible representations of the Poincaré group in terms of mass, spin, and parity.[73] But for the needs of particle physics, this axiomatic definition of the particle concept is *too* general. The definition does not establish reference to individual particles or scattering events. It applies at the type level, not at the token level. The referential gap is filled by associating the symmetries with conservation laws according to Noether's theorem and applying the conservation laws to individual scattering events. This application, however, cannot avoid making implicit use of Heisenberg's generalized correspondence principle. For example, spin is a quantity without classical correlate. In the last analysis, its measurement is associated with *some* measurement of

---

[73]Wigner 1939. Such a definition also works for the non-relativistic Galilean group.

an angular momentum or a magnetic moment which couples to the spin and which is interpreted in terms of Heisenberg's analogies with the corresponding classical quantity.

4. **Superselection Rules.** In quantum mechanics, mass, charge, and spin commute with all other observables. Hence they have classical behavior. According to quantum mechanics in Hilbert space, they only occur in eigenstates, that is, their quantum states do not give rise to superpositions. This fact is expressed by a superselection rule which selects the physically realizable states.[74] The superselection rules are closely related to the symmetries of a quantum theory, and to Wigner's particle definition in terms of the irreducible representations of symmetry groups. They select the states of the observables which are characteristic of the particle type and which correspond to the permanent, unchangeable properties of a quantum process. However, as bridge principles they have limited validity. On the one hand, the extension of the classical concepts of charge and mass to their quantum analogues is once again based on Heisenberg's generalized version of the correspondence principle. Only the latter furnishes the physical interpretation of the concepts of mass and charge in the quantum domain, by telling us which quantum states are physically realisable in Hilbert space and which are not. On the other hand, the superselection rules are no longer valid in quantum field theory. In particular, the quark states of the current standard model of particle physics violate the superselection rules for mass and (electroweak) charge. According to the Salam–Weinberg theory (SWT) of electroweak interactions, the mass and flavour eigenstates of the quarks are not identical. The SWT contains the Kobayashi–Maskawa matrix which expresses superpositions of the quark states. This so-called *quark mixing* is an essential, well-tested part of the current standard model of particle physics.[75]

5. **Dimensional Invariance.** Dimensional invariance is a very general symmetry principle. It claims the invariance of a physical theory under a change of the system of units, say, a change from the cgs system to the Planck length, Planck mass and Planck time. The assumption that all physical theories are dimensionally invariant is a very powerful heuristic assumption of classical physics. It gives rise to the heuristic method of dimensional analysis[76] as well as to current attempts to construct theories for a physics at the Planck scale.[77] The underlying theoretical assumption is very weak. It is the Archimedean axiom of the theory of real numbers, according to which, for any two real numbers a, b with $a < b$,

---

[74] See Streater and Wightman 1964, 5.
[75] See Nachtmann 1990, 361–368.
[76] See Appendix B.
[77] See Callender and Huggett 2001.

there is a natural number n such that $n \times a > b$.[78] Hence, any unit a may serve to measure any quantity b. Obviously, the Archimedean axiom and with it the assumption of dimensional invariance is the most general unifying principle of all physics, as long as physical theories make use of real numbers. It makes it possible to construct the *scales* of physical quantities from the size of the universe down to the Planck scale. However, even though the principle of dimensional invariance is *much* more general than the principle of correspondence, it must be emphasized that the former cannot do without the latter. As in the case of symmetries and superselection rules, an inter-theoretical relation is needed which establishes a correspondence between the physical concepts of the classical and quantum theories at the different scales.

The unifying principles discussed so far only work in *combination* with the correspondence principle. Together, they suggest how to extend the classical physical quantities into the quantum domain. Beyond the validity of the correspondence principle, they suggest how to construct new quantized quantities such as spin or parity which are defined in terms of symmetries and which couple in some way to the hitherto known quantities. Concerning the extension of length, time, and mass scales to the subatomic domain, all of these principles work in the *top–down* direction. They help us to explain the quantum domain in terms of classical physical quantities and symmetries. But they do not give *bottom–up* explanations of the classical world in terms of quantum concepts. The only principle which works the other way round is decoherence, that is, the assumption that in a classical thermodynamic environment the entangled quantum states decohere:

6. **Decoherence.** Decoherence means that in a macroscopic environment with infinitely many degrees of freedom, the interference terms of an entangled quantum system rapidly dissipate into the environment. They are damped away due to thermodynamic effects. Even though the theory of decoherence *assumes* a classical thermodynamic world (rather than *predicting* the emergence of a classical world), it explains how radiative decays as well as measurement results may have definite results.[79] But it explains the emergence of definite events in a classical world only at the statistical level and only *for all practical purposes*, that is, not as a matter of principle. Decoherence does not let the superpositions of entangled quantum states disappear. The superposed states remain forever superposed, even though their amplitudes become *very* close to zero within a very short time. Decoherence does not give rise to any collapse of the wave function. And it does not give any hint of a physical mechanism that explains *which* individual measurement result is finally obtained.

---

[78]This was already emphasized in Hilbert's famous article on the axiomatic method; see Hilbert 1918, 149. See also Appendix A.

[79]See Giulini et al. 1997.

Nevertheless, it establishes unity between the classical and the quantum domain at the probabilistic level. In particular, it explains why the quantum mechanics of scattering is compatible with classical statistical methods of data analysis.[80]

To sum up, the correspondence principle and the symmetry principles of physics together give a top–down explanation of the *physical properties* of subatomic particles, scattering processes, and subatomic structure. But the only bridge principle with *referential* import is the correspondence principle. The usual probabilistic interpretation of quantum mechanics only establishes reference at the ensemble level. The same is true of the only bottom–up principle, namely the assumption that decoherence happens in a thermodynamic environment. It explains nothing but the macroscopic ensemble behavior. Only the correspondence principle claims that the quantum mechanics of scattering refers to individual particle tracks and subatomic scattering centers. Where correspondence breaks down, that is, in the relativistic domain, the physical interpretation of individual particle tracks and scattering events is only supported by symmetries, conservation laws, and superselection rules. But all of these principles work only for the particle *type*. They do not refer at the token level, that is, to *individual* particles and measurement results. In addition, the principle of dimensional invariance supports the physical practice of extending the familiar length, time, and mass scales down to the Planck scale, giving rise to heuristic dimensional considerations about physical systems at whatever scale. But these scales are constructed and extended *in classical terms*. Hence, Bohr's dictum still stands that the measurement results of quantum physics have to be expressed in classical language.

5.5 The Scales of Physical Quantities

The scales of physical quantities are expressed in terms of length, time, and mass. The unity of the scales is constructed in classical terms even though the classical concepts break down stepwise in the quantum domain.

How is it possible that the construction of the length, time, and mass scales from the subatomic to the cosmological domain does not give rise to contradiction? The length scale covers the size of the universe, the size of this sheet of paper and the distance of the quarks within a proton or neutron. The size of the universe is obtained from models of general relativity (above all the *big bang* model) plus many kinds of astrophysical data. The size of this sheet of paper is measured with a ruler. The distance of the quarks within

---

[80]In first order perturbation theory, the quantum theory of scattering predicts *incoherent* scattering. Thus to first order, decoherence is guaranteed. For the higher orders of perturbation theory, decoherence means that the interference of scattering at *different* atoms is rapidly damped away.

the nucleon has been measured from lepton–nucleon scattering in high energy physics as well as predicted from models of quantum chromodynamics. Similarly, the mass scale embraces the mass of electrons, billiard balls, and black holes. Electrons are subject to quantum electrodynamics, the motions of billiard balls to classical mechanics, black holes to general relativity. According to Kuhn, each theory generates its own world view. If we adopt this philosophy in face of the current theoretical pluralism of physics, we have to conclude that the scales of physical quantities span a fragmented world *if* they are able to span a world at all.

In constructing the length, time, and mass scales without despairing of a fragmented world, physicists employ the unifying principles discussed above. But the latter are not strong enough to bring the axiomatic, referential, and operational aspects of physical concepts completely together. In the quantum domain, these aspects fall apart. The axiomatic foundations of a quantum theory do not give rise to operational concepts which refer at the level of individual measurement results. However, the unifying principles and the operational foundations of length, time, and mass *together* make the concepts of different parts of the scales sufficiently overlap. In this way, *some* unity of the scales is established.

1. The *axiomatic* unity of the scales is based on general unifying principles. Above all, the principle of dimensional invariance and the related Archimedean axiom establish the unity of the scales.[81] The Archimedean axiom provides a very weak axiomatic basis. Any physical theory based on the theory of real numbers is expected to satisfy it. The axiom is obviously not sufficient for extending such a theory to quantum operators and operator-valued measures.[82] Nevertheless, it is indispensable for establishing the physical interpretation of quantum mechanics in terms of length, time, and mass, and derived physical quantities based on them. In addition, symmetries and invariances are powerful axiomatic tools.[83] They establish a general framework of theory construction and several pleasant features of a unified physics. The associated conservation laws such as energy–momentum conservation give rise to axiomatic definitions of general concepts which satisfy superselection rules. They are formal keystones in the construction of the language of physics.

2. The *operational* unity of the scales is obviously *not* due to a unique axiomatic basis. Each measurement method comes with its own axiomatic foundations. But in all empirically accessible parts of the scales there are several independent measurement methods which have overlap. Therefore, the scale of a quantity can be operationally defined from a *chain of*

---

[81] See Appendices A and B.

[82] In addition, the generalized correspondence principle discussed in Sect. 5.4 is employed. It makes the correspondence-conforming standard identification of classical and quantum observables according to Scheibe 1999, 224–225 possible.

[83] Wigner 1939, 1979.

measurements. For example, the length scale is established by a chain of measurements performed with measuring devices such as star parallax, geodesic instruments, ruler, micron screw, microscope, electron microscope, or particle accelerators. Thus, the scale of a quantity is a *cluster concept* in the sense of Ellis' theory of measurement.[84] The empirically accessible part of a scale is a cluster of operational concepts. It stems from all empirical and experimental operations which give rise to measurements with unique numerical representations and overlapping numerical ranges. In the overlap of the ranges, any two measurements of the same phenomenon must give approximately the same results. This condition can be made precise in such a way that is necessary and sufficient for defining a scale.[85] It is closely related to the redundant features of the measurement theory of particle physics discussed in Sect. 5.2. Overlap means redundancy.

3. The remaining conceptual gaps in the scales are *referential*. As far as possible, they are closed by means of the correspondence principle. Heisenberg's generalized version suggests interpreting quantum concepts in correspondence with the classical particle and wave pictures. Hence, the correspondence principle suggests to which kind of entity a measured mass, length, or time may be attributed: to the particle-like (or wavelike) quantum phenomena corresponding to the classical description of particle tracks, scattering centers, interference phenomena and so on. Hence, what about quantum concepts without correspondence? For them, the only remaining reference is the realism of properties suggested in Sect. 1.6: a quantum concept (or a number in the scale of the corresponding quantum observable) refers to the physical properties of the respective quantum phenomenon. The carrier of these properties is nothing but the quantum phenomenon itself, say, a particle track on a bubble chamber photograph or the interference pattern of the double slit experiment. The way in which quantum concepts refer can only be understood in *contextual* terms. Quantum concepts refer to the properties of quantum phenomena which occur by means of a given experimental setup in a given physical context. In order to establish the referential unity of the length, time, and mass scales, this contextual account of physical properties of classical objects and quantum phenomena is sufficient.

The sloppy construction of the length, time, and mass scales is empirically most successful. Many modern technologies rely on them. In particular they give rise to the worldwide standardization of precision measurements to atomic units. According to Hilbert's famous paper on the axiomatic method, this fact indicates that the Archimedean axiom is empirically valid.[86] This means, strikingly enough, that the current pluralistic structure of modern

---

[84] Ellis 1968.

[85] Ellis 1968, 41–42.

[86] See Hilbert 1918, 149, and Appendix A.

physics is at odds with an axiom of measurement theory which is hitherto known to be empirically adequate. Axiomatic measurement theory has a larger extension than the incompatible axiomatic structures of current physics at a small, a medium, and a large scale. Indeed, physicists trust more in the Archimedean axiom than in the axiomatic structure of current physics. In order to obtain constraints for the construction of a unified quantum gravity, they extend the length, time, and mass scales down to the Planck scale. This may finally turn out to be problematic. In any case, the idea of a quantum gravity is hardly compatible with the validity of the Archimedean axiom *below* the Planck scale.

Even though the scales are empirically successful they are obviously based on construction. The construction unifies the measurement methods which make up their operational basis. One might argue that a constructivist view of science sufficiently explains the empirical success of this construction. There are matters of fact, however, which constructivism cannot explain. The scales turn out to be empirically adequate when tested by independent measurement methods. Constructivism can by no means explain the extent to which the length, time, and mass scales can be coherently constructed without giving rise to contradiction.

Indeed, Putnam's famous miracle argument[87] can now be employed in order to defend a realistic interpretation of the scales. If the scales did *not* represent the physical properties of natural kinds, the coherence of the cluster of operational concepts underlying the scale of a quantity would be a miracle. This version of the miracle argument implicitly makes two claims. First, the construction of the scales commits one to believe in natural kinds underlying the physical properties. Second, the empirical success of the construction indicates that the construction is not arbitrary but agrees with the actual structure of the empirical phenomena. At a subatomic scale, the natural kinds underlying the physical properties are the physical *phenomena* that are measured. In order to avoid the problems of classical realism, however, they should not be considered to be quantum *objects* in the sense of isolated entities. Thus, here the miracle argument should be understood in terms of a realism of physical properties, the carriers of which are *quantum phenomena* rather than subatomic substances on their own. The resulting realism of properties is an empirical (or critical) realism in Kant's sense, as already explained in Sect. 1.6. The extent to which such a realism of the properties of quantum phenomena is finally compatible with the traditional view of natural kinds, however, remains an intriguing question.

It is indeed in the spirit of Putnam's position of internal realism[88] to bring Kant's views into play at this point. On the one hand, electrons, photons,

---

[87] See Putnam 1975, 73: "The positive argument for realism is that it is the only philosophy that doesn't make the success of science a miracle." See also Smart 1963, 39.
[88] Putnam 1980, 1990.

quarks, and so on are not conceivable in traditional (Aristotelian) terms of (individual) substances (or particulars). From an empiricist view, they are nothing but bare collections of conserved quantities that may appear and disappear contingently. On the other hand, from a Kantian point of view it is *our* concept of substance that makes it possible to conceive of them as such collections, just as *our* construction of the scales is a condition for the possibility of subatomic physical experience. Only the extension of the classical length, time, and mass scales into the quantum domain makes the measurements of subatomic physics possible. Neither is this extension empirical in a simple sense, nor is it only due to construction. But the semantics of quantum concepts is and remains problematic. Hence, the following question is at stake: To what extent is the extension of the classical quantities into the quantum domain semantically consistent?

5.6 Questions of Semantic Consistency

The scales of physical quantities establish a certain semantic unity of physics. Their construction makes it possible to attribute the same kinds of physical properties to physical phenomena at all scales. This semantic unity is expressed in the language of classical physics, as Bohr emphasized. However, it is substantially *weaker* than the unity an all-embracing axiomatic theory would establish. The axiomatic, operational, and referential aspects of the quantum concepts involved at the subatomic scale do not come together. The operational meaning of the axioms of a quantum theory is given in terms of the probabilistic interpretation. Nevertheless, the single events which make up a quantum ensemble are subject to measurement laws with classical foundations. Hence, the axiomatic and the operational meaning of quantum concepts fall apart as far as these concepts refer to individual events and systems.

From an axiomatic point of view, these three semantic aspects of physical concepts should agree. If they do, the axiomatic and operational concepts have the same empirical models. In this case (which is obviously *not* given for quantum concepts), Einstein's dictum holds that the theory tells us what can be measured.[89] Einstein's dictum is an intuitive way of expressing that a theory is *semantically consistent*. Here, semantic consistency is obviously not meant in the logical sense of having a formal model. It means rather that the measurement methods which give rise to the empirical models of a theory are themselves models of it.[90]

---

[89]See above, end of Sect. 5.1, and the dispute between Heisenberg and Einstein on this issue reported in Heisenberg 1969, 91–92.

[90]Semantic consistency implies that there are no independent measurement methods. As long as a theory is still under test, this gives rise to fatal theory-ladenness, i.e., to a vicious circle of measurement. Therefore, semantic consistency

This intuitive account agrees with the definitions of semantic consistency given by von Weizsäcker and Mittelstaedt. According to von Weizsäcker, *semantic consistency* means that a theory agrees with the pre-theoretical understanding of the phenomena to which it applies.[91] Here, the term 'pre-theoretical understanding' needs to be made more precise. For Bohr, the pre-theoretical understanding of quantum theory can be expressed in *plain language*, that is, in terms of the familiar classical concepts. In this Bohrian sense, a physical theory is semantically consistent if and only if its axiomatic content may be expressed in classical terms. This requirement is obviously too strict. It is more reasonable to identify the pre-theoretical understanding with the operational concepts which make it possible to test the predictions of a theory. Taken in this sense, the pre-theoretical understanding of a theory stems from the independent measurement methods which give access to the empirical (or experimental) domain of the theory. Then, a theory is semantically consistent if and only if it is consistent with the measurement theory of its empirical domain. In this case, its operational concepts belong to its models. Hence, semantic consistency means that the operational concepts of a theory derive from its axioms, as required by Einstein's dictum.

Mittelstaedt defines semantic consistency in exactly this way. For him, the language of a physical theory and the language in which the respective measurement results are expressed relate like an object language and the respective meta-language in the sense of Tarski's semantics.[92] According to this Tarskian view of the language of physics, semantic consistency means that the meta-language can be embedded in the object language. For the language of physics, this means exactly that the operational concepts underlying the measurement methods which serve to test a theory finally turn out to derive from the axioms of that theory.

In addition, Mittelstaedt showed that quantum mechanics is semantically consistent at the probabilistic level but semantically inconsistent at the level of individual measurement results and physical systems. The operational meaning of quantum concepts is based on the probabilistic Born–von Neumann interpretation of quantum mechanics. The latter, however, derives from the axioms of quantum mechanics without measurement (i.e., without von Neumann's projection postulate[93]) in a model of infinitely many independent individual quantum systems.[94] So far, so good with semantic consistency. But the well-known non-objectivation theorems regarding quantum mechan-

---

must only stand at the end of theory formation. For a theory under test, it has to be avoided. See the case study discussed in Sect. 3.5.

[91] See von Weizsäcker 1985, 514: "Semantic consistency of a theory means that its pre-theoretical understanding with the help of which we interpret its mathematical structure satisfies the laws of the theory." My translation.

[92] Mittelstaedt 1986; 1995.

[93] See von Neumann 1932.

[94] Mittelstaedt 1997a,b; 1999.

ical measurement results still hold.[95] They show that quantum mechanics is *not* semantically consistent at the level of individuals since a quantum mechanics of measurement is by no means able to explain how an individual measurement result occurs. Therefore, quantum mechanics *per se* does by no means refer to individual systems. Indeed, the operational concepts of quantum physics only agree at the probabilistic level with the axiomatic concepts of quantum mechanics. At the level of individual systems, the operational concepts are not quantum concepts, whereas the axiomatic concepts do not refer. Hence, von Neumann's *projection postulate* is employed to bring axiomatic quantum mechanics down to the earth of individuals.

The measurement methods discussed in this chapter are much more complicated. But they thoroughly confirm these well-known formal results. The measurement methods of particle physics are designed to overcome the *axiomatic* incommensurability of classical concepts and quantum concepts at an *operational* level. This is a bit awkward since the operational concepts of experimental particle physics apply to individual position measurements, particle tracks, and scattering events, whereas the axiomatic concepts of the quantum mechanics of scattering do *not* refer to any individual event or track. Indeed, physical practice overcomes this difficulty with striking empirical success. It proceeds by constructing unified scales of operational quantities where there is no unified axiomatic theory. In particle physics, the basic operational quantities are length, time, and mass, but the derived quantities of momentum and energy give rise to a scale which is nevertheless important. The data analysis of scattering experiments is based on the patching together of classical measurement laws (which apply to individual particle tracks) with quantum corrections (which only apply at the probabilistic level).

For the analysis of particle tracks, this works out as follows. At the *probabilistic* level, the quantum mechanics of scattering is semantically consistent with the outcome of classical measurement results, while at the level of *individual* measurement results it is *not*. This was a crucial conclusion of the energy loss case study given above in Sect. 5.3. In first order scattering theory, coherent scattering is negligible, and hence, the individual measurement points of a particle track are only due to incoherent scattering. Thus, they occur independently, without any additional decoherence mechanism (which, likewise, only works at the probabilistic level). In first-order scattering theory, there is pre-established harmony between the quantum mechanics of a particle track and the classical picture that the individual measurement points are generated independently of each other. Nevertheless, to apply Bethe's *probabilistic* calculation of the energy loss of charged particles in matter to *individual* particle tracks is only justified if the errors produced by this method are corrected at the probabilistic level of many particle tracks.[96]

---

[95]See Mittelstaedt 1997a,b; 1998a; 1999. See also related earlier results, in particular von Neumann 1932.

[96]See Sect. 5.3.4.

Hence, quantum physics is semantically inconsistent at the level of individual events, but this can be coherently handled at the probabilistic level. Formal results tell us that quantum mechanics does not refer to individual systems but only to statistical ensembles. This claim is confirmed by the heterogeneous measurement theory of particle physics. Only classical and quasi-classical measurement laws refer to individual quantum systems. Genuine quantum laws do not. The reference of genuine quantum laws to individual quantum systems can only be established in the following ways:

(i)   by means of the experimental context, i.e., the concrete experimental device in which a quantum phenomenon occurs;

(ii) at the (probabilistic) level of *many* subsequent independent individual measurement results which are obtained in one and the same experimental context and belong to one and the same quantum ensemble;

(iii) in terms of the superselection rules for conserved quantities such as mass, charge, or spin and the corresponding conservation laws according to which individual scattering events are analyzed;

(iv) using Bohr's correspondence principle or Heisenberg's generalized version of it.

These four ways contribute to making quantum theory semantically consistent *for all practical purposes*, i.e., in a non-formal, pragmatic sense. An application of quantum theory to individual systems can only be obtained by taking into account (i) the experimental context, (ii) repeated measurements of one and the same system, (iii) superselection rules, or (iv) by employing the generalized correspondence principle in order to derive quantum models of subatomic event sequences, charge distributions, and so on. In general, two or more of these ways of obtaining concrete quantum phenomena, respectively individual quantum systems, come together.

In particular, only the experimental context (and our ways of conceiving of it in classical terms) makes it possible to talk in a sloppy way of *quantum objects*. Quantum theory alone does not admit of constructing objects in the sense of individual spatio-temporal systems with position and temporal duration. Abstracting particle properties such as mass, charge, or spin from their experimental or environmental context means dispensing with spatio-temporal objects and keeping nothing but *bundles of dynamic properties* of the respective kinds. These bundles are made up of those magnitudes that can be measured dispersion-free at the same time in *any* experiment. Bare quantum 'objects' are just bundles of properties which underlie superselection rules and which exhibit non-local, acausal correlations. Somehow these bundles of dynamic properties propagate through an experimental context, respecting the conservation laws of mass–energy, charge, and spin (as well as Einstein's causal condition that signal transmission is only possible within the light cone). They seem to be Lockean empirical substances, that is, collections of empirical properties which constantly go together.[97] However, they are only individuated by the experimental apparatus in which they are measured or the concrete quantum phenomenon to which they belong. Regarding reference, their 'trans-phenomenal' identity as well as their non-local features remain mysterious. They can only be individuated as context-dependent quantum *phenomena*. Without a given experimental context, the reference of quantum concepts goes astray. In this point, Bohr is absolutely right up to the present day.

In the measurement of quantum phenomena, both quantum and classical concepts are employed. Even the probabilistic expectation values of quantum observables are expressed in terms of the classical quantities length, time, and mass. According to Bohr's (authentic) version of the Copenhagen interpretation, i.e., the complementarity view of quantum mechanics,[98] the latter serve to give the abstract concepts of a quantum theory physical meaning. This confirms Bohr's view of quantum mechanics according to which the correspondence principle makes it possible to express quantum phenomena in the language of classical physics. This physical meaning, however, is not expressed in terms of empirical observations, that is, in the empiricist's observational vocabulary. Rather, it is expressed in terms of *properties*, namely the classical physical quantities. And it was Bohr who emphasized that it is not possible to construct quantum objects out of these properties.[99]

Thus, the (generalized) correspondence principle is indispensable for extending the scales of physical quantities into the quantum domain. It is a meta-theoretical rule about inter-theoretical relations. However, it establishes very peculiar inter-theoretical relations. It gives quantum mechanics concrete models which are quasi-classical individual systems. To put it in other terms, it is a bridge principle which makes it possible for quantum mechanics to take a grip on its empirical applications. But these are (approximate) models of the theories of classical physics rather than observations in an empiricist sense. This bridge principle gives rise to a common measure for concepts that are incommensurable in the sense of Kuhn's incommensurability thesis. In terms of abstract measurement theory, it fills the axiomatic and referential gaps between distinct operational concepts. In a certain sense, the generalized correspondence is at once *more* and *less* powerful than other unifying principles. It gives rise to a partial semantic unity of physics where there is no

---

[97]Locke 1689, Book II, chapter XXIII. See also Sects. 1.6 and 6.3. As to the question of the carriers of these properties, there remain only metaphysical concepts which are at odds with the idea of Lorentz invariance – as long as no Kantian point of view is adopted, according to which it is in the end *our* concept of substance that makes it possible to conceive of them as such bundles or collections of properties.

[98]See below Sect. 7.2 and Howard 2002.

[99]Bohr 1948, 315.

axiomatic unity. Together with axiomatic theories, it makes up the backbone of the architectonics of current physics.[100]

At this point, the following general conclusion about the interpretation of quantum theory can be drawn. Quantum theory in general (including the current relativistic quantum field theories of high energy physics) has *no realistic interpretation* which establishes reference to individual systems. Strictly speaking, up to the present day *there is no quantum theory of individual systems*. Only by means of superselection rules, the generalized correspondence principle, and so on, does quantum physics apply to individual systems. Quantum theory *on its own* commits one to an *ensemble interpretation*.[101] Indeed, this is just another way to state that in a strict sense quantum mechanics is only semantically consistent at the probabilistic level. The current metaphysical interpretations of quantum mechanics are at odds with relativistic principles and do not suggest any way out.

---

[100] As far as I can see, this non-empiricist view of correspondence as a bridge principle is compatible with Nancy Cartwright's way of criticizing the empiricist bridge principles (Cartwright 1983, 135–134 and 144–145). However, it is in opposition to her later views about a *completely* disunified physics, as proposed in Cartwright 1999.

[101] I am indebted to Reinhard F. Werner for drawing my attention to this point, in contradistinction to my earlier views expressed in Falkenburg 1995, 196–198. According to my present understanding of quantum physics, the application of a quantum theory to an individual system *always* makes tacit use of the generalized correspondence principle.

6 Metamorphoses of the Particle Concept

The 20th century history of the particle concept is a story of disillusion. It turned out that in the subatomic domain there are no particles in the classical sense. Neither the atoms, nor their constituent parts, nor the causes of the particle tracks observed in the cloud chamber, nor the pointlike or extended scattering centers within some target matter are classical particles. And it will turn out in the course of this chapter that a generalized concept of quantum particles is not tenable either. Particles are experimental phenomena rather than fundamental entities.

Nevertheless, the particle concept was not given up. After the quantum revolution physicists still speak of particles. However, now they do so in various senses. On the one hand, the classical particle model is still used. It is indispensable for measuring the events and particle tracks registered by Geiger counters, nuclear emulsions, bubble chambers, and more recent particle detectors. The classical measurement theory of subatomic particles was extended and corrected in the course of the quantum revolution, but its classical core was kept. On the other hand, several quantum successors of the classical particle concept were introduced. They should *not* be confused with the classical particle concept. They result from several meaning shifts towards quantum particles.

These meaning shifts of the particle concept are neglected in the philosophy of physics. They give rise to several conceptual ambiguities. Today, the particle concept is no longer unique. In addition to the classical particle concept, there is a family of quantum particle concepts. The task of this chapter is to analyze this conceptual family and the relations between its members. Philosophers should be warned at this point. Current physical practice relies on the usual probabilistic interpretation of quantum theory. In order to understand the many facets of what is called a particle today, one should do the same. Therefore, in the following I neither join the philosophical discussion of the foundations of quantum mechanics nor the recent debate concerning the ontological aspects of quantum field theory.[1] I just want to clarify what physicists mean, after the quantum revolution, when they talk of particles.

---

[1] See Brown and Harré 1988, Auyang 1995, Teller 1995, Cao 1999, Kuhlmann et al. 2002.

In comparison to the classical particle concept (CP) (Sect. 6.1), the rise of quantum mechanics and the light quantum hypothesis brought two semantic shifts. These resulted in the particle concepts (QM) and (LQ) (Sect. 6.2). Their empirical basis is an operational particle concept (OP) which is metaphysically modest though not abstinent (Sect. 6.3). Current philosophy of language claims that meaning is use. At the axiomatic level, physicists use the classical particle concept *plus* as many particle concepts as there are quantum theories. They are even using more than this. In particular, the concepts of *field quanta* (FQ), group theory (GT), *virtual particles* (VP), and *quasi-particles* (QP) have to be discussed. All these quantum concepts have in common that they attribute non-local or wavelike features to particles (Sect. 6.4). They also give rise to new views about the constituent parts of matter (MC). Here, the quark model in particular deserves attention (Sect. 6.5). What do all these meanings have in common? Virtual particles apart, the *independence, discontinuity, and localizability of a collection of conserved quantities* remain the general hallmarks of particles (GP). But in a strict sense this only holds for the non-relativistic domain. The underlying entity has the hallmarks of a field, if it makes sense to ask for such an entity at all (Sect. 6.6).

Hence, after the quantum revolution only an *informal* particle concept remains. This concept is more than a mere *façon de parler*. But it is metaphysically more modest than the mereological and causal particle concept associated with classical physics. It has an operational basis which stands in precise relations to the current quantum theories, and it has some typical features, amongst them *statistical independence*.

6.1 Classical Particles

Literally, 'particle' means 'small part of a whole'. The original meaning of the particle concept is mereological rather than causal. It is based on the part–whole relation between material substances and their microscopic constituent parts.[2] Ancient atomism assumed that matter consists of movable, absolutely impenetrable ultimate parts which make up all material things. Dating back to Descartes and Newton, these particles or matter constituents were conceived as tiny mechanical bodies (corpuscles). Until the end of the 19th century, it was assumed that they are subject to the laws of classical dynamics. They were ideally described as mass points. The classical particle concept brings the ideas of ancient atomism and the mathematics of point mechanics together. Classically, the parts of matter behave like tiny celestial bodies which move along deterministic trajectories. According to this, every particle has a unique trajectory which is completely determined by external forces.

---

[2]See Appendix E.

Hence, classical particles have the physical properties of mechanical bodies. Their spatio-temporal properties are position, (non-relativistic) velocity, and acceleration, all parameterized by time. Their dynamic properties are inertial mass m_i, electric charge q_e, gravitational mass m_g (the charge of gravitation which agrees numerically with the inertial mass), and derived quantities such as momentum $\boldsymbol{p}$ and energy E. A simple version of classical mechanics is constituted by Newton's three axioms. More elaborate versions are based on the Lagrange or Hamilton equations. Any of these axiomatic systems gives rise to the typical properties of classical particles which one may state by the following list of informal predicates:

(CP) Classical particles are

(MQ)	carriers of *mass m* and *charge q*,
(INDEP)	*independent* of each other,
(POINT)	*pointlike* in interactions,
(CONS)	subject to *conservation laws*,
(LOCAL)	*localized*,
(DET)	*completely determined* by the laws of mechanics,
(TRAJ)	moving on *trajectories* in phase space,
(INDIV)	spatio-temporally *individuated*,
(BOUND)	able to form *bound systems*.

The properties (MQ), (POINT), and (LOCAL) together express the fact that a classical particle is a *mass point*. They are typical of a particle ontology. In contradistinction to classical particles, fields and waves are non-local. The properties (LOCAL), (TRAJ), and (INDIV) are typical of the *classical* particle concept. They are weakened, replaced, or skipped in the quantum successors. (DET) is common to classical mechanics and classical field theories. In all quantum theories, it is replaced by probabilistic determination.

Property (INDEP) is crucial for the classical particle concept and its generalizations, as will be seen in the following. It means that particles are primarily considered to be interaction-free, or uncorrelated, and hence statistically independent. To be more exact, independent particles are in unbound states and they are considered to be (approximately) non-interacting, except by collisions. Hence, they only have inertial motions and the contingent initial conditions of their motions are statistically uncorrelated. In classical mechanics this means that the positions and momenta of particles are uncorrelated. This model gives rise to the kinetic theory of an ideal gas and Boltzmann's *Stoßzahlansatz*. In general, due to forces the particles are not *really* independent. However, the very concept of force presupposes the interaction-free case as a specific possibility. Indeed any linear physical dynamics presupposes that there may be non-interacting entities with uncorrelated initial conditions of their motions. In the most general sense,

(INDEP) Independence means that particles

(UNCOUP)   *may* be in *non-interacting* or uncoupled states, and

(UNCORR)   their *initial conditions* are *statistically uncorrelated*.

The properties (INDEP), (POINT), and (CONS) are *shared* by classical point mechanics, quantum mechanics, and all classical and quantum field theories. They are to a certain extent meta-theoretical. The same is true of the dynamic properties (MQ). The list of dynamic properties may vary, but it is always closely related to the symmetries and conserved quantities of physics. (POINT) means that particles interact locally and (CONS) is associated with their symmetries. In order to avoid confusion by introducing too many particle properties, I keep the labels (INDEP), (POINT), and (CONS) throughout this chapter, even though the meaning of these particle properties will change in the transitions to quantum mechanics and quantum field theory.

However, it has to be noted that the precise meaning of these meta-theoretical predicates depends on the theoretical context. The meaning of (INDEP) is affected by the transition to quantum theory in various ways, as will be discussed in the course of this chapter and summed up in Sect. 6.6. In classical mechanics, (POINT) states that a body may be replaced by its center of gravity. But in classical field theory, (POINT) means that there is no action-at-a-distance. Only pointlike interactions are compatible with Einsteinian causality, that is, with the requirement that signalling is only possible inside the light cone and on its surface, but not over spacelike distances. (CONS) includes the conservation of momentum, energy, and angular momentum. The conservation laws are closely related to the group of symmetries of the underlying dynamics which is the Galilean group here. In a relativistic theory, they are associated with the Poincaré group, giving rise to momentum–energy conservation, i.e., the invariance of the squared momentum–energy 4-vector. In addition, they give rise to the sum rules which underlie the models of bound systems and the associated constituent models of matter. The theoretical transformations of the particle property (BOUND) are investigated in more detail in Sect. 6.5.

In the subatomic domain, the classical trajectories (TRAJ) are *empirically underdetermined by the phenomena of subatomic physics*. The better the spatial resolution of a particle detector, the better the observable particle tracks are resolved into single measurement points, showing that no continuous trajectory is measured, but only discrete positions. This empirical underdeterminacy of the particle trajectory is by no means accidental. It does not depend on the choice of the particle detector but on the very possibilities of measuring position with a macroscopic device. As Born already pointed out in his seminal paper on the probabilistic interpretation of quantum mechanics, it is impossible in principle to pursue the complete trajectory of a scattered particle in a subatomic scattering process. Classical physics makes untestable assertions about subatomic events whereas quantum mechanics does without them:

> Here the whole problem of determinism comes up. From the standpoint of quantum mechanics, there is no quantity which in any *individual case* causally fixes the consequence of a collision; but also experimentally we have so far no reason to believe that there are some inner properties of the atoms which condition a definite outcome for the collision. Ought we to hope later to discover such properties (like phases of the internal atomic motions) and determine them in individual cases? Or ought we to believe that the agreement of theory and experiment – as to the impossibility of prescribing conditions for a causal evolution – is a pre-established harmony founded on the non-existence of such conditions?[3]

Hence, the trajectory which connects the measurement points of a particle track belongs to the *non-empirical* features of the classical particle concept. It is an axiomatic classical concept without operational counterpart. From an empiricist point of view, it belongs to metaphysics, even though the *data model* of a particle track on a bubble chamber photograph results from the spatio-temporal concatenation of the successive position measurements. The data model of a particle track *is* already a classical trajectory. In the sense of Quine's ontological commitments,[4] it commits us to a classical particle. This data model is *not really* compatible with quantum theory, even though the measurement errors of position and momentum along a track are usually more than 10 orders larger than the indeterminacy predicted by Heisenberg's uncertainty relation. It is well known today that this classical data model is only empirically adequate at a macroscopic scale and for subatomic particles of low energy. The higher the particle energy, the more the classical data model goes astray.[5] A particle track is not caused by a classical particle. It is due to subatomic interactions which are subject to the quantum mechanics of scattering.

6.2 The Shift to Quantum Particles

With the development of quantum theory, the classical particle concept was given up, but several informal ways of speaking about particles remained. Up

---

[3]Born 1926a, 51; translation from Wheeler and Zurek 1983, 54. Here, Born identifies causality and determinism. This identification was attacked by Cassirer in 1937. However, it became influential in the discussion of quantum mechanics.

[4]Quine, *On What There Is*.

[5]See Sect. 5.3.

to the present day, physicists call the entities of atomic, nuclear, and particle physics 'particles'. To give only one textbook example:

> Today we know of the existence of a whole series of elementary particles. Examples are the electron (e), the photon (γ), the proton (p), the neutron (n), the pi-meson or pion (π), the K-mesons (K), and the hyperons (Λ, Σ, etc.). Of these, the electron and the photon have been known for the longest time. At the time when the electron was discovered (Thomson 1897), there was still a general belief in the validity of the classical physics in which waves and particles were totally separate concepts. It is really no more than a historical accident that the electrons were first observed in an experiment in which they behaved like classical particles [...] In the case of light quanta, the situation was exactly the reverse [...] The revolutionary idea of Einstein – the hypothesis of light quanta, in which light is also assumed to possess a particle nature (Einstein 1905) – at first met with great scepticism. It took approximately twenty years before the γ-quantum was accepted as a 'particle' [...] In the meantime, with the rise of quantum mechanics, it had been realized that particles and waves cannot be regarded as strictly separate phenomena, but must instead be considered as two aspects of the same phenomenon.[6]

Here, the term 'particle' is used in several *different* meanings. To name just four of them:

- elementary particles (electrons, photons, protons, etc.),
- classical particle,
- light quantum,
- particle aspect of a phenomenon which also has a wave aspect.

Apparently, the particle concept originally associated with classical point mechanics was not completely given up with the quantum revolution. It only obtained new meaning. To be more exact, it obtained *several* new meanings, giving rise to semantic confusion in the dialogue between physics and philosophy. Within the physics community, however, no one complains about such confusion. Of course, behind this lack of semantic discomfort stands a firm realistic belief in the existence of subatomic particles, *whatever they may be*. This belief is associated with a strong conviction that the reference of the term 'particle' remains at least stable to *some* extent when it is applied to quantum phenomena. What should be thought of this realistic belief in the existence of subatomic particles? This question can only be answered by analyzing the semantic shifts of the particle concept in detail.

The break with the classical particle concept took place on two fronts, giving rise to *different* meaning shifts in the particle concept. (1) In atomic physics the results of Rutherford's scattering experiments demanded an

---

[6]Nachtmann 1990, 3.

atomic model which could no longer be expressed within classical physics. This happened just at the time when the first particle trajectories caused by α-rays were observed in the cloud chamber. Bohr's phenomenological atomic model introduced quantization rules into the classical model of a bound system of charged particles. They were at odds with classical electrodynamics. Only the quantum mechanics of the hydrogen atom re-established consistency. (2) Einstein's light quantum hypothesis extended the domain of the particle concept to radiation, that is, to the area of electrodynamics. Relativistic kinematics provided this application with a measurement theory, describing the trajectory of a massless particle on the light cone and its energy loss due to the collision with a massive particle. The Compton effect proved that this application is empirically adequate.

6.2.1 Matter Waves

In Bohr's atomic model[7] as well as in non-relativistic quantum mechanics, the *mereological* particle concept went on being applied to matter constituents. Bohr's atomic model is still based on a classical particle concept. Quantization conditions are added to it in an *ad hoc* manner, in order to prevent the unstable electron orbits predicted by Maxwell's electrodynamics. Later, the subatomic parts of matter were described by *quantum mechanics* instead of classical point mechanics. The quantum mechanics of the hydrogen atom replaces the classical particle orbits by the bound solutions of the Schrödinger equation. An electron bound by the Coulomb potential is now described in terms of a *stationary wave*. In this way, the classical trajectory is replaced by Schrödinger's wave function. The classical property of being local is now restricted to sharp position states of the wave function, that is, to a wave packet.

Schrödinger's corresponding realistic interpretation of the wave function did not succeed. The wave packet of a free particle rapidly develops towards a plane wave. Hence localized states turned out not to be stable. In order to derive the quantum mechanics of scattering and the probabilistic interpretation of the wave function, Born applied Schrödinger's wave mechanics to asymptotically free stationary waves.[8] According to the probabilistic interpretation of Schrödinger's wave mechanics, quantum mechanical particles seem to be like particles in the former (i.e., classical) sense, *except* with regard to locality and individuation:

(QM) Quantum mechanical particles are

(MQS)   carriers of *mass m*, *electric charge q*, and *spin s*,

(INDEP)   *independent* of each other,

---

[7] Bohr 1913b.

[8] Born 1926a,b.

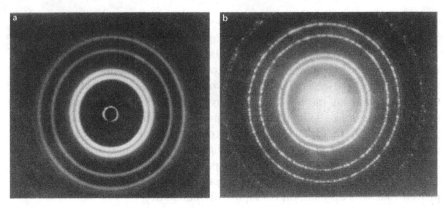

Fig. 6.1. Matter waves. Electron diffraction (**b**) compared to diffraction of gamma rays (**a**) (Raether 1957, 443)

(POINT)	*pointlike* in interactions,
(CONS)	subject to *conservation laws*,
(LOCPD)	*localizable* by a particle detector,
(PROB)	*probabilistically determined* by the Schrödinger equation,
(WAVE)	in states that superpose and interfere,
(UNPQ)	unsharp in momentum p and position q according to the *uncertainty relation* $\Delta p \Delta q \geq h/2$,
(PAULI)	not spatio-temporally individuated but only *distinguished by their quantum states*, according to Pauli's exclusion principle,
(BOUND)	able to form *bound systems*.

This meaning change is *intensional*.[9] And it is serious: particles are no longer local. Classical locality (LOCAL) is weakened to the *mere possibility* (LOCPD) that only becomes actual through a particle detector that makes a position measurement. This is related to Heisenberg's uncertainty relation (UNPQ) which gives rise to unsharp properties, that is, properties defined at a probabilistic level, in the sense of the width of the statistical distributions of the results of repeated position and momentum measurements.[10] This liberalization of classical locality makes the particle concept context-dependent. The property of being local now depends on the experimental context.[11] The

---

[9] According to modern semantics, the intension of a concept is its conceptual content (Frege's 'sense'), whereas the extension is its domain, i.e., the class of objects to which it refers (Frege's 'reference').

[10] See Busch et al. 1991, 127–131; Busch et al. 1995, 3 and 59–60.

[11] I avoid talking of dispositions or propensities, since they usually commit one to an ignorance interpretation of quantum theory which I cannot defend. Bohmian

classical trajectory has been replaced by Schrödinger's wave function, dispensing with classical spatio-temporal individuality (INDIV) and determinism (DET). According to quantum theory, subatomic particles do not have a trajectory which individuates them. Due to Heisenberg's uncertainty relation, they cannot be described in terms of spatio-temporal coordinates and momentum–energy at the same time. (This point was mainly emphasized by Bohr, giving rise to his complementarity view of quantum physics.[12]) This also affects property (INDEP), the way in which quantum particles may be considered to be independent of each other. On the one hand, due to Heisenberg's uncertainty relation statistically uncorrelated measurement results come into play. On the other hand, quantum mechanics affects the statistical behavior of particles. Electrons are fermions. They obey Pauli's exclusion principle (PAULI) and Fermi statistics. A many-particle system has to be described by an antisymmetrized wave function. This is related to the spin 1/2 of electrons which enters the list (MQS) of dynamic properties. Only the features (POINT) and (CONS) of the classical particle concept remain unaffected.

However, the particle concept (QM) still shares the *extension* of (CP). It refers to subatomic matter constituents with mass and charge. Since these are the very features of classical particles which make it possible to *measure* them, the operational content of the particle concept is also untouched. The *causal* particle concept also escapes more or less unaffected, up to this point.[13] Hence, as far as non-relativistic massive charged particles are concerned, it seems unproblematic to consider (QM) as the legitimate successor of (CP) in making the mereological and causal particle concept precise.

6.2.2 Light Quanta

Einstein's light quantum hypothesis gives rise to an *extensional* meaning change. It bridges the classical distinction of matter and radiation by extending the particle concept to radiation phenomena. This step was decisive. It resulted in using the term 'particle' for all field quanta of whatever quantum field theory. The particle picture of light and the name 'photon' for the light particles were accepted once the light quantum hypothesis had been extended to relativistic kinematics and the energy conservation in single scattering processes had been demonstrated in the Compton effect.[14] Relativistic

---

quantum mechanics tries to keep locality, but at the price of implementing action-at-a-distance into the theory, that is, skipping the property (POINT) (which is a *sine qua non* for Einstein causality). This is at odds with the principles of relativity.

[12] For Bohr's complementarity philosophy, see Meyer-Abich 1965; Jammer 1966, 345–361, and 1974, Chap. 4–7; Scheibe 1973, Chap. I; Murdoch 1981; Folse 1985; Honner 1987; Faye 1991; Chevalley 1991; Falkenburg 1998; Pringe 2006.

[13] At least in the domain of non-relativistic quantum mechanics. See Mott 1929 and the discussion in Sect. 5.3.1.

[14] See Sect. 3.2.2.

kinematics brings about a differentiation of the concept of mass. It distinguishes the constant rest mass m_0 and the mass $m(v)$ of a moving particle that depends on velocity. A rest mass 0 and an inertial mass $h\nu/c^2$ is assigned to the photon. Applying relativistic kinematics to the light quantum hypothesis resulted in a *double* extensional meaning change of the particle concept:

(i) A rest mass 0 may now be attributed to an uncharged particle described by the energy–momentum 4-vector of relativistic kinematics. Here, the particle concept is extended to massless particles that only propagate on inertial trajectories, on the light cone.

(ii) Particles are considered to be the constituents of light, in just the way Newton speculated in the *Queries* of his *Opticks* about the atomic nature of light.[15] According to the relativistic equivalence of mass and energy, this particle is associated with an inertial mass. In this way, one arrives at the following informal concept of particles in the sense of light quanta or photons, as they were conceived around 1923:

(LQ) Light quanta (photons) are

(E)	*massless* quanta of *energy* $E = h\nu$,
(DISC)	*discontinuous*, i.e., they come in quanta,
(INDEP)	*independent* of each other,
(POINT)	*pointlike* in interactions,
(CONS)	subject to *conservation laws*,
(BOSE)	*indistinguishable*, giving rise to Bose–Einstein statistics,
(PROB)	*probabilistically determined* by classical radiation theory,
(WAVE)	in states that superpose and interfere,
(NLOC)	non-local,
(LOCPD)	*localizable* by a particle detector.

The relation between (NLOC) and (LOCPD) requires comment. According to their theoretical description, photons are *non-local*. Since they are described by a certain wave vector $\boldsymbol{k}$ or frequency ν, they do not exist in local states,[16] even though they may be measured by a particle detector and therefore considered to be localizable *for all practical purposes*, in accordance

---

[15]See Newton 1730 vs. Einstein 1905.

[16]Indeed this is true of *any* relativistic particle or field state; see Clifton and Halvorson 2002. And this is stated by a theorem that derives from very general assumptions about relativistic quantum fields, independently of all operational considerations about the impossibility of localizing particles with lower precision than their Compton wavelength without effects such as pair production.

with (LOCPD). It should be noted, however, that (LOCPD) is an *operational* property that does not have any direct counterpart in the theoretical description of photons, neither around 1923 nor in modern quantum field theory.

Consequently, in the photon concept only the classical particle properties (INDEP), (POINT), and (CONS) survived. However, their meaning has changed. (POINT) has become a *field property*, meaning only that there is no action-at-a-distance or that the photon couples locally to other particles or fields. (CONS) is associated with the Poincaré group of relativistic kinematics. The independence property (INDEP) has further changed. Again, statistically uncorrelated measurement results come into play, but in addition indistinguishability and the associated Bose–Einstein statistics (BOSE) have to be taken into account.

Now, *all* subatomic particles are subject to quantum theory whereas *some* of them are massless and do not form bound systems. The classical light properties wavelength or frequency are taken over to the quantum theory of light, giving rise to wave–particle duality.[17] Indeed this produced a *further* extensional meaning shift. Classical locality is generalized to *discontinuity* (DISC). Photons are best characterized by their *discontinuous* nature. They are discrete energy quanta. This point was emphasized by Bohr at the beginning of his famous Como lecture by stating the

> quantum postulate, which attributes to any atomic process an essential discontinuity, or rather individuality, completely foreign to the classical theories and symbolized by Planck's quantum of action.[18]

According to this, any quantized property that comes in multiples of $\hbar$ may count as a particle, whether it is localizable or not. (It should be noted that here Bohr does not differentiate between the integer and half-integer multiples of $\hbar$ in terms of which angular momentum and spin are expressed in quantum mechanics and the quantization of energy quanta in terms of $h\nu$.) Perhaps this is indeed the most general modern particle concept. However, it even applies to the harmonics of a string in classical mechanics.

Today, particle physicists agree more or less about two realistic assumptions. *First*, the particle concept refers to discontinuous phenomena. *Second*, these phenomena underlie the laws of a quantum theory. In addition, they agree that at the operational level quasi-classical measurement laws are needed in order to capture these phenomena quantitatively. Taking the Compton effect into account, it is admitted that all subatomic particles are subject to a *measurement theory* which describes a *classical non-relativistic or relativistic particle with rest mass* $m \geq 0$.

Beyond this basic agreement, the particle concept is used liberally today. The furthest-reaching consequences had the generalization suggested in

---

[17]See next chapter.
[18]Bohr 1928, 580.

Bohr's Como lecture, namely to take the discontinuity of quantum phenomena into account. Since the energy quanta of a radiation field are measured by a local device, they also count as particles in an operational sense.

6.3 The Operational Particle Concept

Hence, the current particle concept has two roots. The first lies in the experiments of atomic, nuclear, and particle physics and in a measurement theory which is based on the classical particle concept. From the first e/m measurement of the electron up to the present day, it is justified by the experimental phenomena and by the theoretical assumptions involved in their data analysis. It is primarily based on the quasi-classical properties of observable particle tracks, on the time-of-flight measurements performed in high-energy scattering experiments at a particle accelerator, etc. The second root is Einstein's light quantum hypothesis, which is also based on experimental phenomena such as the photoelectric effect and the Compton effect. Therefore, the following operational definition of a particle is very general:

(OP) Operationally, particles are

    (MESQ)   collections of *mass m, energy E, spin s, charge q,*

    (LOCPD)   *localizable* by a particle detector,

    (INDEP)   *independent* of each other.

These properties stem from the quantum concepts (QM) and (LQ) explained above. In (MESQ), the classical property of having mass and charge is generalized in such a way that massive particles and light quanta are likewise included. The spin of massive charged particles can be measured, e.g., in the Stern–Gerlach experiment. In (MESQ), the metaphysical expression 'carrier' for the owner of these properties is replaced by the operational concept of a *collection* of properties. A metaphysical carrier of properties can obviously *not* be measured. In (LOCPD), the classical property of being local is generalized in the sense of the operational possibilities of localizing quantum mechanical particles and the photon. This generalization is still based on a firm belief that there *is* an entity which *may* be localized by means of a particle detector. The independence property (INDEP) is now related to the occurrence of uncorrelated particle detections, granting the statistical independence of the single events which make up an ensemble.

One may ask whether particles in the operational sense do not *reduce* to such events, to the mere *clicks* in the Geiger counter or another particle detector. They do *not*. The concept is theory-laden. It includes the realistic belief mentioned above. Operationally, particles are conceived as collections of certain dynamic properties which under well-defined experimental conditions always go together. Otherwise, it would not make any sense to perform experiments with subatomic particles. Experimental results must be reproducible. Only stable collections of dynamic properties can give rise to reproducible experiments. Hence, the operational particle concept deals with Lockean empirical substances, that is, collections of empirical properties which constantly go together[19] or bundles of properties which repeatedly appear together. In a chaotic world where no such bundles of properties show up and endure, it would be impossible to practise the experimental method successfully. Metaphysically, such Lockean empirical substances are *more* than mere non-reproducible isolated events but *less* than carriers of properties. (Moreover, they may well be interpreted in terms of a Kantian scheme of substance, according to which it is in the end *our* own concept of substance that makes it possible to conceive of them as such bundles or collections of properties.) The particle concepts of the next section also make use of this unavoidably metaphysical though modest way of stating stable dynamic properties of a kind of particle.

Unfortunately, there is no unambiguous axiomatic counterpart of this metaphysically modest particle concept. Quantum theory gives rise to wave–particle duality. Only in some quantum states are the particles of the definitions (QM) and (LQ) (approximately) localized. But the localized states are *not* preferred. *Any* quantum dynamics prefers the *non*-local states. Without repeated position measurements, an unbound quantum mechanical state develops towards a plane wave. And without a macroscopic measuring device which is itself local, *nothing* is localized and no reidentification of the same kind of particle in subsequent measurements is possible.[20] Things are even worse. Light quanta are relativistic. But according to relativistic quantum theory, localization in a strict sense is impossible.[21] Any relativistic quantum theory is at odds with the very possibility of local states. Hence *strictly speaking*, in the relativistic domain the operational particle concept does not have *any* axiomatic counterpart. And this is *not* only due to the ubiquitous quantum measurement problem but to a major mismatch between relativistic quantum theory and the concept of a local state. Given this problem, it has to be stressed that the particle concept of subatomic physics is *primarily operational*.

Any attempt to understand the particle concept exclusively in axiomatic terms gives rise to conceptual confusion. When philosophers face the generous usage of physicists regarding the term 'particle', such confusion almost automatically arises. Non-experts in quantum theory only know about classical particles. They identify particles with the local entities individuated by their trajectories. Philosophers of quantum theory most often do not have detailed knowledge of the experimental foundations and operational content of the

---

[19] Locke 1689, Book II, beginning of Chapter XXIII.

[20] See Sect. 6.6.

[21] See Clifton and Halvorson 2002; see also Sect. 6.6.

particle concept. Obviously, they focus on locality as the predominant particle property.[22] Most physicists, however, keep their corpuscular talk *without*
adhering to a hidden mechanistic world view or a classical particle ontology.
Indeed, the particle tracks and scattering events recorded by particle detectors provide them with rather good reasons for keeping the term 'particle'.
Up to the present day, they speak of *particle physics*. But the corresponding particle concept is partially operational, partially based on the axioms of
some quantum theory.

As pointed out in the last chapter, the core of the incommensurability
problem in the transition from classical to quantum physics is the mismatch
between the operational, axiomatic, and referential aspects of quantum concepts. This mismatch arises as follows. The data analysis of any high energy
physics experiment forces physicists to analyze individual particle tracks and
scattering events in quasi-classical terms. But from an axiomatic point of
view, the operational basis of quantum mechanics and quantum field theory
is probabilistic. Given that only classical point mechanics deals with individual particles, and given the unresolved quantum measurement problem at the
level of individual particle detections, the mismatch is unavoidable. Nevertheless, the operational particle concept (OP) builds *some* semantic bridges
where the theoretical concepts (CP), (QM), and (LQ) fail to fit in with each
other. If this is neglected, one remains in the dark as to how the classical
particle concept and its various quantum successors are related.

6.4 More Quantum Particles

Today, quite a number of informal meanings of the term 'particle' are floating
around. Without a unified theory of fundamental particles and their interactions there can be no general agreement about how to use the term 'particle'.
Beyond the operational particle concept (OP) given above, this is only clear
in the context of a specific theory. Axiomatically, particles are the entities to
which certain terms of an axiomatic theory refer. The *kind* of entity is specified by the respective theoretical description. In this sense, one might then say
that a particle concept is *implicitly defined* in the sense of Hilbert's axiomatic
method. It is implicitly defined by the axioms of classical point mechanics,
by a quantum theory, or by a related theoretical model or structure.

In addition it is usually required that an axiomatic particle concept have
some empirical correspondence with the operational particle concept (OP).
The belief in the *existence* of a certain kind of particles depends on the empirical support for their theoretical description, i.e., on the existence of an

---

[22]See the characterization of a particle theory which Redhead sends ahead of
his discussion of the philosophical problems of quantum field theory: "A particle
theory attributes to certain individuals (the particles) a variety of properties. These
properties will include space-time location." Redhead 1988, 10. According to this,
the current theories of elementary particles are not particle theories.

(approximate) empirical model. That is, the theoretical description must fit in with the collections of dynamic properties found in reproducible kinds of particle tracks and scattering events. This correspondence makes the difference between physical particles and mere mathematical tools of calculation.

Hence, the nature of a particle is specified in terms of the axioms of some dynamics. This typically embraces a state space, a set of state functions, and an equation of motion. Three examples of such implicit definitions of the particle concept in terms of theoretical structures have already been mentioned. They give rise to the informal definitions (CP), (QM), and (LQ). A classical particle is subject to the axioms of classical point mechanics. A quantum mechanical particle is subject to the axioms of Hilbert space, the Schrödinger equation, and the projection postulate for the states in Hilbert space. These definitions are well-known. The textbooks of theoretical physics give them by declaring that the entities of the dynamics under investigation are particles (*if* there is any such declaration at all). For example, the famous textbook by Landau and Lifschitz only introduces the particle concepts of classical mechanics and quantum mechanics in footnotes.[23] Our third example was based on a rudimentary quantum theory of radiation, namely Einstein's light quantum hypothesis. In 1905, respectively 1916, it was based on the Planck–Einstein relation $E = h\nu$, the laws of relativistic kinematics (in particular momentum–energy conservation), and the laws for the radiative transitions in Bohr's atomic model. A full axiomatic definition of the light quantum can only be given in the framework of quantum field theory.

But here, the terminological ambiguities begin. In the context of quantum field theory, *several* informal particle concepts are used. With the same right a quantized field or its field quanta may be considered to refer to particles. Quantum field theory deals with the creation and annihilation of field quanta. Real field quanta are physical particles in the sense of corresponding to the operational particle concept (OP). In addition, there are the so-called *virtual particles*. Their ontological status is unclear. They are involved in the perturbation expansion of the interactions of quantum fields, that is to say, they are formal calculational tools and they *cannot* be measured in accordance with the operational particle concept (OP). However, they give rise to collective effects that *can* indeed be measured. The quasi-particles of condensed matter physics in turn result from collective effects in solids. They are formal tools of calculation, too, but they are formal analogues of *real* field quanta and they do indeed *have* an operational meaning, as will be shown below.

In order to make this range of uses of the term 'particle' clearer, I now discuss the following concepts: (1) the *field quanta* of relativistic quantum field theory (FQ), (2) Wigner's particle definition in terms of *group theory* (GT), (3) *virtual particles* (VP), and (4) the *quasi-particles* of condensed matter physics (QP) which are non-relativistic analogues of relativistic field quanta. Ironically, in the sense of Bohr and Heisenberg's correspondence principle

---

[23] See Landau and Lifschitz 1987, 1, note 1 and Landau 1988, 2, note 2.

none of these concepts correspond to classical *particles*. As far as they have *any* correspondence to classical models of entities, it is to classical *fields*. However, in the domain of relativistic quantum fields and their field quanta, any correspondence breaks down and the referential claims of classical realism become highly problematic. Nevertheless, all of these concepts except the virtual particles may (approximately) correspond to particles in the operational sense.

6.4.1 Field Quanta

The identification of particles with field quanta results from the generalization of Einstein's light quantum hypothesis to the quanta of *any* quantum field theory. The first fully-fledged quantum field theory was quantum electrodynamics. Its foundations were laid by Dirac's 1927 paper *The Quantum Theory of the Emission and Absorption of Radiation*.[24] The next step was Dirac's theory of the electron which was originally developed in the framework of relativistic quantum mechanics. In addition to the usual positive-energy solutions for the electron, the Dirac equation predicted wave functions corresponding to negative-energy solutions. Later, they were associated with the positron, the antiparticle of the electron. The positron has the mass and spin of the electron, but the opposite charge. Particles and antiparticles annihilate. The corresponding non-conservation of particle number provoked the step from relativistic quantum mechanics to quantum field theory.

Quantum field theory applies to matter and radiation likewise. Quantized matter and radiation fields are only distinguished by their spin values, the corresponding field equations, and the associated Fermi or Bose statistics of the field quanta. Field quanta of half-integer spin (fermions) obey Pauli's exclusion principle and Fermi statistics. Field quanta of integer spin (bosons) obey Bose–Einstein statistics. The respective fields are described by the Klein–Gordon equation (spin 0), the Dirac equation (spin 1/2), and the Maxwell equations (spin 1). Many textbooks on quantum field theory identify the quanta of these fields with particles. In their famous textbook, Bjorken and Drell generalize the concepts of the electron and the positron in this way. According to this, the field quanta which belong to positive or negative eigenvalues of the charge operator are called particles and antiparticles, respectively.[25]

The usual textbook approach to quantum field theory follows the heuristics suggested by Heisenberg's generalized correspondence principle.[26] Quantum field theory is introduced by quantizing a classical field theory. The classical field observables are replaced by operators and certain commutation rules are defined for them. This gives rise to Heisenberg-type uncertainty relations which express non-classical features of the field observables.

---

[24] Dirac 1927.

[25] Bjorken and Drell 1965, end of Chap. 12.5.

[26] Heisenberg 1930a, 78, and 1930b, 105; see the discussion in Sect. 5.4.2.

For a spin 0 particle or field, the quantization procedure starts from either a classical scalar field or the Hamiltonian of the quantum mechanical harmonic oscillator. The latter quantization procedure has been called *second quantization* for the following formal reason. Quantizing the wave function of the quantum mechanical harmonic oscillator or quantizing the oscillations of a classical scalar field yields exactly the same formal results, namely the solutions of the quantized Klein–Gordon equation. Second quantization works as follows. First, the classical harmonic oscillator is quantized. In the second step, the resulting Schrödinger equation of the harmonic oscillator is transformed to the Heisenberg picture and quantized again. Since the formal results are the same in both cases, a quantized classical field is formally equivalent to a quantum mechanical many-particle system with infinitely many particles, respectively oscillators, or degrees of freedom. The Maxwell field is quantized as follows. The classical radiation field is developed into field modes of a certain momentum p or angular frequency $\omega = 2\pi\nu$, which are expressed in terms of the wave vector $k = p/\hbar$. They are then quantized in terms of creation and annihilation operators a_k^+, a_k for field quanta of energy $E = h\nu_k$, which belong to the field mode of wave number k.

Generally, the field quanta of a quantized field are characterized in terms of the wave number $k = \hbar p = h/\lambda$ and a set of values of quantized dynamical properties such as mass, spin, and eventually charge. The field quanta are regarded as the smallest units of these properties. As collections of such dynamic properties, they obviously correspond to the operational particle concept (OP).[27] The creation and annihilation operators a_k^+, a_k obey the following commutation relations:[28]

$$[a_k, a_k^+] = 1 \ , \tag{6.1}$$

$$[a_k, a_k] = 0 \ , \tag{6.2}$$

$$[a_k^+, a_k^+] = 0 \ . \tag{6.3}$$

For fermions or bosons, $[a_k, a_{k'}^+]$ means the anticommutator $[a_k a_{k'} + a_{k'} a k]$ or the commutator $[a_k a_{k'} - a_{k'} a k]$, respectively. These operators act on the field modes. The creation operator a_k^+ generates a field quantum of wave number k, the annihilation operator a_k makes a field quantum of the respective field mode disappear:

$$a_k^+ \Psi_{k,n} = \sqrt{n+1} \Psi_{k,n+1} \ , \tag{6.4}$$

---

[27]Leaving aside the localization problem mentioned above.

[28]Any dependence on the spacetime coordinates (which is expressed in terms of δ functions) is suppressed here. For two different wave numbers k, k', the only non-vanishing commutator is (6.1), which then becomes $(a_k, a_{k'}^+) = \delta(k, k')$.

$$a_k \Psi_{k,n} = \sqrt{n} \Psi_{k,n-1} \ . \tag{6.5}$$

Now, the field state Ψ may be developed into a sum of states of field modes $\Psi_{k,n}$ of wave number k and well-defined occupation numbers n (these states form the Fock space):

$$\Psi = \Sigma_{k,n} c_{k,n} \Psi_{k,n} \ . \tag{6.6}$$

The states in this sum are eigenstates of the occupation number operator $N = a_k^+ a_k$:

$$N\Psi_{k,n} = n\Psi_{k,n} \ . \tag{6.7}$$

Hence, if the creation operator a_k^+ acts on the vacuum state Ψ_0, a field mode $\Psi_{k,1_k}$ with one field quantum of wave number k is generated, and the corresponding eigenstate of the occupation number operator is

$$N\Psi_{k,1} = 1\Psi_{k,1} \ . \tag{6.8}$$

It is tempting to identify the eigenstate of the occupation number operator with a collection of field quanta or particles. But this does not work, at least if a classical particle picture is tacitly presupposed. A quantum field does not dissolve into a swarm of particles in a classical sense. Quantum field theory has several features that destroy any obvious correspondence with the classical particle concept. In particular, low-intensity quantum fields have striking non-local and non-classical properties. At the operational level, these properties show up in experiments with low-intensity beams.[29] At the axiomatic level, they are expressed in terms of commutation relations of the field operators. The operators for the electric and magnetic field strengths of a quantized radiation field do not commute, and the phase of the field modes of an occupation number state are not well-defined. Therefore, it is impossible to attribute well-defined field strengths to spacetime points or sharp field intensities to finite spacetime regions.

The field intensity is given in terms of the expectation value of the occupation number operator N. The basic quantum field state is the vacuum. It belongs to the expectation value $\langle \Psi_{k,0} | a_k^+ a_k | \Psi_{k,0} \rangle$ of the operator $a_k^+ a_k$, that is, to an *expected* or *average* number 0 of field quanta. But according to the laws of quantum field theory, the vacuum state *fluctuates* around the occupation number 0. Even though the average number of field quanta may be 0, the spontaneous creation and annihilation of field quanta seems to be ubiquitous, as shown for example by the Casimir effect and polarization experiments with a low-intensity photon beam.[30]

---

[29] See the discussion of such experiments in Sects. 7.3–7.5.

[30] The Casimir effect is the attraction of two metal plates in the vacuum, which is due to the modification of the vacuum state between them; see Itzykson and Zuber 1985, 138–141. In polarization experiments with single photons, two vertically crossed polarizers let no signal pass through, whereas inserting a third, diagonal polarizer makes the signal reappear; see the discussion in Sect. 7.3.3.

Hence, the field quanta do not make up the quantum field as constituents in any usual sense. They contribute to the field intensity in a far less obvious way. The sum of quantized field modes given in (6.6) replaces the states of classical physics and quantum mechanics, namely the trajectories in phase space, electromagnetic waves, and Schrödinger's wave functions. The solutions of the relativistic field equations describe quantized fields in terms of the effects of creation and annihilation operators rather than in terms of an evolution of the underlying wave functions. They make predictions for the emission and absorption of field quanta. Einstein's light quanta are one example of such field quanta. They are subject to the quantized Maxwell equations (spin 1 particles). Other examples are the electrons and positrons of the Dirac equation (spin 1/2) or the pions of the Klein–Gordon equation (spin 0). Generally, field quanta may be qualified as follows:[31]

(FQ) Field quanta are

(MESQ)	collections of *mass m, energy E, spin s,* and *charges q_i*,
(DISC)	*discontinuous*, i.e., they come in quanta,
(INDEP)	*independent* of each other,
(POINT)	*pointlike* in interactions,
(PROB)	*probabilistically determined* by a quantized field equation,
(WAVE)	in states that superpose and interfere,
(NLOC)	non-local,
(LOCPD)	*localizable* by a particle detector,
(COMM)	subject to certain *commutation rules* for anticommutators (fermions) or commutators (bosons),
(NDIST)	*only numerically distinguishable*, giving rise to Fermi or Bose statistics (depending on the spin),
(CONS)	subject to *conservation laws*,
(BOUND)	able to form *bound systems*.

Not only fermions, but also many bosons have the particle-like property (BOUND).[32] The conserved quantities (CONS) depend on the symmetries of the underlying field. They give rise to the group theoretical particle concept (GT) discussed in the next section. The wavelike properties of classical matter or radiation fields are preserved in (WAVE). The particle-like property (DISC) expresses the quantization of the field modes in terms of the quanta

---

[31] For relativistic field quanta, in (MESQ) the *rest mass* of relativistic particles is intended. To the electric charge q, add the generalized charges of the electroweak and strong interactions, i.e., *flavor q_f* and *color q_i*. See Sects. 6.4 and 6.5, concerning the group theoretical particle concept and quarks.

[32] See Sect. 6.5.

of energy $E = h\nu$ or momentum $\boldsymbol{p} = \hbar\boldsymbol{k}$ that are created and annihilated. This stems from the field quantization or the commutation rules for the field operators. For bosons, these rules are given in terms of commutators, for fermions, in terms of anticommutators. Hence, it is crucial to include the spin of the field quanta in the list (MESQ) of intrinsic properties. By fixing the use of commutators or anticommutators, the spin determines the statistical consequences of (NDIST), giving rise to either Fermi or Bose–Einstein statistics. In this way the spin affects the independence property (INDEP), which has substantially changed compared with the corresponding classical property.[33]

The relation between (NLOC) and (LOCPD) is the same as for photons.[34] Since the field quanta are described by a certain wave vector $\boldsymbol{k}$ or frequency ν, they do not exist in local states, even though they may be measured by a particle detector and therefore considered to be localizable *for all practical purposes*, in accordance with the operational property (LOCPD).

The conservation laws (CONS) give rise to non-local correlations of measured field quanta, as the well-known example of EPR (Einstein–Podolsky–Rosen) correlations shows.[35] In addition, the commutation rules (COMM) are responsible for the possibility of extremely non-local correlations. For example, the operators of electric and magnetic field strengths of the quantized Maxwell field do not commute. So *where* is the field intensity? And the phase operators that may be constructed in quantum optics[36] turn out not to commute with the occupation number operator. So how many light quanta does the coherent state of a laser beam contain? Obviously, in many field states *neither* of these quantities has a sharp value. The non-local consequences will be discussed in the next chapter.

One further point has to be emphasized. Only the field quanta of *non-interacting* quantum fields correspond to the *operational particle concept* (OP). The field quanta of free fields are collections of dynamical properties such as energy (or frequency), mass, charge, and spin. In addition, *for all practical purposes* they can be considered to be localizable in particle detectors. Their detection corresponds to the absorption of a field quantum (or to a more complex scattering process, such as the Compton effect). The claim that the field quanta of *non*-interacting particles are *detected* seems paradoxical. However, the interactions of quantum fields are construed in terms of scattering amplitudes, where the incoming and outgoing particles are considered to be *asymptotically free*. (The squared scattering amplitudes have the usual probabilistic meaning.) In addition, the detection processes underlie conservation laws and superselection rules for the dynamical properties of the field quanta. At least with regard to the detected amount of energy,

---

[33] See Sect. 6.1.

[34] See Sect. 6.2.2.

[35] Einstein et al. 1935.

[36] See, e.g., Walls and Milburn 1994, 23–26.

mass, charge, spin, etc., the field quanta behave quasi-classically. In any case, they are *never* individuated by a trajectory.

6.4.2 The Group Theoretical Definition

According to the structure of the state space and the field equations, quantum field theory offers *two* possible successors of the classical particle concept: either the field modes on which the creation and annihilation operators act or the field quanta which are created and annihilated. Both are described in terms of the same collections of dynamic properties. From an *operational* point of view it is an obvious step to identify the particles of current particle physics with the field quanta of quantum field theory. But from an *axiomatic* point of view, the structural properties of the field equations of a quantum field theory are predominant. What matters is the structure of the space of solutions of these field equations.

In this regard, particles are usually identified with the irreducible representations of the Poincaré group. The Poincaré group is the group of spatio-temporal symmetry transformations of a relativistic theory, one of its subgroups being the Lorentz transformations. According to the Noether theorem of mathematical physics, the invariances of a physical theory are associated with certain conservation laws satisfied by the physical quantities of the theory. In this way, the conservation laws for mass, energy, momentum, angular momentum, and charge are closely related to the symmetries of a physical dynamics. In 1939, Wigner showed in a seminal paper that the solutions of any relativistic field equation can be classified according to the irreducible representations of the Poincaré group, in terms of the quantities *mass*, *spin*, and *parity*.[37] This holds for the solutions of *all* field equations, be they quantized or not. Hence it is not specific to a quantum theory at all. It applies quite generally to *any* relativistic theory of non-interacting fields. In addition, it may be extended to *non*-relativistic field equations, i.e., to the Galilean group of non-relativistic mechanics and quantum mechanics:

> It may be mentioned, finally, that these developments apply not only in quantum mechanics, but also to all linear theories, e.g., the Maxwell equations in empty space. [...] The increase in generality, obtained by the present calculus, as compared to the usual tensor theory, consists in that no assumptions regarding the field nature of the underlying equations are necessary.[38]

The choice of a specific relativistic field equation depends on the spin. The Klein–Gordon equation belongs to spin 0, the Dirac equation to spin 1/2, the Maxwell equations to spin 1. (Later, it turned out that parity conservation or

---

[37]Wigner 1939.

[38]Wigner 1939, 151. The 'usual tensor theory' is quantum mechanics in Hilbert space.

non-conservation makes another important difference.) In relativistic quantum field theory, the respective *field quanta* have spin 0, 1/2, or 1. According to Wigner's classification, a given spin value corresponds to the *solutions of the respective field equations.* Hence we face the semantic ambiguity already mentioned above: in the context of quantum field theory, particles may be identified either with the field quanta or with the solutions of the field equations, i.e., fields or field modes. This makes a substantial difference for all field states except the eigenstates of occupation number 1. Added to this conceptual ambiguity is the fact that Wigner's 'particle' definition is not specific to particles at all.

Wigner's paper of 1939 had great heuristic value for later attempts to find an all-embracing relativistic quantum field theory of elementary particles. In particular, it influenced Heisenberg when he tried to construct a unified field theory based on symmetries. He connected the concept of an elementary particle characterized by certain values of quantized properties with Wigner's way of classifying the solutions of relativistic field equations in terms of mass, spin and parity. Heisenberg considered the elementary particles as physical representations of the invariances of his field theory:

> The law of nature defines certain fundamental symmetry operations, like, e.g., translations in space or time. In this way, it determines the frame in which all events can take place. The elementary particles correspond to the simplest representations of these symmetries (in the sense of mathematical group theory).[39]

This characterization fits in with the assumption that the elementary particles are the referents of quantum field theories. Every representation of a symmetry group is characterized by a specific field equation of a quantum field theory. The different representations correspond to the different states of the quantized field described by the equation. The irreducible representations correspond to the basic field modes, that is, to the particles which are considered to be elementary.

Later, other authors connected the particle concept to Wigner's classification of the irreducible representations of the Poincaré group, but without making any difference between the mathematical representations of a symmetry group and the physical particles. For example, von Weizsäcker gives the following sloppy explanation of the relation between symmetries and elementary particles:

> Current elementary particle physics tries to establish the system of the elementary particles by symmetry groups. [...] The existence of

---

[39]Heisenberg 1971, 879. My translation. Even in 1975, Heisenberg still wanted to include the scale invariance of deep-inelastic lepton–nucleon scattering (see Sect. 4.3.2) in this approach; see Heisenberg 1976.

particles is an immediate consequence of special relativity theory; they are the irreducible representations of the Poincaré group.[40]

Wigner's original paper deals with a mathematical problem of group theory. Where the physical applications of this problem are mentioned, the paper deals with quantum mechanical states and fields rather than particles. The term 'particle' does not belong to its vocabulary. One might say that Wigner's definition of the particle concept is due rather to Wigner's followers than to himself. It is indeed a *very* implicit definition of the particle concept (in the sense of Hilbert's axiomatic method). According to it, particles of a given mass, spin, and parity *are* a physical model of mathematical group theory. The following historical reference to Wigner's particle concept of 1939 is therefore a typical legend of the historiography of physicists:

> In 1939 Wigner was very successful in making the definition: an elementary particle *is* an irreducible projective representation of the Poincaré group, $\mathcal{P}$, with mass ≥ 0 and energy ≥ 0, and spin $s \in \{0, 1/2, 1, \ldots\}$. He did not merely say that a particle is well described by such a representation: this would leave the word particle still undefined. Thus a particle *is* a pair $(\mathcal{H}, U_{(m,s)})$ where $\mathcal{H}$ is a Hilbert space, and U is a unitary continuous action of $\mathcal{P}$ on $\mathcal{H}$, obeying $U(a, \Lambda)U(b, M) = \omega U(a + \Lambda b, \Lambda M)$, where $a, b \in \mathbb{R}^4$ are spacetime vectors, and Λ, M are Lorentz matrices, and where (m, s) are the mass and spin.[41]

Here, a particle of given mass m and spin s is defined in terms of an axiomatic structure, that is, a mathematical structure based on the axioms of functional analysis and group theory. The structure consists of a functional space, namely Hilbert space, and a set of Lorentz invariant mappings on this space. This leads to a *very general* meaning of the term 'particle'. It has been extended to the generalized charges q_i of the dynamic or 'internal' symmetries of gauge invariant theories, such as the $U(1) \times SU(2) \times SU(3)$ symmetry of the current standard model of particle physics. In addition, it has been liberalized to the concept of spontaneous symmetry breaking which is involved in the standard model, giving rise to the lepton masses. Since symmetries do not fix the specific shape of the functions in Hilbert space, the concept is very open for future theoretical developments in particle physics. According to this, for example the strings of the current superstring theories (which are conceived of as extended rather than pointlike objects) are particles, too. Informally, the resulting particle definition reads as follows:

---

[40] Von Weizsäcker 1985, 37–38. My translation. Heisenberg and von Weizsäcker refer to *elementary* particles as corresponding to the *irreducible* representations of the respective symmetry groups. This point is taken up again in the next chapter where the parts of matter are discussed.

[41] Streater 1988, 144.

(GT) According to group theory, (elementary) particles

(MESPQ)	are collections of *mass m*, energy *E*, *spin s*, parity *P*, and *charges q_i*,
(REPR)	belong to the *(irreducible) representations of symmetry groups*,
(INDEP)	are *independent* of each other,
(FIELD)	are subject to some *field equations*,
(CONS)	are subject to *conservation laws*.

(MESPQ) expresses the dynamic properties of the operational particle concept (OP) and the field quanta concept (FQ) in terms of symmetry groups. Mass, spin, parity, and some generalized charges classify the solutions of field equations of given spatio-temporal and dynamic (or internal) symmetries. These symmetries are closely related to the conservation laws (CONS) of the kinematic and dynamic properties of the particles and fields. The classical locality property (LOCAL) and its successors (LOCPD), and (DISC) of (QM), (OP), (LQ), and (FQ) have been skipped. According to Wigner's definition, particles are no longer local, and they must be neither localizable, nor discontinuous. The only feature that (GT) has in common with the particle concepts of the preceding sections is that particles are considered to be *primarily non-interacting*. They are conceived as independent of the rest of the world, like the substances of traditional metaphysics. But here, (INDEP) cannot mean statistical independence. This property does not make any sense for the solutions of a field equation. It can only be understood in the sense of being interaction-free or uncoupled.[42] In this way (INDEP) becomes a constraint for the field equations (FIELD). Particles are subject to *uncoupled* field equations. Obviously, this is also true of free electromagnetic waves. In this way, the distinction between particles and fields is neglected.

According to group theory, the only hallmark of particles is (REPR). They fit in with some reducible or irreducible representation of the symmetries of some field equation with regard to its kinematics and dynamic quantities. One may ask whether such an abstract particle concept is *too* liberal. Indeed, it is not directly connected to experimental phenomena. It defines particle *types* of given mass, spin, and parity, and some generalized charges, but it does not bring them down to earth, to individual particle detection or local events in the experiments of particle physics. Its operational content is at the type level rather than the token level. It corresponds to *classes* of individual particle detections, tracks, and scattering events. This is in agreement with the ensemble interpretation of quantum theory but much more abstract than the quasi-classical measurement theory of particle physics. Indeed, (GT) is a

---

[42]This is a *dynamic presupposition* of *statistical independence*. The problem of the various uses of the term *independent* is taken up again in Sects. 6.6 and 8.5.

meta-theoretical concept which serves to classify the kinds of possible *theories* of the phenomena rather than the phenomena themselves.[43]

6.4.3 Virtual Particles

So far, the particle concept has only been discussed in the context of non-interacting quantum fields. With the interactions of quantum fields, several new conceptual problems come into play. Amongst them are the distinction between 'real' and 'virtual' field quanta and the renormalization problem. The latter will only be mentioned at the end of this section, whereas we deal with the former in some detail now.

Obviously, the concept of non-interacting quantum fields is an unrealistic idealization. The empirical basis of particle physics only consists of the observable effects of subatomic interactions, namely position measurements, particle tracks, scattering events, total and differential cross-sections, resonances, jets, etc. Whenever these phenomena are quantitatively explained, the quantum theory of scattering is involved. Quantum field theory meets the phenomena in the cross-sections of high energy scattering experiments. But the precise low-energy predictions of quantum electrodynamics such as the Lamb shift and the gyromagnetic factor of the electron are also calculated within the quantum theory of scattering. Indeed, the quantum field theory of interactions *is* scattering theory. Its crucial quantity is the S-matrix which contains the transition probabilities of scattering processes. Their calculation is complicated. They are calculated within time-dependent perturbation theory. Here, the *virtual particles* come into play. They are called 'virtual' in contradistinction to the 'real' incoming and outgoing particles or field states, in the sense of *not being real*. Their propagators belong to the perturbation expansion of the S-matrix.

The calculation starts from the interaction term $\mathcal{L}'(x)$ of the Lagrange density of coupled fields, such as the coupled Dirac and Maxwell fields.[44] The lowest order of perturbation theory is the Born approximation. It is based on the assumption that the incoming and outgoing particles, respectively the corresponding initial and final quantum field states,

$$|i\rangle = \lim_{t \to -\infty} |t\rangle , \qquad (6.9a)$$

$$|f\rangle = \lim_{t \to +\infty} |t\rangle , \qquad (6.9b)$$

---

[43] See Wigner 1964, Falkenburg 1988, on the meta-theoretical character of symmetries. In the subsequent particle definitions parity is suppressed for two reasons. The *first* is theoretical. It turned out that parity conservation is violated in weak interactions. The *second* is operational. Even though parity is an intrinsic particle property, it cannot be measured for single particles but only at the ensemble level, i.e., from asymmetries in the spatial distribution of *many*-particle detection.

[44] The coupling is based on the idea of gauge invariance which cannot be explained here. An analysis of this concept is given in Lyre 2004.

are identical with non-interacting quantum fields. In the Born approximation, the interaction Lagrangian is given by the following expression corresponding to the lowest order Feynman diagrams:

$$\langle f|\boldsymbol{S}|i\rangle_{\text{Born}} = \delta_{fi} - \text{i} \int \text{d}t' \left\langle f \left| \int \text{d}^3x \mathcal{L}'(\boldsymbol{x},t) \right| i \right\rangle . \qquad (6.10)$$

The elements of the S-matrix are the *scattering amplitudes* of the respective quantum scattering processes. The scattering amplitude is related to the differential cross-section of a particle reaction as follows:

$$\left(\frac{\text{d}\sigma}{\text{d}(q^2)\text{d}E} \right)_{\text{QFT}} \sim f(p_1^i, \dots , p_n^i; p_1^f, \dots , p_m^f) |\langle f|\boldsymbol{S}|i\rangle|^2 . \qquad (6.11)$$

The two magnitudes agree up to a flux factor which is not given here. The kinematical factor f depends on the momenta $p_1^i, \dots , p_n^i; p_1^f, \dots , p_m^f$ of the n incoming particles and m outgoing particles of the individual particle reactions. The incoming and outgoing particles or quantum field states count as *real*. This means that they have *operational* meaning. They correspond to the quantum states which are prepared and measured in scattering experiments. (At least, they do *for all practical purposes*.)

The scattering process itself is a *black box*. The beam of incoming particles is prepared in a well-defined quantum state, the tracks and corresponding quantum states of the outgoing particles are measured, but what happens inside is unobservable. Mathematically, the scattering is described in the lowest orders of perturbation theory: in the Born approximation, possibly to second order, and on rare occasions to some higher order. The perturbational expansion reconstructs what goes on inside the black box in terms of so-called *virtual processes*. It describes the emission and absorption of virtual field quanta. The propagators of the virtual field quanta are mathematical components of a quantum theoretical superposition. Operationally, it is by no means possible to resolve them into single particle contributions. They are nothing but the mathematical contributions to an approximation procedure: like the harmonics of the oscillations of a mechanical string, the Fourier components of a classical electromagnetic field, or the cycles and epicycles in Ptolemy's planetary system. Formally, they are quantum analogues of the straight lines in the classical model of diffraction, along which the wave fronts of spherical waves propagate. The classical wave fronts propagate according to the laws of geometrical optics. At a given distance the propagated amplitudes sum up in constructive or destructive interference to the observed intensities of the interference fringes.

Feynman modelled the interaction of quantum fields in the spirit of the last example. What goes on inside the *black box* of the scattering of quantum fields is reconstructed in terms of propagators. Mathematically, these propagators are Green's functions which have to be integrated and summed up *before* calculating the probability from the scattering amplitudes. The propagators describe the propagation of virtual field quanta. In the classical analogue they correspond to the propagation of wave fronts (or non-quantized field amplitudes). Quantizing the classical field is once again formally analogous to the second quantization of quantum mechanical wave functions. Hence, the propagators that contribute to the interaction of a quantum field give rise to a quantized version of the path integral formalism of ordinary quantum mechanics.

So far, so good. Up to this point, no mysteries are involved and no conceptual confusion arises. The effect of the scattering of quantum fields is expressed in terms of quantization of the path integral formalism of ordinary quantum mechanics. The calculation of the S-matrix is lengthy and tedious.[45] It consists in summing up the amplitudes of several possible processes for the emission and absorption of virtual field quanta. In order to do so, it is much more convenient to express the field operators in the *momentum representation* rather than the spacetime coordinates. (The change from spacetime coordinates to the momentum representation is made by a simple Fourier transform. However, this is like switching from Ptolemy's system to Copernicus' world view. Things get much simpler from a more abstract point of view.) The momentum representation of quantum fields and their scattering no longer admits of any spatio-temporal idea of what happens during the scattering of quantum fields. Only the abstract idea of the probability of the emission and absorption of field quanta associated with a certain momentum–energy transfer remains. The next step is to simplify the calculation by translating the sums over different propagators into the formal language of Feynman diagrams. Each of the propagators which are summed up to the scattering amplitude corresponds mathematically to a line in a Feynman diagram. Each Feynman diagram symbolizes one of the virtual processes which contribute to the scattering amplitude. Hence, there are well-defined rules of formal correspondence between the mathematical formalism of the perturbation expansion and the iconic representation of the emission and absorption of *virtual* field quanta. There is nothing more to Feynman diagrams.

The propagators inside the *black box* only contribute to the perturbation expansion of interacting quantum fields. To talk about them in terms of *virtual particles* is a mere *façon de parler*. Some quantum field theory textbooks use this expression whereas others avoid it.[46] Feynman himself describes *vir-*

---

[45]In order to describe physical states, it includes mathematical tricks such as normal ordering and time ordering of the operator products; see, e.g., Nachtmann 1990, 108–110, Itzykson and Zuber 1985, 110–111 and 123.

[46]Nachtmann 1990 combines both attitudes. On the one hand, he emphasizes that within the limits of the so-called uncertainty relation between energy and time, non-interacting charged electrons may emit and re-absorb virtual photons (Nachtmann 1990, 99). On the other hand, he points out that the individual contributions to the scattering amplitude ('elementary processes') are fictitious: "One must be on one's guard, however, not to ask when, or how many, elementary processes take place in

tual processes, virtual field quanta, virtual photons, etc., in his papers and lectures[47] but he does talk about virtual *particles* in his popular talks and books.[48] So which particle properties do these virtual particles still have? One might characterize them as follows:

(VP) Virtual particles are

(MESQ)	collections of *mass m, energy E, spin s,* and *charges* q_i,
(DISC)	*discontinuous,* i.e., they come in quanta,
(PERT)	part of the *perturbation expansion of the S-matrix,*
(WAVE)	in states that superpose and interfere,
(PATH)	propagate along trajectories that are summed up according to the path integral formalism of quantum mechanics,
(NIND)	*not independent,* i.e., they belong to interactions,
(NLOC)	*neither local nor localizable* by a particle detector,
(NCON)	allowed to violate *energy conservation,* i.e., they may be *off mass shell* during the interaction time, according to the so-called uncertainty relation for energy and time, $\Delta E \Delta t \geq \hbar/2$.

To talk about virtual particles in the sense of (VP) obviously stems from dealing with virtual *field quanta,* and calling them *particles* in the sense of the particle concept (FQ). But in contradistinction to all the particle concepts introduced before, virtual particles are not conceived as independent. They essentially belong to interactions. Hence, they are neither localizable, nor independent, nor separable, nor do they underlie the usual law of energy conservation.[49] In particular, the theoretical particle-like property (PATH) should not be confused with the trajectories (TRAJ) of classical particles at all. The latter stand on their own, the former under an integral which runs over infinitely many possibilities. (PATH) and (WAVE) rely on equivalent

---

a particular experiment. This would be a completely meaningless question, since the decomposition of the reaction into elementary processes is a purely theoretical aid for calculating the transition amplitudes." Nachtmann 1990, 126.

[47] Feynman 1949a,b, 1961.

[48] Feynman 1985; 1987, 10.

[49] It has to be noted that these are statements about virtual particles that correspond to the inner lines of Feynman diagrams and *not* about the virtual *processes* that correspond to the *vertices* of the Feynman diagrams. The latter have to be considered in a different way. In addition, it is crucial that the virtual particles may only be *off mass shell* for a time $\Delta t = \hbar/2\Delta E$. The lifetime of particles is another important particle property which I neglect throughout this book. It is taken into account for example in the particle definition of Bogoliubov and Shirkov 1959, 5. However, it is a probabilistic property.

formal grounds. For all these reasons, the concept of a virtual particle on its own does not have any operational meaning. Therefore, this *façon de parler* provokes conceptual confusion as soon as one insinuates that virtual particles are *physical*, i.e., on a par with the real field quanta or the incoming and outgoing physical particles of a scattering experiment. Virtual particles only occur in superposed contributions to the perturbation expansion. They are called *particles* in the sense of *field quanta* which are created and annihilated during the interaction process, but they are called *virtual* in the sense of having *no operational meaning*. The concept of a virtual particle on its own only admits of an instrumentalist interpretation. Virtual field quanta are nothing but formal tools in the calculation of the interactions of quantum fields.

This does not mean, however, that the perturbation expansion of the *S*-matrix in terms of virtual particles is *completely* fictitious. The virtual processes described in terms of the emission and absorption of virtual particles contribute to a scattering amplitude or transition probability. Hence, *infinitely many* virtual particles *together* may be considered to cause a *real collective effect*. In *this* sense, they obviously *have* operational meaning. What is measured is an *S*-matrix element or the probability of a transition between certain real incoming and outgoing particles. The transition probability stems from *all* virtual field quanta involved in the superpositions of the relevant lowest and higher order Feynman diagrams.

In the low energy domain it is sometimes even possible to single out the contribution of one single Feynman diagram to the perturbation expansion. There are even several well-known high precision measurements to which mainly one Feynman diagram or the propagator of a virtual particle corresponds. This is demonstrated in particular by the best high precision tests of quantum electrodynamics, the measurement of the hydrogen Lamb shift and the $(g-2)/2$ measurement of the gyromagnetic factor g of the electron or muon. Dirac theory alone incorrectly predicts the fine structure of the hydrogen spectrum (no splitting of the levels $S_{1/2}$ and $P_{1/2}$ for $n=2$) and a gyromagnetic factor $g=2$ for the electron or muon. Measurements reveal the Lamb shift of the hydrogen fine structure and the anomalous magnetic moment of the electron.

The anomalous difference $(g-2)/2$ between the prediction of the Dirac theory and the actual magnetic moment was measured with high precision from the spin precession of a charged particle in a homogeneous magnetic field. The next order quantum electrodynamic correction stems from a single Feynman diagram which describes electron self-interaction. Here, theory and experiment agree at the level of 1 in 10^8, with a tiny discrepancy between theory and experiment in the eighth digit.[50] In such a case, the experiments are *for all practical purposes* capable of singling out the real effect of a single

---

[50]Lohrmann 1992, 109. Taking several higher orders into account does not change this result; see Perkins 2000, 41–42.

Feynman diagram (or virtual field quantum). The case of the Lamb shift is similar. Here, the next order of perturbation theory gives a correction based on two Feynman diagrams, namely for vacuum polarization and electron self-interaction. The correction shows that only 97% of the observed Lamb shift can be explained without the vacuum polarization term. A textbook on experimental particle physics tells us therefore that the missing 3% are "a clear demonstration of the actual existence of the vacuum polarization term."[51] Any philosopher should counter that this is *not really* the case. The virtual field quanta involved in this term cannot be *exactly* singled out.

Hence, the above conclusions remain. Virtual particles are formal tools of the perturbation expansion of quantum field theory. They do not exist on their own. Nevertheless, they are not fictitious but rather produce *collective effects* which can be calculated and measured with high precision. The Lamb shift in the hydrogen spectrum or the anomalous magnetic moment of the electron are such collective effects. Another collective effect of virtual particles is the mass and charge renormalization to which the higher orders of perturbation theory give rise. The renormalization procedure results in the so-called 'physical' particles, that is, in collections of units of mass and charge which *differ* from the mass and charge of the respective non-interacting particles, the so-called *effective mass and charge*. Formally, this collective effect is due to vacuum polarization. Several physicists reify it by saying that the 'naked' electron is 'dressed' by a 'cloud' of virtual particles. This ontological *façon de parler* neglects the fact that this 'dressing' only takes place in interactions. The virtual particles always belong to such collective effects. In this point they are like the Fourier components of electromagnetic radiation or the epicycles of Ptolemy's planetary system. In several regards, however, they also are similar to the quasi-particles of condensed matter physics.

6.4.4 Quasi-Particles

The quasi-particles of condensed matter physics are excitations of a macroscopic many-particle system. Like virtual particles, they do not come on their own but belong to collective effects. However, here the relation between the cause and the effect is *the other way round*. Virtual particles do *not* have separable effects. A superposition of many of them generates a collective effect, namely the contributions to a scattering amplitude up to the nth order. Quasi-particles, however, *are* the collective effects of all charges or nuclei within a macroscopic atomic lattice such as a crystal. Under certain conditions they are even separable and localizable. Quasi-particles are charged or uncharged energy quanta which propagate through a solid and interact with each other *as if* they were single particles, in a sense that obviously needs to be qualified. They result from a collective excitation of *many* subatomic

---

[51]Lohrmann 1992, 108–109; my translation.

constituents of a solid, where *many* means a collection of the order of 6×10^{23} electrons and atomic nuclei.[52]

The basic theory is the quantum mechanics of a many-particle system. The constituents of a solid are electrons and atomic nuclei. They are particles in the sense of the quantum mechanical particle concept (QM). It is not obvious how the phenomenological properties of the different kinds of solids derive from this theory. The solids are classified in metals and various kinds of crystals distinguished in terms of chemical binding on the one hand, conductors, semiconductors, and insulators on the other hand. Many aspects of the behavior of conductors, semiconductors, and insulators can be described by means of good quantum mechanical approximation methods. The simplest approximation procedure is the Hartree–Fock theory, also called the self-consistent field or mean-field approximation. It aims at describing the behavior of one electron within the field generated by all the other charges of the constituents of a solid. This gives rise to the well-known band structure of solids (which for example underlies the model of a transistor). Here, it is assumed that a single electron moves independently in the field of all the other electrons and nuclei of the solid. This approximation is based on the collective effects that give rise to the appearance of the kinds of excitations called quasi-particles. In general, such effects are so large that they cannot be treated as perturbations. Therefore, the Hartree–Fock Hamiltonian does not usually give rise to a correct description of the ground state of a solid, i.e., the quantum state of lowest energy. The best result that can be achieved in this way is to give a surprisingly good approximation of the ground state energy. However, the most important quantity of a solid is not this energy but the energy difference between the ground state and the lower excited states.[53] It determines many macroscopic phenomenological features of a solid, in particular its magnetic properties.

In order to calculate this crucial energy difference, the methods of quantum field theory are very helpful. There is a far-reaching formal analogy between quantum field theory and the many-particle theory of solid state physics. The analogy is based on the fact that the so-called second quantization (the formal quantization of the many-particle Schrödinger wave function) and the quantization of a classical field give the same formal results. Since the starting point is a non-relativistic field equation, the result is a non-relativistic analogue of quantum field theory. Even in the relatively simple Hartree–Fock approach the analogy is very helpful. It permits one to get rid of the complicated antisymmetrized many-electron wave functions of a solid and replace them by the number representation of field modes of a quantum field in Fock space.[54] Electrons are only numerically distinguishable and obey Pauli's exclusion principle. Hence, only the occupation of the quantum

---

[52]For the following, see Anderson 1997.

[53]Anderson 1997, 97–99.

[54]See Anderson 1997, 15–28.

states of the electrons matters. The formalism is based on exactly the same creation and annihilation operators and commutation rules as given above for field quanta. Since electrons are fermions, the anticommutator has to be taken. Any electron is represented by a state $c_k c_k^+ \Psi_k$ that belongs to the occupation number 1 in the Fock space. According to this approach, independent or approximately uncoupled electrons have the properties of a Fermi gas.

Quasi-particles enter where the Hartree–Fock approximation becomes too crude. They are due to *collective excitations* of the electrons and/or nuclei of a solid. Collective excitations concern the solid as a whole, a collection of the order of 6×10^{23} electrons and atomic nuclei. This means that they are nothing but *macroscopic quantum states* of a solid. Most important are the lower excited states mentioned before. They are called *elementary excitations*. In a first approach they can be defined as analogues of field quanta, in terms of creation and annihilation operators q_k^+, q_k for quanta of a given momentum k:

> This, then, is a preliminary definition we can make of an elementary excitation: *An elementary excitation of momentum k is that operator which creates the lowest excited state of a particular type of momentum k from the ground state.*[55]

It has to be noted that Anderson calls the definition 'preliminary' because the quasi-particle concept is based on an *approximation procedure* and in a second step the interactions *between* quasi-particles have to be taken into account.[56] The excited states are "of a particular type" regarding the various *kinds* of excitation. These may be due to adding an electron, the slow vibrations of the ionic atom cores of a solid, spin fluctuations, optical excitations, etc. The lowest excited energy states which result from these different elementary excitations may differ substantially. They are calculated in perturbation theory. The approximation procedure is analogous to the perturbation expansion of a quantum field theory, including the use of Feynman diagrams. To first order, each of them is considered independently. The calculation shows that the elementary excitations have *approximately* particle properties in the sense of the quantum particle concepts of the preceding sections. In particular, they have (approximately) a quantized charge and/or spin, they are fermions or bosons, they behave *independently*, and they have some effective mass m_e. For their approximate independence and their effective dynamic properties they are called *quasi-particles*.

The *effective mass* m_e of a quasi-particle is interaction-dependent. It is associated with the energy E of a quasi-particle given by the energy difference between its state and the ground state of the solid. It is due to a collective effect that is analogous to mass renormalization in quantum field theory. As in quantum field theory, physicists like to say that quasi-particles are 'dressed'.

---

[55] Anderson 1997, 102. Anderson's momentum k is the quantity called $p = \hbar k$ in the other parts of this book.

[56] See below.

In solid state physics, this ontological *façon de parler* seems to be better justified since the effective mass is due to the interactions within a solid. (However, this point is debatable. The analogy between the field quanta of a quantum field and the quasi-particles of solid state physics induces a formal analogy between the vacuum state of a quantum field and the ground state of a solid. Therefore one might argue in turn that the vacuum of quantum field theory admits of a stronger realistic interpretation than suggested at the end of the preceding section.)

There are different kinds of quasi-particles. Some of them are localizable, others not. In particular, the following three quasi-particle concepts have to be distinguished:[57]

1. **Free Electronic Charge Carriers.** These result from adding an electron to a solid. They inherit their charge and spin from this electron, but their effective mass depends on their velocity. In a metal, the additional electron 'disappears' in the partially filled band of conduction valence electrons, i.e., its momentum distributes within the solid, being carried away by the band electrons and giving rise to a vanishing excitation energy. In a semiconductor or insulator the wave function of the added electron develops in a different way. After a certain time, a wave packet remains. The wave packet is centered around a central momentum value k_0 (which corresponds to the excitation energy) and has a width Δk. It is then localized within a region of size $1/\Delta k$ around some point r_0. This wave packet approximately carries the charge and spin of the electron. Hence, it behaves like a fermion.

2. **Phonons.** Like other density waves such as magnons or plasmons, these result from collective excitations of the nuclei. Phonons result from the slow vibrations of the ionic cores mentioned above. They are stationary density waves and as such they do not inherit any dynamic properties from physical particles. In particular, they carry no charge and no spin, being approximately bosons. Phonons behave like the harmonics of sound waves (hence their name). In many solids they are not localized. However, in crystals they may even be approximately localized at impurities and imperfections.

3. **Excitons.** Excitons occur in insulators. They are due to optical excitation. An electron of an ionic atom core in a crystal such as GaAs (which is of great importance for the telecommunications industry) is excited from a predominantly arsenide p state to a predominantly gallium s state. The excited electron can move in the p-band through the crystal. Due to Coulomb interaction it repels the s-electrons of the atoms it passes. Therefore, it draws its corresponding hole along behind it in the s-band. Such an electron–hole pair is a bound state of an electron and a hole that gives rise to a compound quasi-particle of charge 0, of a certain effective mass, and with bosonic behavior. Like free electric charges and

---

[57]See Anderson 1997, 102–104.

phonons, they may be localized under certain circumstances. Spin waves (which give rise to the magnetic state of a solid) and the Cooper pairs of superconductivity are similar to excitons.

There are many other kinds of quasi-particles that give rise to more conceptual differences and that cannot be mentioned here. To all of them, however, particle-like properties are attributed. In general, quasi-particles are excitations of a solid which *behave approximately like particles* even though they *are not particles*. To the momentum, charge, spin, and effective mass which they may carry, further particle-like features may be added. They do not behave only approximately as if they were independent of the rest of the solid. In several regards, they also behave approximately independently of each other. They superpose in such a way that they give approximately linear contributions to the energy of a solid, they behave to a high degree like fermions or bosons, and they interact as if they were particles. Along the lines of the preceding particle concepts, one may characterize the quasi-particles as follows:

(QP) Quasi-particles are (to a very good approximation)

(ESQM)	collections of *energy E, spin s, charge q*, and some *effective mass m*,
(DISC)	*discontinuous*, i.e., they come in quanta,
(PROB)	*probabilistically determined* by the many-particle Schrödinger equation or the Hamiltonian of a solid,
(COLL)	*collective excitations* of the ground state energy of a solid,
(INDEP)	*independent* of each other,
(WAVE)	in states that superpose and interfere,
(UNKR)	unsharp in momentum k and location r according to the *uncertainty relation* $\Delta k \Delta r \geq h/2$,
(COMM)	subject to certain *commutation rules*, i.e., commutators for bosons and anticommutators for fermions,
(NDIST)	*only numerically distinguishable*, giving rise to Fermi or Bose statistics (depending on the spin),
(CONS)	subject to *conservation laws*.

This list combines the properties of the particle concepts (QM) and (FQ). Quasi-particles behave like quantum mechanical particles or non-relativistic field quanta. Some of them have the electron charge and spin, others have charge and spin zero, others are bound systems of excitations. Many quasi-particles may even occur in localized states, in accordance with Heisenberg's uncertainty relation: (UNKR) is completely analogous to (UNPQ) of (QM), setting limitations on the momentum width Δk and the region Δr within which quasi-particles are found to be localized in measurements. Hence, in various regards they look like particles. The 'quasi' in their name is obviously due to the property (COLL). Quasi-particles are fake entities. They look *as if* they would come on their own but indeed they do not.[58] They emerge from collective excitations. This gives rise to an effective mass which is not fixed (it depends on the internal interactions of the solid as well as on external fields) and to the fact that the quantum particle properties (INDEP), (COMM), and (NDIST) are only approximately realized.

Since the quasi-particles only behave *approximately* in this way, their interactions have to be taken into account in a *second* step. They give rise to scattering and polarization effects. For example, a charged quasi-particle may lose a quantum k of its momentum to the core vibrations, or vice versa. Such an energy transfer is identical to the emission or absorption of a phonon. It can be described in terms of corresponding creation and annihilation operators q_k^+, q_k. There are many possible interactions of this kind. They are again calculated with the tools of quantum mechanical scattering theory. This gives rise to a perturbation series which is indeed in many regards analogous to the calculation of the S-matrix elements of interacting quantum fields. In particular, propagators, Green's functions, and Feynman diagrams for several possible scattering processes are involved. They superpose like the virtual processes or Feynman diagrams of quantum field theory. As in quantum field theory, these interactions also include contributions from the creation and annihilation of *virtual* quasi-particles. The analogy with quantum field theory goes further. Analogously to the non-empty vacuum of quantum field theory and its effects on the charge and mass of the field quanta, there is a feedback between the interactions within a solid and its ground state. There are always interactions between the vibrations of the ionic cores and the electrons of the valence band of a solid which *cannot* be taken into account in the approximation methods sketched above. They affect the difference between the lowest excited states and the ground state of a solid as follows:

> Thus a great deal of the electron–phonon interaction must be thought of as already included in the definition of the phonon excitation itself, just as the electrons which are being scattered are not really single independent electrons, but quasi-particles.
>
> The problem is remarkably like that of renormalization in field theory; the 'physical' quasi-particles and phonons we see are not the same at all as the 'bare' particles we can simply think about, and their experimental properties – energy, interactions, etc., – include

---

[58]Indeed, they are called quasi-particles for the same reasons as the particle tracks of Mott's model of 1929 may be called quasi-classical. According to Mott's model, quantum particles in the sense of (QM) cause sequences of position measurements which *look like* completely determined classical trajectories, even though they *are not*.

contributions from the cloud of disturbances surrounding the bare particles.[59]

This gives rise to effects which mean that the 'preliminary definition' of the 'elementary excitation' no longer holds. Due to the many possible scattering processes, the quasi-particle states are not *really* eigenstates of the Hamiltonian of the whole system. One way of dealing with this problem is the 'full renormalization' approach,[60] which is once again similar to renormalization in quantum field theory.

Up to this point, I have explained the definition of quasi-particles, its foundations, its advantages, and the technical problems it raises. I did this at some length for philosophers. Indeed it may seem to them somewhat long-winded. However, from the standpoint of the physicist, the explanation given above is rather superficial and the real, technical problems are neglected. Nevertheless, I must leave the physical problems in abeyance now and come finally to the philosophical questions. These obviously concern the *existence* of quasi-particles.

Are the quasi-particles of condensed matter physics as fictitious as the virtual particles of a single Feynman diagram in quantum field theory? Obviously *not*. In contradistinction to the former, they *come on their own as quantized appearances*, even though they are due to collective excitations. And it is not only due to the mathematical approximation methods of condensed matter physics that they enter the stage. Their calculation explains details of the energy band structure of a solid which the Hartree–Fock approach cannot explain. In addition, it predicts details concerning the charge structure of a solid with impurities and imperfections. All this can be investigated in experiments. And it is most successful in predicting the phenomenological macroscopic behaviour of solids, in particular the magnetic states or superconduction. As far as the quasi-particles are localized in a crystal, they exist in the same sense as the spatial structure of subatomic scattering centers in the non-relativistic domain of the scattering energy. They contribute to the quasi-classical charge distribution within a solid in the same sense as the squared amplitude of the many-particle wave function of atoms gives rise to a quasi-classical charge distribution (i.e., via the generalized correspondence principle[61]). This is particularly true of the unsharply localized wave packets

---

[59] Anderson 1997, 116. In the following, Anderson mentions that the analogy ends at the divergences of quantum field theory, which fortunately do not occur when calculating the interactions of quasi-particles. Anderson calls only the free electronic charges (type 1) quasi-particles, but not the above quasi-particles of type 2 and 3. I also call them quasi-particles in accordance with the use of the term in current condensed matter physics. This use is justified since all three types have in common that they are approximately independent, quantized, localizable, and due to collective effects.

[60] Anderson 1997, 120.

[61] See Sects. 4.3 and 5.4.2.

of charged quasi-particles or of phonons which clump around an impurity, making a localized target for the scattering of charged quasi-particles.

According to the analogy with quantum field theory, quasi-particles are on a par with the *real* field quanta of some given observable mass or energy. They are *real energy states* which underlie energy conservation, in contradistinction to the property (NCON) of the virtual particle concept (VP). Otherwise, it would make no sense at all to define *virtual* quasi-particles on the lines of (VP), in order to complete the analogy between the interactions of real field quanta and quasi-particles.

In the context of the philosophical debate on scientific realism, it was argued that quasi-particles counter Hacking's reality criterion: "If you can spray them, they exist."[62] It was claimed that quasi-particles do *not exist* even though it is possible to *spray* them, given that one may use them as markers, etc., in crystals even though they are fake entities. However, this line of reasoning misses the crucial point that quasi-particles are *real* collective dynamic effects of the constituents of a solid. As quantized energy states, they *exist*. They do *not* of course exist as *absolutely* independent entities, such as the classical particles were supposed to be. I agree that in this sense they are fake entities. But they exist as *relatively* independent collections of energy, charge, and spin. (Physicists like to regard them as *carriers* of these quantities. This, however, is a misleading fallback to the classical metaphysics of substance.) Due to the technical problems mentioned above, the quantitative prediction of their precise excitation energy may be bad. But this does not affect their existence as discrete energy states of a solid which may sometimes even be localized.

Indeed, quasi-particles are as real as a share value at the stock exchange. The share value is also due to a collective effect (as the very term indicates), namely to the collective behavior of all investors. The analogy may be extended. It is also possible to 'spray' the share value in Hacking's sense, that is, to manipulate its quotation by purchase or sale for purposes of speculation. Its free fall can make an economy crash, its dramatic rise may make some markets flourish. And the crash as well as the flourishing may be local, i.e., they may only affect some local markets. But would we conclude that the share value does *not exist*, on the sole grounds that it is a *collective effect*? Obviously, share values as well as quasi-particles have another ontological status than, say, Pegasus. Pegasus does not exist in the real world but only in the tales of antique mythology. But quasi-particles exist in real crystals, as share values exist in real economies and markets. Indeed, both concepts have a well-defined operational meaning, even though their *cause cannot* be singled out by experiments or econometric studies. But the same is unfortunately true of the cause of an electron track.

Therefore, the alleged 'home truths' which entity-realists only admit about real entities are slippery and delusive. If it is a home truth that elec-

---

[62]Gelfert 2003, against Hacking 1983, 22–25.

trons exist as entities in their own right (i.e., in the sense of substances-on-their-own),[63] and if this may be taken as a case against the existence of quasi-particles, all the worse for the philosopher's home truths.

6.5 The Parts of Matter

So far, the particle property (BOUND) of giving rise to bound systems has not been discussed. According to the current constituent models, matter consists of molecules, molecules of atoms, atoms of electrons and nuclei, nuclei of nucleons, i.e., protons and neutrons, and protons and neutrons of quarks and gluons. The sense in which these matter constituents form bound systems also changed with the transition from classical mechanics to quantum theory. The quantum mechanical particle concept (QM) gives rise to a generalized account of matter constituents, (MC), given below in Sect. 6.5.1. It is based on conservation laws, sum rules, and the dynamic properties of the particle concepts discussed in the preceding sections. They make it unproblematic to consider matter as made up of atoms and the latter of electrons, protons, and neutrons. Even though these subatomic matter constituents are indistinguishable, the sum rules for their mass and energy, spin, charges, and momentum give them a well-defined and testable operational content.[64]

But, what about the quarks? According to the current standard model of particle physics, the nucleons (protons and neutrons) are supposed to be composed of quarks and gluons. In comparison to the higher-level constituent models of matter the quark model has several peculiar features. Until the rise of the quark model the electron was associated with the smallest unit of electric charge. Quarks are supposed to carry fractional charges, namely $\pm 1/3$ or $\pm 2/3$ of this charge unit. Unlike the electrons, they have never been detected as free particles in unbound states. They are not localized as charged particles (or collections of mass and fractional charge) in the sense of the operational particle concept (OP). But they are considered to be *real* field quanta which belong to the incoming and outgoing physical states of subatomic scattering process and the corresponding S-matrix elements. Hence, the applicability of the particle concepts of this chapter to the quarks needs further investigation.[65]

6.5.1 Matter Constituents Generalized

Let us start with classical atomism and the bound systems of classical mechanics before we proceed to quantum mechanical bound systems. Traditionally, atomism means the assumption that all material things consist of small-

---

[63]See Gelfert 2003, referring to Hacking 1983, 22–25 and 265.

[64]See Sect. 6.5.1 and Appendix E.

[65]See Sect. 6.5.2 and Appendix E.

est indivisible constituents. Ancient atomism can be traced back to Democritus and Epicurus. At the beginnings of modern science, it was taken up by Gilbert and Galileo. Later, atomism was defended by Boyle, Newton, Locke, and Boscović. Until the late 19th century the atoms were inaccessible to experimental methods. They were most controversial objects of metaphysical debates such as the Leibniz–Clarke debate or, two centuries later, Mach's opposition to Boltzmann and Planck's atomism. Traditional atomism embraces two assumptions which one should strictly distinguish from each other. The *first* is that matter consists of discrete microscopic constituent parts. This gives rise to a theory of the constituent structure of matter. The *second* assumption is that there are ultimate constituent parts which are atoms in a literal sense (the Greek word $\alpha\tau o\mu o\varsigma$ means *indivisible*). But the search for the ultimate foundations of physics is no more than a regulative principle of theory formation. Here, we only deal with the first assumption.

Since ancient natural philosophy, atoms were thought to be extended and impenetrable. Extension is a spatial predicate, and impenetrability a dynamic predicate. A constituent model of matter should explain both properties. Since the 18th century, all atomistic models of matter have explained the spatial extension of matter in terms of a physical dynamics. Kant and Boscović laid the foundations for this by explaining the dynamical structure of matter in terms of attractive and repulsive forces. These forces act between the parts of matter in such a way that pointlike matter constituents may form an extended bound system.[66] Classical point mechanics makes this idea precise as follows.

The state space of a classical N-particle system is the Cartesian product of the N one-particle state spaces, the phase space $\mathbb{R}^{6N}$. The state function of the system is a trajectory, $\varphi : \mathbb{R} \to \mathbb{R}^{6N}$. It is unambiguously fixed by the dynamic equations of the N particles. For every instant t it coordinates the $3N$ positions x_k^i and the $3N$ momenta p_k^i (where $i = 1, \ldots, N$; $k = 1, 2, 3$):

$$\varphi : t \mapsto \left(x_1^1(t), \ldots, p_3^N(t) \right) . \tag{6.12}$$

The N-particle state factorizes into one-particle states. They are obtained by projecting the trajectory $\varphi(t)$ onto the one-particle spaces. In ordinary spacetime the one-particle states are disjoint, that is, they do not intersect.

The standard examples for classical bound systems are the solar system and Rutherford's atomic model. The respective binding forces are gravitation and the Coulomb force. The spatial extension of a bound system is finite if the one-particle trajectories are periodic, it is the volume surrounded by them. In addition, the description of a bound system includes *sum rules* for

---

[66]Surprisingly, this idea did not have any consequences for the 17th century debate on atomism. Kant's second antinomy of pure reason is still exclusively expressed in terms of a spatio-temporal part–whole relation. See Kant 1781/87, A 434–437/B 462–465; Falkenburg 2000, 227–239.

the conserved dynamic quantities of the system and its constituents. In particular, the momentum P of the system results from the N particle momenta p^i according to $P = \sum_i p^i$ and the mass M and charge Q of the system sum up from the individual particle masses and charges according to $M = \sum_i M^i$ and $Q = \sum_i Q^i$ $(i = 1, \ldots, N)$, respectively.

Quantum mechanical particles in the sense of the particle concept (QM) also form bound systems. In order to describe atoms, the classical Coulomb potential of the atomic nucleus is put into the Schrödinger equation for the wave function of the electrons. The one-particle solution gives rise to the quantum mechanics of the hydrogen atom. The quantum mechanics of a bound many-particle system is only required for the purposes of atomic physics beyond hydrogen (starting with helium), molecular physics, nuclear physics, and condensed matter physics.

The state space of a quantum mechanical N-particle system is the the tensor product of the N one-particle Hilbert spaces $\mathcal{H}^i$, the Hilbert space $\mathcal{H}^N = \mathcal{H}^1 \otimes \mathcal{H}^2 \otimes \ldots \otimes \mathcal{H}^N$. The corresponding state function is the (totally antisymmetrized) quantum mechanical N-particle wave function Φ^N. It is obtained from the one-particle states $\phi_{k_i}^{(i)} \in \mathcal{H}^i$ by tensor product formation and antisymmetrization (for fermions) or symmetrization (for bosons):

$$\Phi^N = \sum_{k_1, k_2, \ldots, k_N} c_{k_1, k_2, \ldots, k_N} \phi_{k_1}^{(1)} \otimes \phi_{k_2}^{(2)} \otimes \ldots \otimes \phi_{k_N}^{(N)} \pm \text{permutations}. \quad (6.13)$$

The spatial extension of the bound system is explained from the probability density $|\Phi^N(r)|^2$. As discussed in Sects. 4.3 and 5.4, a generalized correspondence principle permits one to interpret this probability density as a quasi-classical charge distribution. Due to (anti-) symmetrization, Φ^N does *not* factorize into well-defined one-particle states. The constituents of a bound quantum mechanical N-particle system are *entangled*. It is not possible to individuate them spatio-temporally. They are only numerically distinct.

Now, fermions and bosons have to be distinguished. Fermions are particles of half integer spin, such as the electron, proton, and neutron. Bosons are particles of integer spin, such as the photon, the α-particle, or the unionized helium atom. According to Pauli's exclusion principle, the dynamic states of fermions are distinguishable in terms of their quantum numbers. For this reason it is possible to define dynamic one-particle states of a bound quantum mechanical system in Fock space. They are simply eigenstates of the occupation number operator.[67] For bosons this is not the case. Many of them may occupy the same quantum state, giving rise to coherent collective behavior as in laser light or in the Bose–Einstein condensate. The photons and many other bosons do not form bound systems. Nevertheless, there *are* bound systems of bosons. For example quantum chromodynamics postulates bound systems of gluons, the particles which give rise to the binding forces

---

[67]See Sect. 6.4.4.

between quarks. In addition, any bound quantum mechanical system that consists of two fermions is a boson.[68] Hence, the different statistical behavior of fermions and bosons does not give rise to a major distinction in terms of matter constituents and interaction quanta.[69]

In order to identify the 'single' particles which belong to a bound quantum mechanical system (or an entangled system in general), one has to destroy, uncouple, or disentangle the system by detecting one of its parts. This is not only true of the notorious EPR correlations in an entangled 2-photon system[70] but also for bound quantum mechanical systems. There are no non-destructive measurements of the position of an electron within the atom. Only at the level of whole atoms has it been possible to localize individual parts of matter at the surface of crystals, by means of an electron microscope. What is observed here is once again the quasi-classical charge distribution to which a quantum mechanical many-particle system gives rise in a macroscopic environment.

However, a quantum mechanical bound system underlies conservation laws just like a classical bound system. Its conserved quantities are the energy, momentum, angular momentum, spin, and charge of the bound system and its constituent parts. They give rise to sum rules for the respective magnitudes of an atom and its constituent parts. These sum rules take into account other physical laws such as Einstein's mass–energy relation $E = mc^2$. In particular, the following sum rules hold for the momenta P, p^i, the masses M, m^i and binding energy E_B, and the electric charges Q, Q^i for an atom or atomic nucleus and its n constituent parts $(i = 1, \ldots, n)$:

$$P = \sum p^i , \tag{6.14}$$

$$M = \sum M^i - \frac{1}{c^2} E_B , \tag{6.15}$$

$$Q = \sum Q^i . \tag{6.16}$$

Classical bound systems and coupled quantum systems therefore have the following properties in common. The spatio-temporal structure of a bound system is explained in terms of the dynamics of its constituents. The system and its constituent parts are related by well-defined conservation laws and sum rules. They apply to the dynamic quantities *mass, energy, spin*, the *generalized charges*, and the *momentum*. Indeed, such sum rules have a well-defined

---

[68] This is in particular true of the excitons that may occur in solids.

[69] Redhead 1988, 15–16, brings several arguments for and against taking the distinction of fermions and bosons as a distinguishing mark of matter constituents or their interactions.

[70] Einstein et al. 1935.

operational content and they establish well-defined *part–whole relations* for the constituent models of matter. In atomic and nuclear physics, they are predominantly based on mass–energy and charge which are conserved according to (6.15) and (6.16).[71] Even though quantum particles lack spatio-temporal individuality, this part–whole relation is well-defined and testable. For the quark model in addition the momentum sum rule becomes crucial. But before turning to this, let us express the properties of matter constituents in terms of another informal particle concept:

(MC) Matter constituents are

    (MESQ)   collections of *mass m, energy E, spin s*, and *charges q_i*,

    (BOUND)   able to form *bound systems*,

    (SUM)   subject to *sum rules* which derive from the conservation laws for mass, spin, generalized charges, and the momentum,

    (PART)   dynamic parts of a spatio-temporally extended whole.

The route to the quark model involves such sum rules (SUM) and their relations to the symmetry groups of the strong and electroweak interactions. Strict conservation laws also apply to angular momentum and spin. They are associated with the representations of the symmetry groups $O(3)$ and $SU(2)$, respectively. The corresponding sum rules stem from the quantum mechanical algebra of angular momentum. In 1932, Heisenberg defined analogous algebraic sum rules for the so-called *isospin*.[72] Isospin was conceived as a symmetry associated with the charge independence of the strong interactions of the proton and neutron. This was the first *dynamic* symmetry of subatomic particles, in contradistinction to the kinematic symmetries of the Poincaré group.

6.5.2 The Quark Model

Later, such dynamic or *internal* symmetries were associated with the concept of gauge invariant quantum field theories.[73] Today, they are related to the generalized charges *flavor* and *color* of the current *standard model* of particle physics. This in turn is associated with the symmetry group $U(1) \times SU(2) \times SU(3)$ and based on the quantum field theories of the electroweak and strong interactions, the Salam–Weinberg theory (quantum *flavor* dynamics) and

---

[71]It is possible to express this in terms of a mereological sum of disjoint dynamic parts of matter. See Appendix E.

[72]See Pais 1986, 423–425.

[73]Lyre 2004 explains what the textbooks of particle physics neglect.

quantum *chromo*dynamics.[74] Today, the electroweak interactions of quarks are associated with the generalized charge *flavor*. In addition, it is assumed that each of the flavors *up, down, strange, charm, bottom, top* comes in three *colors*. In accordance with the collection of dynamic properties (MESQ) of the above particle concept (MC), the standard model attributes to the quarks the following masses, spins, and charges:

masses: m_f (where f = u, d, s, c, b, t, are the six *flavors*)

spin: $s = 1/2$

charges: $q_e \in \{\pm 1/3, \pm 2/3\}$ (electric)

             $q_f \in \{u, d, s, c, b, t\}$ (flavor)

             $q_c \in \{red, blue, green\}$ (color)

The quark model arose step by step. In its various stages it came to satisfy more and more particle criteria. Nevertheless, in comparison to the higher-level constituent models of matter the quark model has retained some peculiar features up to the present day.

Its first version came in 1964. The quarks were associated with the irreducible representations of the $SU(3)$ symmetry of some unknown quantum field theory of strong interactions. (Expressed in terms of the current standard model, the hadron $SU(3)$ of the 1964 quark model was based on the first three *flavors* of the electroweak interactions, i.e., the *up, down*, and *strange* quark, and not on the SU(3) *color* symmetry of the strong interactions.) In the preceding years, this symmetry group was very successfully applied to the hadrons. It explained the exploding *particle zoo* of the 1950s and early 1960s by bringing systematic order into a multitude of kinds of particle tracks and scattering events found in experiments and cosmic ray data.[75] The increasing number of heavy particles (hadrons) and the collections of dynamic properties typical of them turned out to exhibit a structure corresponding to the representations of the symmetry group $SU(3)$. In 1961, the hadrons known around 1960 were classified according to multiplets corresponding to the representations of the $SU(3)$ symmetry of conserved quantum numbers such as isospin and strangeness. The corresponding conservation laws and sum rules for the isospin and other quantized magnitudes served to analyze the particle tracks and scattering events obtained from cosmic rays or at particle accelerators.

---

[74]The Salam–Weinberg theory has a complicated symmetry structure. In order to explain the non-zero quark and lepton masses, the concept of broken symmetry and the related Higgs boson are introduced. The existence of the Higgs boson, however, has not been experimentally confirmed up to the present day.

[75]This was explanation by unification (Friedman 1974) rather than by a deductive approach. The respective measurement methods are sketched in Sect. 3.3.3.

Once the symmetry $SU(3)$ had been associated with an attempt to *explain* the hadron multiplets, the classification gave rise to the quark model of 1964. It was assumed that the multiplets result from the composition of quarks of identical spin and parity, approximately the same mass, and different (fractional) electric charges. That is, the kinds of *hadrons* found in the analysis of particle tracks and scattering events were interpreted in terms of an approximate model of certain *reducible representations* of $SU(3)$. The early quark model explained this symmetry in terms of *three quarks* which corresponded to the *irreducible representations* of $SU(3)$. Hence, in the early quark model the part–whole relation between the quarks and hadrons was only based on the distinction of irreducible and reducible representations of a symmetry group. The constituent parts and the wholes made up from them were associated with the irreducible, respectively reducible representations of $SU(3)$. Hence, at that time the quarks were only considered to be particles in the sense of the group theoretical concept (GT). Their only operational basis was the conservation laws (CONS) and corresponding sum rules (SUM) of some generalized charges. What is more, they had to be attributed such odd dynamic properties as fractional electric charges. Therefore, on the basis of symmetries alone the quark model was not accepted. The nice group theoretical quark concept *lacked any operational basis* in the sense of the operational particle concept (OP). There was *no measurement of such particles*. All attempts to detect isolated fractional charges have been fruitless.

The first empirical evidence for the existence of pointlike nucleon constituents was found in 1968. The theoretical basis for this observation was the chain of models of pointlike scattering and form factors in the non-relativistic and relativistic domain explained in Sects. 4.3–4.4. The electromagnetic *structure functions* $W_{1,2}(q^2, \nu)$ of deep inelastic electron–nucleon scattering (which are related to the electromagnetic form factors of the proton or neutron) depend on two magnitudes: the 4-momentum transfer q (or its square) and the relative energy transfer ν. These express the relevant experimental knowledge about the dynamic structure of the nucleon. The crucial experiment in 1968 showed that the nucleon structure functions did not depend on both magnitudes independently, but only on their ratio. The measured structure functions showed scale invariance (invariance under change of energy scale), in straightforward analogy to Rutherford scattering.[76]

In 1969, Bjorken and Feynman developed the *parton model* of the nucleon in order to explain the experimental results. Later, it was further elaborated to the *quark–parton model* which identifies the pointlike nucleon constituents found in 1968 with the quarks of the 1964 model. According to both models, the hadrons are made up of either two or three quarks. Mesons (hadrons

---

[76]Bloom et al. 1969, Breidenbach et al. 1969; both reprinted in Cahn and Goldhaber 1989. In 1968 too, one of the phenomenological features of the measured cross-section was large-angle scattering, in perfect analogy with the backward scattering discovered in Rutherford's laboratory.

with a mass which is small compared to the nucleon mass) are made up of a quark and an antiquark. Baryons (hadrons with a mass which is at least of the order of the nucleon mass) are made up of three quarks. For any hadron, the fractional charges of the quark constituents sum up to an integer multiple of the electric charge. The most familiar hadron is the nucleon, i.e., the constituent of the atomic nucleus (the proton or neutron, unified by the concept of isospin). The other kinds of hadrons were either found in cosmic rays or generated in the scattering experiments of high energy physics. The breakthrough of the current standard model of particle physics came in 1974 when the J/Ψ resonance was detected. To it, the conserved quantity *charm* was attributed. It had been predicted as a fourth quark *flavor* within an extended quark model.[77]

Heisenberg still could not accept the quark model in 1976, even though he had opened up the way to it by conceiving of the isospin as a first dynamic or *internal* symmetry. He interpreted the *scaling invariance* of the nucleon structure functions in terms of *just another symmetry* of the elementary particles (or the all-embracing field theory he was in search of). In his famous essay *Was ist ein Elementarteilchen?* he criticized the underlying view of the part–whole relation. Following Kant's *second antinomy of pure reason*, he emphasized that in the last analysis the quark model is still based on interpreting the part–whole relation in *spatial* terms. He argued that such a view is problematic for two reasons. *First*, such intuitive terms do not apply to subatomic particles. *Second*, he claimed that it must be possible to decompose a composite system into constituent parts of a rest mass which is very large in comparison to the energy required to decompose the system.[78] His first objection was not really convincing. It neglects the dynamic structure of the constituent models of physics. As in any other model of a bound system, in the quark model the part–whole relation is understood in terms of dynamic quantities and sum rules.

His second objection was more serious. Relativistic quantum theory predicts the creation of particles (or real field quanta) from the energy transfer of a scattering experiment. Protons have a rest mass of about 1 GeV but the energy transfer in deep-inelastic lepton–nucleon scattering is typically of the order of 100 GeV. In the experiments which measure the nucleon structure functions, this energy transfer gives rise to a hadron shower, that is, to the production of *many* particles which are *much heavier* than the proton.[79] Thus, in the very scattering experiments which gave rise to the quark–parton model the proton (or neutron) was not decomposed into detectable constituent parts. Rather, new particles were created out of it which could

---

[77]See Riordan 1987, 210 and 294–321; Pickering 1984, 184 and 253–279.

[78]Heisenberg 1976, 5.

[79]The energy of this hadron shower is one of the quantities of a scattering event on which the measurement of a differential cross-section is based.

not be regarded as its constituents. Hence, Heisenberg's objection was still related to the missing operational basis of the concept of fractional charges.

The solution of the problem, now generally accepted, is the *confinement hypothesis*, a phenomenological hypothesis which was first suggested on theoretical grounds. It is related to the dynamic property *color* which was attributed to the quarks as an additional quantum number for the reason that otherwise Pauli's exclusion principle would be violated. According to this, in the quark model each quark *flavor* comes in three *colors*. The latter are the generalized charges of the strong interactions, giving rise to the name 'quantum *chromo*dynamics'. The bound states of quarks, however, only come in singlets according to the quark model and the phenomenological structure of the hadron multiplets. Therefore, in order to explain the non-existence of free or unbound quarks it was proposed that no quark states except the 'colorless' (singlet) states are stable. The potential $V(r)$ of the quark interactions is phenomenologically described by the following expression:[80]

$$V(r) = -\frac{4}{3}\frac{\alpha_S}{r} + kr , \qquad (6.17)$$

where the first term on the right side is a Coulomb-like term (dominating at short distances) and the second term expresses a constant force (dominating at greater distances). According to (6.17), the quarks are approximately uncoupled within the nucleon, i.e., in their bound states. Hence, the confinement hypothesis suggests that quarks have the following *paradoxical dynamic behavior*: they are *asymptotically free in bound states*, whereas the potential $V(r)$ which ties them together, or their *binding energy, rises with increasing distance r*.[81] But crucial deductive gaps remain between the quarks and the nucleons. It is still debated whether quark confinement is really explained from quantum chromodynamics. In particular, up to the present day, it has *not* been explained how quark confinement gives rise to the dynamics of bound systems.

The confinement hypothesis has enormous explanatory power. It explains the peculiar *jet events*, that is, single scattering events with two or more well-separated bundles of particle tracks which look like jets and belong to hadron showers. They are explained as follows. Due to the exceptional kind of force which gives rise to confinement, the distance of bound quarks increases with the energy of their state. Therefore, with increasing scattering energy of a collider experiment the probability for the following event sequence increases. A quark–antiquark state with a large internal distance is created. The energy of the state is so large that another quark–antiquark pair can be created. The

---

[80]See Perkins 1987, 20. α_S is the coupling constant of the strong interactions, k is a constant.

[81]In 2004, D.J. Gross, H.D. Politzer, and F. Wilzcek received the Nobel prize for developing the basic ideas of quantum chromodynamics, as a theory with asymptotic freedom.

latter quark recombines with the former anti-quark and vice versa, giving rise to two new bound states of lower energy. These appear as two unstable mesons coming out of the former quark–antiquark state, decaying into two spatially separated hadron showers which are sharply centered and directed in two opposite directions.[82] From a causal analysis of the individual tracks making up each shower or jet, the quark *flavors* of each such pole of the primary quark–antiquark state can sometimes be unambiguously identified. In this case, all possible decay modes of all possible quark–antiquark states except one can be significantly excluded, so that the remaining kind of process *must* be the one which actually happened and caused the observed individual jet event. In this way, a few years ago, experimental evidence for the *top* quark was announced, making the quark model empirically complete.[83]

Indeed, the causal analysis of such an unambiguous jet event may be considered as an *observation in the generalized sense* suggested in Sect. 2.5. The jet event is unambiguous if and only if it is clear that it does not belong to the background or *noise* of events of some other kind. (There are not many such significant events, but this does not matter. The main point is that such events are *reproducible*.) For such *gold-plated events*, both parts of the condition for a reliable observation are fulfilled: *First*, the data analysis of the particle tracks of the secondary hadrons shows that the quark content of the initial quark–antiquark state can be unambiguously determined. This means that the measured magnitudes of each of the two jets can be unambiguously related to a particular kind of quark of mass m, spin s, and flavor q_f. *Secondly*, the impressive computer picture of such a jet event is obtained from the record of a concrete scattering process which happened at a given moment in a concrete collider experiment, giving rise to a perfect causal story about what was going on inside the interaction zone. The story relates each jet unambiguously to an individual quark, which according to a well-established theory was one of the two constituents of a compound quark–antiquark system.[84]

---

[82]Three-jet events result from similar processes. Here, in analogy to the bremsstrahlung of quantum electrodynamics an additional higher-order effect of quantum chromodynamics is assumed, namely the creation of some gluon state which decays into a third hadron shower or jet.

[83]Abe et al. 1995; Abachi et al. 1995. See also Liss et al. 1997; Perkins 2000, 134–138.

[84]However, it may still be argued on Heisenberg's line of reasoning that in these jet events one only observes the mesons which stem from the fragmentation of the primary quark–antiquark pair, and *not* the constituent parts of the latter. According to this argument, only a quark–antiquark *pair* goes through as the cause of the individual causal story. See Fox 2006, Chap. 19. This point is obviously related to the concept of *distance* involved in the confinement hypothesis, a concept which Heisenberg surely would not have accepted due to its intuitive, heuristic, and analogical character.

Even though quarks never come on their own but only in bound states, they are considered to be dynamic parts of the nucleons in the sense of matter constituents (MC). The integral charge and the isospin of the nucleons result from the sums of the electric and *flavor* charges of three quarks, in accordance with simple sum rules (SUM) for the generalized charges of the quarks and nucleons. In addition, there are sum rules for the momentum. The structure functions $W_n(q^2, \nu)$ can be expressed in terms of weighted sums of the momentum distributions of quarks. The weights in these weighted sums have probabilistic meaning. They are nothing but squared scattering amplitudes from incoherent scattering of the probe particles at the different quarks inside the nucleons.[85] Indeed, they express the probability that a given quark q_f (or a given kind of quark *flavor*) within the nucleon carries (or is associated with) the relative amount $x = p_f/P_{\mathrm{nucl}}$ of the nucleon momentum, where x is the scaling variable of the structure functions.[86] In this way, the structure functions of deep-inelastic lepton–nucleon scattering are interpreted in terms of quark momentum distributions which belong to three out of the six quark *flavors*, namely the *up*, *down*, and *strange* quarks.

But some peculiar features of the quark–parton model still remain. For example, the *mass* and *flavor* eigenstates of the quarks are not identical. (This is due to CP violations. In the standard model this so-called *quark mixing* is expressed by the Kobayashi–Maskawa matrix.) And at very high scattering energies, scaling violations in the structure functions and the corresponding momentum sum rules indicate that there must be *more partons* than the three expected (so-called valence) quarks. According to the current version of the quark–parton model, gluons give rise virtual sea quarks (i.e., quark–antiquark pairs) and to other higher-order collective effects which add to the dynamic nucleon content.[87] Both effects show the wavelike or field-like nature of the nucleon constituents which is indeed nothing peculiar in the quantum domain. But a really striking feature is that according to the confinement hypothesis the quarks are *approximately independent inside the nucleons*. In contradistinction to the higher-level matter constituents, they are assumed to be *asymptotically free* in their *bound states*. Hence, even though they always belong to compound systems, they have approximately the (quantum) particle property (INDEP) of independence. The more one tries to uncouple them from their asymptotically free states, the harder together they stick, finally giving rise to fragmentation.

---

[85]The quarks are fermions. They are distinguished by their quantum state in terms of their three kinds of charges. This far it makes sense to talk of *different* quarks inside the nucleons. They are localized *inside* the nucleon according to the spatial resolution of the scattering experiment (see Sect. 4.5 and below).

[86]See Perkins 1987, 171–181.

[87]See Perkins 2000, 162–192; Povh et al. 1999, 107–111; and Sect. 8.3.

6.6 What Kinds of Particles Remain?

The quantum revolution shifted the original meaning of the term 'particle' to several kinds of quantum particles. But the classical particle concept has not been completely abandoned. Very large molecules or macroscopic dust particles still belong to the approximate models of the concept (CP). Even though in the case of macromolecules this may be a crude idealization, they behave classically. However, the borderline between the classical and the quantum particle concepts is fuzzy. Fullerenes, that is, C_{60} molecules, pass a double slit and generate an interference pattern as if they were electrons or photons.

Today, the particle concept is much more differentiated than it was in the early days of atomic physics. The term 'particle' is no longer fixed but has many uses which depend on the theoretical and experimental context. Bearing in mind the list of particle concepts developed above, we may ask: What are the predominant features of a particle, after all? Is there any rationale behind the bewildering uses of the term 'particle' in current physics? Or to put it in other words: Is there *any* general particle concept which may be considered as a 'rational generalization' of the classical particle concept, say, on the lines of Bohr's complementarity philosophy and Heisenberg's suggestion of an analogical use of the particle concept?

Physicists use the term mainly as an *informal* expression of ordinary language which denotes several physical phenomena and concepts. 'Particle' does not only refer to the phenomena of atomic, nuclear, and particle physics, giving rise to the operational particle concept (OP). It also refers to the mathematical tools of various quantum theories, such as the concepts of wave functions in (QM), photons (LQ) and other field quanta (FQ), virtual particles (VP), and quasi-particles (QP). In the case of real field quanta and quasi-particles, the relation to particle-like empirical phenomena and the operational particle concept is far from being obvious, but it is still there. Only in the case of virtual particles has this operational content been lost. Here, the particle talk just tells a fairytale about what goes on *virtually* within the *black box* of the scattering (even though this is a precise story with full internal mathematical rigor and an empirical happy ending lying in well-predicted transition probabilities). The other extreme is the group theoretical particle concept (GT) which is no longer specific to particles but applies to the solutions of *field* equations.

These various uses show *family similarity* in Wittgenstein's sense. But the aggregation of this family of particle concepts is by no means arbitrary. In biological families, the similarity of the family members is based on heredity. Analogously, the similarity between the members of the 'particle' family is based on some common heritage. All of them stem in one way or another either from the classical particle concept or from the associated measurement laws. They are still tied together in terms of the familiar physical quantities *mass and/or energy* and eventually *charge*. There are more relationships. The quantum members of the family are tied together by introducing in addition the spin and by quantizing the dynamic properties except mass. Operationally, these quanta correspond to the particle detections or position measurements of the operational particle concept. The latter states that the dynamic properties can be localized and that the localizations are independent of each other. All particle concepts (except virtual particles) share the property of being *independent* or uncoupled, and statistically uncorrelated.

Independence is a crucial particle property.[88] It grants the statistical independence of uncorrelated particles, even though the statistics change in the transition to the fermions and bosons of quantum theory. As for the matter constituents which make up bound systems, it turns out rather surprisingly that the *quarks* are *asymptotically free* and *independent* in their bound states. According to this, they are particles. The group theoretical particle concept (GT) includes independence too, even though only in the sense of applying to uncoupled fields. (GT) states that particles come in terms of the dynamic properties of mass, spin, parity, and some generalized charges and relates these dynamic properties to the symmetries and conservation laws of the theories behind the various particle concepts. As for localizability, group theory neither includes nor precludes it.

Hence, we have the following chain of generalization of the classical particle concept:

$$(CP) \rightsquigarrow (QM), (LQ) \rightsquigarrow (FQ), (QP) \rightsquigarrow (GT). \qquad (6.18)$$

On the left there are classical particles. On the right there are fields. In the middle are four members of the quantum family. (Their exact relations are not specified here, but obviously field quanta and quasi-particles are generalizations of light quanta and quantum mechanical particles.) The virtual particle concept is not amongst them. It is the only family member which really deviates. Virtual particles are neither localizable, nor independent of each other, and nor do they underlie energy conservation. Therefore they are not usually considered to belong to *physical* particles. One might be inclined to think the same of quasi-particles. But in contradistinction to virtual particles, they behave in many regards approximately like physical particles. For all these reasons, one should *not* include virtual particles in a concept of *physical particles*; but one *should* include *quasi-particles*. Hence, we might differentiate a *general* particle concept and a concept of *physical* particles. Making this distinction, the following properties seem to be typical of particles:

- **(GP) Generally, particles** are:

  (MESQ)   collections of *mass m, energy E, spin s*, and *charges* q_i,

  (DISC)   *discontinuous*, i.e., they come in quanta.

---

[88]Its various meanings, which have not been analyzed here, will be discussed in the general conclusions of this book (Sect. 8.5).

- **(PP) Physical particles** are in addition:

    (INDEP)   *independent* of each other,

    (CONS)   subject to *conservation laws*, ... and (hopefully) ...

    (LOCPD)   *localizable* by a particle detector.

The properties (MESQ) (or some shorter or longer list of dynamic properties) are shared by *all* particle concepts discussed in this chapter. (DISC) is shared by all *quantum* particle concepts as a theoretical property which is expressed in multiples of Planck's constant or of some elementary charge. Classical particles, however, also come in certain quanta of mass and charge. Only in quantum field theory does (DISC) acquire axiomatic status. Here, it is based on the commutation rules for the creation and annihilation operators.

The virtual particle concept (VP) is restricted to (MESQ) and (DISC). The concepts of *physical* particles in addition share at least (INDEP) and (CONS). These four kinds of property together express a great deal of the meta-theoretical, algebraic structure which underlies current physics and the heuristics of theory formation. (MESQ) and (CONS) are closely related to the unifying principles discussed in the last few chapters, in particular dimensional arguments, symmetries, and superselection rules.[89]

In (MESQ), particles are considered to be *collections* rather than carriers of these properties. This is on the lines of the operational particle concept. A metaphysical carrier of properties cannot be measured. Operationally, particles are conceived as collections of certain dynamic properties which go constantly together under well-defined experimental conditions. In the generalized particle concept proposed here, they are still conceived as Lockean empirical substances or bundles of properties. This concept is metaphysically modest, but not abstinent. Even a liberalized particle concept *must* be metaphysical. It is based on the realistic belief that there *is* an entity which appears as a stable bundle of properties in the phenomena. It has been emphasized in Sect. 6.3 that this is a metaphysical presupposition of the operational particle concept. Only stable collections of dynamic properties can give rise to reproducible experiments. This assumption involves the realism of properties suggested in Sect. 1.6, but it is even stronger. It claims that the dynamic properties of subatomic particles are tied together to property bundles with the status of natural kinds.[90] Obviously, this is a weakened version of the traditional metaphysics of substance. Due to the loss of spatiotemporal individuality and the indistinguishability of all quantum particles, the metaphysical carriers of the properties are cancelled, but some metaphysical glue is left which makes them stick together. This is an obstinate residue of the traditional concept of substance in particle physics.

---

[89] See Sects. 4.2 and 5.4.3.

[90] A related ontological approach is trope ontology. See, e.g., Seibt 2002.

However, this metaphysical glue is *inherent in the empirical basis of the particle concept* pointed out in Chap. 2. This is reflected by (LOCPD), a predicate that comes in dispositions even though they do not really work in the quantum domain.[91] Here, the classical property of being local is generalized to localizability, taking into account the operational possibilities of localizing electrons, photons, and other quantum particles by means of particle detectors. Here again the underlying idea is that there *is* a something which is localized by a position measurement. This idea is supported by the conservation laws (CONS), above all energy conservation. Operationally, the particle behind a track is *nothing but* the repeated localization of conserved dynamic quantities. This is *a lot* since we cannot but interpret the repeatability as indicating an underlying entity. But it is not *too* much. The particle concept may well be considered to play the role of Kant's schematized concept of a substance. According to Kant, the substances of physical science have to be interpreted in terms of conserved quantities.[92]

Another indispensable relic of the traditional metaphysics of substance is the *independence* property (INDEP). In classical mechanics, it implies that particles are entities on their own, similar to the way in which Descartes or Leibniz conceived quite differently of substances. To my knowledge, this crucial *independence* property, its various aspects, and its quantum restrictions have been completely neglected in the philosophical discussion of the particle concept. On the one hand, it is closely related to the statistical properties of several kinds of particles. On the other hand, in its broader sense of being uncoupled it is a meta-theoretical bridge between the concepts of particle and field.[93] In the classical particle concept (CP) of Sect. 6.1, the independence property was stated as follows. In the most general sense,

(INDEP) Independence means that particles

| (UNCOUP) | *may* be in *non-interacting* or uncoupled states, and |
| (UNCORR) | their *initial conditions* are *statistically uncorrelated*. |

In classical mechanics, the second aspect of independence derives from the first. Indeed, the property of independence is also closely related to the constructive features of the experimental method explained in Sect. 2.2 as well

---

[91] ... because dispositions are usually conceived of as local causal powers, whereas quantum phenomena and their causes are non-local.

[92] See Kant 1786, A 116 (Akad. 4.541–542) and von Weizsäcker 1971, 383–404.

[93] One might think that the pointlikeness of interactions is another meta-theoretical bridge between particle and field theories, i.e., the property (POINT) of the particle concepts discussed above. However, its meaning depends too drastically on the underlying theories. It involves very distinct concepts of locality and causality. This point is taken up in the next section. Independence is much easier to clarify in meta-theoretical terms, namely in terms of being interaction-free and showing uncorrelated statistical behavior.

as the goals of causal and mereological analysis. But (UNCOUP) and (UN-CORR) are more than mere construals. If the independence of physical objects and systems were *not* realized to a certain degree in empirical reality, the experiments of physics would not be possible and neither would the concepts of a physical dynamics be tenable.

In the quantum domain, however, together with the classical particle concept the independence property breaks down stepwise, too, and with the quantum particle concepts it becomes differentiated. Primarily, particles are considered to be non-interacting or uncoupled. This assumption (UNCOUP) is crucial for the group theoretical particle concept (GT), *without* having any statistical counterpart (UNCORR). Conversely, in the operational particle concept (OP) independence *only* means that the single particle detections are uncorrelated in the sense of (UNCORR), indicating the statistical independence of the single events that make up a quantum ensemble. But this is not all. In quantum theory, *both* aspects of the classical independence property eventually become problematic. At first glance, the opposite seems to be the case since the results of a quantum measurement are statistically independent if the measurement is not repeatable. But in addition there are various restrictions on (UNCOUP) and/or (UNCORR):

1. Classical statistics has to be replaced by Fermi and Bose statistics. This means that there are fewer possible quantum states than classical states. In addition, the antisymmetrization or symmetrization of the quantum mechanical many-particle wave function of a bound system gives rise to a dynamic effect, namely the so-called exchange energy contribution to the energy of a many-particle system.

2. In the case of quantum entanglement, due to conservation laws under certain experimental conditions, particle detections are no longer uncorrelated. A well-known example are the EPR correlations,[94] others are the quantum erasers in the recent *which-way* experiments discussed in Sect. 7.5.

3. Beyond first order perturbation theory, quantum field theory is in need of renormalization, for the very reason that the concept of independent, non-interacting particles is no longer tenable.

Nevertheless, the quasi-particles of condensed matter physics demonstrate how, due to collective effects in solids, new quasi-independent particle-like phenomena show up. Obviously, they underlie the above restrictions of (UNCOUP) and (UNCORR), too.

Localizability is crucial for the operational content of the particle concepts (CP), (QM), (LQ), (FQ), and (GT). It is *violated* by the quarks and in general in the relativistic domain. However, the quark model as well as the quasi-particle concept (QP) show that today localizability is no longer considered to be a *necessary* particle criterion. In the case of the quarks the

---

[94]Einstein et al. 1935, Aspect et al. 1982.

sum rules of the dynamic properties and momentum, the pointlike scattering behavior in a certain energy domain, and the causal stories associated with jet events are considered to be sufficient. In the case of the quasi-particles even a sharp momentum state suffices. The structure of quasi-particle states shows that in the quantum domain *discontinuity is ranked higher than locality or localizability*. Phonons and other quasi-particles are discontinuous in the same way as photons. They are *localized in momentum space*. Only under certain circumstances do quasi-particles become approximately local in a spatial region. Similarly, photons only become approximately local in a particle detector. Nevertheless, due to the independence property and the conservation laws which give rise to sum rules they may be considered to be physical particles. And under certain boundary conditions they exist in *approximately* localized states or (in the case of quarks) as dynamically discontinuous constituent parts of localizable bound systems.

As already mentioned, *any quantum theory prefers momentum states*. Without interactions, quantum particles are localized in momentum space rather than in spatio-temporal coordinates. Position measurements are interactions with a particle detector. Without them, a quantum mechanical state develops towards a plane wave. And without a local macroscopic measuring device, no reidentification of the same kind of particle along the subsequent measurements of a particle track is possible.[95] In addition, the localization of particles in a finite spacetime region is at odds with relativistic quantum theory.[96]

Surely subatomic physics deals with discontinuous, independent, localizable collections of mass-energy, spin, parity, and some generalized charges. But there is no all-embracing theory of them. The underlying entity (*if* there is one) has the hallmarks of a field rather than a particle. But one should also be *very* cautious with regard to any ontology of non-interacting quantum fields. As far as I can see, the commutation rules of the field operators preclude any obvious physical interpretation of the quantum field itself. What remains is the usual probabilistic interpretation of quantum theory. It applies to the S-matrix, that is, *to interactions alone*.[97] In addition, the *collective* effects of virtual particles remain. Their contributions to the dynamic structure of matter constituents can be measured.[98] The underlying quantum field is non-local because its relativistic structure is incompatible with local states.

---

[95] See Sect. 6.6.

[96] See Clifton and Halvorson 2002.

[97] This and other operational features of quantum field theory favor a relational ontology; see Falkenburg 2002a. At this point, the discussion of the ontological aspects of quantum field theory should begin; see Brown and Harré 1988; Auyang 1995; Teller 1995; Cao 1999; Kuhlmann et al. 2002.

[98] This becomes important in the quark–parton model at high scattering energies, where so-called sea quarks and gluons contribute to the spin and momentum of the proton and neutron.

All the particle concepts discussed in the preceding sections are based on approximations. The virtual particle concept is even *part* of an approximation method. All the other particle concepts include independence. But no physical object or system in the universe is *really* independent. The independence assumption is an approximation which in the last analysis has to be corrected. As the renormalization problems of quantum field theory *and* condensed matter physics show, for many problems this only works to the first order of perturbation theory. However, field theory does not do the full job either. Relativistic fields do not have local states. Perhaps the operational and axiomatic plurality of current physics commits one to ontological pluralism. At this point wave–particle duality needs to be reconsidered.

7 Wave–Particle Duality

Today, many physicists use the expressions 'particle' and 'field' more or less synonymously. This is an obvious consequence of the transformations of the particle concept discussed in the last chapter. Particle physics started from the classical particle concept and arrived at a group theoretical concept which deals with the conserved dynamic quantities of fields. The quantum particles discussed in Sects. 6.2 and 6.4 are neither particles nor fields in a classical sense. They have wavelike and particle-like features. Quantum mechanical systems are described in terms of plane waves and more or less sharply localized wave packets. Quantized fields may be either in a sharp number state or in a state of well-defined phase but not in both states at once. For photons, the coherent states with minimum unsharpness of occupation number and phase are closest to the description of a classical field.

The ambiguous spatio-temporal and dynamic features of quantum particles are often expressed in terms of *wave–particle duality*. Whereas some philosophers of science may think that this is obsolete, in quantum physics wave–particle duality is quite a current concept. Several physicists even like to use the term 'wavicle' in order to denote subatomic particles and their wavelike features. But this is not helpful. It conceals the precise meaning of wave–particle duality instead of revealing it.

Wave–particle duality is an informal concept. It does not belong to any specific quantum theory. Rather it expresses the way in which many physicists talk about the spurious referents of *any* quantum theory. Hence, it is a meta-theoretical concept which serves to interpret the current quantum theories, or a philosophical concept. But it belongs to physical practice. The task of this chapter is to investigate the semantics of this concept. Its function in physical practice is peculiar. It serves to link the axiomatic, operational, and referential aspects of quantum concepts which fall apart, given that the wave functions of quantum theory only have probabilistic meaning. In a certain sense, the concept of wave–particle duality bridges the gap between the probabilistic interpretation and the individual subatomic processes which happen in the measurement devices. We will see that this is partially due to Born's probabilistic interpretation of the quantum mechanical wave function and partially due to Bohr's complementarity interpretation of individual quantum phenomena.

The semantics of wave–particle duality developed with quantum physics. The following versions of the concept have to be investigated. The formal and operational grounds of wave–particle duality are Einstein's light quantum hypothesis, de Broglie's matter waves, and the corresponding quantum phenomena (Sect. 7.1). After the rise of quantum mechanics, more sophisticated versions of wave–particle duality came into play. Those related to Born's probabilistic interpretation of quantum mechanics, Heisenberg's uncertainty relations and Bohr's principles of correspondence and complementarity became most influential (Sect. 7.2). Born's account of probability waves and particle detections is used in physical practice up to the present day. A typical pragmatic attitude distinguishes waves and particles as follows. In the experiments of quantum physics, waves are prepared but particles are detected. This view is based on an asymmetry between the experimental operations of preparing and detecting quantum systems. The asymmetry is discussed for two striking experiments with single photons, namely the interference of two lasers which generate only one photon (Pfleegor–Mandel experiment, 1967) and the subsequent polarization of single photons in varying superpositions (Sect. 7.3). In recent quantum optics, the famous double-slit thought experiment of the Bohr–Einstein debate and Feynman's light microscope have been reconsidered. In 1991, Scully, Englert, and Walther proposed a double-slit experiment with path information (which-way experiment) and quantum eraser. Their idea was to store path information by means of quantum entanglement, against Einstein's recoiling slit and Feynman's claim that it is impossible to measure the path through the double slit without any substantial momentum transfer, finding a way round Heisenberg's uncertainty principle. This gave rise to a debate about the question of whether Bohr's complementarity principle is more fundamental than Heisenberg's uncertainty relation. As a way out of the debate, so-called duality relations were developed (Sect. 7.4). Recent which-way experiments have realized Scully's proposal of an interference experiment with path information and quantum eraser in several ways: employing a Mach–Zehnder interferometer (Ou et al. 1990), an atom interferometer with standing light waves (Dürr et al. 1998a,b, Dürr and Rempe 2000a), and a double slit with quarter-waves and circularly polarized light (Walborn et al. 2003) (Sect. 7.5). After all, it turns out that the concept of wave–particle duality is *not* obsolete, whereas the traditional metaphysical ideas about microscopic particles fail. In particular, a causal analysis of the experiments discussed in this chapter shows that the *causal* particle concept is no longer tenable. One of the reasons is that there are particle detections for which it is impossible to reconstruct an unambiguous causal story. The other reason is that there is no longer an unambiguous concept of causality but a *plurality* of causal concepts (Sect. 7.6).

7.1 Light Particles and Matter Waves

Historically, wave–particle duality traces back to Louis de Broglie. He was in search of a conceptual synthesis of particles and waves that unifies matter and light. In his thesis of 1923, he advanced the hypothesis that matter has wavelike properties which are completely analogous to the particle aspects of light.[1] Whereas the particle aspects of light were considered to be well-confirmed since the proof of the Compton effect in Bothe and Geiger's recoil experiment,[2] the wave aspects of matter were not yet. However, several physicists still attempted to avoid these hypotheses. In particular, the Bohr–Kramers–Slater (BKS) theory of the Compton effect deserves attention. It avoided the light quantum hypothesis at the price of dispensing with energy conservation for individual processes.[3] Indeed, the existence of light particles and matter waves was only established in the context of many experiments and tentative theoretical approaches. Nevertheless, de Broglie's hypothesis of matter waves was a decisive step towards quantum mechanics. It was taken up in Schrödinger's wave mechanics and Born's probabilistic interpretation of the squared amplitude of the wave function. After the rise of quantum mechanics, in 1927, Davisson and Germer observed interference fringes in an electron diffraction experiment, finally giving empirical support to the hypothesis of matter waves. In the same year Bohr expressed wave–particle duality in terms of complementarity, in his famous Como lecture.[4]

The historical development shows that wave–particle duality was put forward on the following axiomatic and operational grounds. Before the rise of quantum mechanics, the only formal bases of wave–particle duality were Einstein's light quantum hypothesis, its integration into relativistic kinematics, and de Broglie's hypothesis of matter waves. The crucial laws were the Planck–Einstein[5] relation $E = h\nu$ and the Einstein–de Broglie[6] relation $p = \hbar k$. These laws relate classical particle properties to classical wave properties. They relate energy to frequency and momentum to wavelength. Today, they belong to quantum mechanics and quantum electrodynamics. Therefore, they may be regarded as an uncontroversial formal basis of wave–particle duality up to the present day.

Operationally, wave–particle duality was confirmed by the particle-like and wavelike experimental phenomena to which the relations apply. For light,

---

[1] See Wheaton 1983, 286–301.

[2] See Sect. 3.2.2 and Wheaton 1983, 279–281, but also Greenstein and Zajonc 1997, 23–35, and my remarks at the end of Sect. 3.2.2.

[3] Bohr et al. 1924. Bohr worked on this theory *after Einstein had received the Nobel prize for the light quantum hypothesis*. See Jammer 1966, 183–187; Wheaton 1983, 281; Beller 1999. The role of the BKS theory in the development of Bohr's thought is investigated in Pringe 2006.

[4] Bohr 1928. See Sect. 7.2.2.

[5] Planck 1901, Einstein 1905.

[6] Einstein 1917, de Broglie 1923.

the Compton effect proved the particle properties predicted by relativistic kinematics, even though the classical diffraction phenomena indicate light waves. For matter, Thomson's e/m measurement and the particle tracks observed in the cloud chamber indicated that cathode rays or α-rays are due to massive charged particles, whereas Davisson and Germer observed a diffraction pattern when they sent an electron beam through a thin crystal of nickel.[7] Hence, the empirical basis of wave–particle duality is the experimental proof that particle tracks or interference fringes may be generated from the same kind of radiative source or particle beam. Which kind of phenomenon appears, a particle-like or a wavelike phenomenon, depends on the experimental devices. In a cloud chamber or a bubble chamber a beam of electrons or α-particles generates particle tracks and scattering events. In a Geiger counter γ-rays or high-energy photons generate a sequence of clicks. In a photomultiplier low-intensity light generates a sequence of electric pulses. In the Compton effect, a photon gives a recoil to an electron, loses energy, and comes out with lower frequency and longer wavelength. But at a double slit, the same kinds of charged particles or radioactive rays or photons are diffracted and give rise to the observation of interference fringes on a screen.

All statements about wave–particle duality have the empirical content that the way in which quantum systems behave depends on the way in which they are measured. Under certain experimental conditions electrons or photons behave in a particle-like way, and under others they behave in a wave-like way. The Planck–Einstein and Einstein–de Broglie relations $E = h\nu$ and $p = \hbar k$ relate these particle-like and wavelike phenomena in terms of physical quantities which can be measured.

7.2 Wave–Particle Duality in Quantum Mechanics

In quantum mechanics and quantum electrodynamics, these physical quantities are expressed in terms of observables which correspond to operators in Hilbert space, Fock space, and so on. These operators act on wave functions or field modes and they obey commutation rules such as Heisenberg's uncertainty relations. Quantum mechanics associates the de Broglie wavelength of massive particles with Schrödinger's wave function. In quantum electrodynamics the photons are considered to be the field quanta of the free Maxwell field, where the momentum or energy of a photon is associated with the wave number of the respective field mode. Both theories express the wave aspects of electrons or photons in axiomatic terms of wave functions, and the particle aspects in operational terms of individual measurement results. In physical practice, the only link between the axiomatic level of wave functions and the

---

[7]Davisson and Germer 1927. For an early visualization of wave–particle duality see Fig. 1.2 in Sect. 1.5, where the magnetic deflection of electron diffraction is shown.

operational level of individual particle detections is the usual probabilistic interpretation of quantum mechanics, quantum electrodynamics, and so on.

Therefore, as a *physical concept* wave–particle duality should primarily be explained in terms of the probabilistic interpretation of quantum theory. The foundations for doing so are laid in Born's seminal papers on the probabilistic interpretation of quantum mechanics. Born, Bohr, and Heisenberg, however, added several other ideas. These ideas not only gave rise to the never-ending philosophical debate on quantum mechanics *beyond* the probabilistic interpretation, but they also became influential within the scientific community of physicists. For the ways in which wave–particle duality may be understood, Bohr's concept of complementarity and Heisenberg's explanation of it in terms of classical analogies are crucial.

7.2.1 Born's Probability Waves

A strictly operational view of quantum theory is based on Born's probabilistic interpretation and von Neumann's extension of it to quantum mechanical expectation values.[8] According to the probabilistic interpretation of quantum mechanics, Schrödinger's wave function Ψ is nothing but a probability amplitude. Its square $|\Psi|^2$ is a probability density. It predicts the probability distribution of the individual measurement results. In the case of position measurements, it predicts the spatial distribution of particle detections. The probabilistic interpretation of quantum field theory is completely analogous. Quantum field theory only comes down to the earthly context of experimental results in terms of the S-matrix. The S-matrix elements are nothing but the amplitudes of transition probabilities.

When Born suggested the probabilistic interpretation of Schrödinger's wave function Ψ, however, he was still puzzled about the relation between the wavelike and particle-like features of quantum phenomena. In his seminal paper on the probabilistic interpretation of quantum mechanics, he obviously had the classical concepts of particles and waves (or fields) in mind when he expressed wave–particle duality as follows:

> The guiding field which is represented by a scalar function Ψ [...] spreads according to Schrödinger's differential equation. Energy and momentum, however, are transferred as if corpuscles were really flying around.[9]

Here, Born characterizes Schrödinger's wave function Ψ in terms of a ghost-like particle-guiding field or pilot wave. According to Jammer, this stemmed from Einstein's view of the relation between electromagnetic waves and the light quanta:

---

[8]Von Neumann 1932.

[9]Born 1926b, 803. My translation. Here, 'guiding field' is the literal translation of 'Führungsfeld'.

Einstein regarded the wave field as a kind of 'phantom field' [*Gespensterfeld*] whose waves guide the particle-like photons on their path in the sense that the squared amplitudes (intensities) determine the probability of the presence of photons or, in a statistically equivalent sense, their density.[10]

In addition, Born considers the energy and momentum transfer to be analogous to the propagation of particles along trajectories. These ideas were later taken up and literally interpreted in Bohm's hidden variable approach.[11] However, the results of the paper quoted above do not support such speculations. According to them, Ψ itself is meaningless, there are no particle trajectories, and determinism is at stake. Only the squared amplitude $|\Psi|^2$ has physical meaning. Since the probability amplitude Ψ is a wave, it was called a *probability wave*:

> [...] as Born pointed out, the "waves" which the wave–particle account portrays as spreading through space are "probability waves"; the square of the amplitude of the wave at any point in space gives the probability of finding the "particle" there.[12]

Born derived this interpretation of Schrödinger's wave function in the model of a scattering process. In this model, Ψ is the mathematical description of a diffracted wave, and $|\Psi|^2$ turns out to express the probability of detecting particles of a given momentum and at a given scattering angle. In this model, wave–particle duality is implemented as follows. The model is about the scattering of charged particles at atoms. (It corresponds to classical Rutherford scattering.[13]) The incoming particles are described by a plane wave which is diffracted at an atom. The scattering results are described in terms of the outgoing particles which are detected at some scattering angle. In this way, the scattering of particles is modelled by the diffraction of a wave. The wave propagates according to the Schrödinger equation, and it predicts the probabilistic distribution of the scattered particles. In the limit of infinitely many particle detections, $|\Psi|^2$ has the empirical meaning of a relative frequency or particle counting rate. Born's paper does not clarify the exact relation between the concepts of probability and relative frequency. But it is absolutely clear about the point that $|\Psi|^2$ is the theoretical correlate of the empirical relative frequency of the individual scattering results or measurement outcomes.

In contradistinction to this physical meaning of $|\Psi|^2$, the ideas of the guiding field and the propagation of corpuscles are fictitious. Operationally, one has to forget all speculations about pilot waves or particle trajectories.

---

[10] Jammer 1974, 41.

[11] Bohm 1952.

[12] Hughes 1989, 302, following the 'latency' interpretation Margenau 1954.

[13] See Sects. 4.2–4.4.

They cannot be *measured*. This corresponds to the probabilistic structure of quantum mechanics, as Born already emphasized in his preliminary communication on the quantum mechanics of scattering:

> From the standpoint of our quantum mechanics there is no quantity which in any *individual case* causally fixes the consequence of a collision; but also experimentally we have so far no reason to believe that there are some inner properties of the atoms which condition a definite outcome for the collision.[14]

On grounds of the probabilistic interpretation, wave–particle duality is the duality of probability waves and particle detections. From an operational point of view, the particle-like and wavelike properties of electrons or photons are indeed very distinct. In *individual* subatomic position measurements, particle tracks, and scattering events, only the *particle-like* properties of electrons, photons, and other subatomic particles can be measured. The *wavelike* properties only show up in a *probabilistic ensemble* of many particle detections. For a large number of position measurements, quantum waves and their squared amplitudes predict the relative frequencies of particle detections. Hence, the wavelike properties of electrons or photons are measured at the ensemble level, and the particle-like properties at the level of the individual measurement results. In order to observe the diffraction pattern behind a double slit or a thin crystal, one needs *many* particle detections. They may stem either from a high intensity beam of laser light, electrons, etc., or from a low intensity beam in a diffraction experiment which runs over a very long time.

Therefore, Born's probabilistic interpretation of the quantum mechanical wave function suggests the following view of wave–particle duality. *Operationally, there are particles. Axiomatically, there are fields and waves. The waves predict the probability distribution of the particles.* In scattering experiments, there are individual particle detections. In quantum mechanics or quantum field theory, the scattering is described in terms of wave functions or quantized field modes. The two meet in the cross-section of the scattering.

This view may also be expressed in terms of the informal particle concepts developed in the last chapter. The detection of particles gives rise to the operational particle concept (OP) of Sect. 6.3. The axiomatic correlates are various ways of describing quantum states in terms of wave functions, field modes, field operators, etc., which give rise to the various quantum particle concepts of Sects. 6.2 and 6.4. In terms of the informal predicates which make up these particle concepts, wave–particle duality simply tells us that quantum particles are

---

[14]Born 1926a, 51; translation from Wheeler and Zurek 1983, 54. See the full quotation in Sect. 6.1.

(LOCPM)   *localizable* by a position measurement, and

(WAVE)   in states which superpose and interfere.

The gap between the particle-like and wavelike properties of quantum systems is bridged by the probabilistic interpretation. Operationally, the quantum waves correspond to a large number of particle detections. However, the founders of the Copenhagen interpretation of quantum mechanics did *not* explain wave–particle duality in this way. They introduced it in a heuristic way which traces back to the Planck–Einstein and Einstein–de Broglie relations $E = h\nu$ and $p = \hbar k$ and to the application of Heisenberg's uncertainty relations to individual subatomic scattering processes.

7.2.2 Bohr's Complementarity View

Bohr's Como lecture of 1927 expressed wave–particle duality in terms of complementarity. This concept became a cornerstone of the Copenhagen interpretation of quantum mechanics.[15] In order to understand what it means, we have to take a step back to the operational content of the Planck–Einstein and Einstein–de Broglie relations $E = h\nu$ and $p = \hbar k$. They were confirmed by quantum phenomena such as particle tracks and energy–momentum conservation in the Compton effect *versus* the diffraction patterns generated by light rays and electron beams.

Bohr called such phenomena *complementary*. His complementarity view of quantum mechanics is based on the correspondence principle.[16] According to this, complementary quantum phenomena correspond to *classical* concepts or models of the phenomena which *mutually exclude and complement each other*. Bohr thought that a full understanding of quantum phenomena is only possible in terms of corresponding classical concepts and models. According to him, the experimental conditions under which particle tracks and scattering events are observed are complementary to the conditions under which interference fringes appear.

Bohr's concept of complementarity changed over time. From the very beginning, however, it embraced several conceptual levels.[17] Only in his later writings did Bohr make it explicit that mutually exclusive experimental devices generate complementary phenomena. This seems to be the last concep-

---

[15] According to Howard 2002, however, what is today called the Copenhagen interpretation is due to Heisenberg, whereas Bohr's own view is just the complementarity view.

[16] See Meyer-Abich 1965, Falkenburg 1998, Pringe 2006.

[17] A careful analysis is found in Scheibe 1971, 29–35. See also Meyer-Abich 1965; Murdoch 1981; Folse 1985. Greenstein and Zajonc 1997 give a short systematic account in the context of quantum optics.

tual stage.[18] In the famous Como lecture, Bohr introduced the concept as follows:

> The very nature of the quantum theory thus forces us to regard the spacetime coordination and the claim of causality, the union of which characterizes the classical theories, as complementary but exclusive features of the description [...].[19]

Later in the Como lecture, Bohr derives this complementarity of spacetime coordination and causality from wave–particle duality in the sense of Sect. 7.1. The derivation begins with the Planck–Einstein and Einstein–de Broglie relations $E = h\nu$ and $\boldsymbol{p} = \hbar\boldsymbol{k}$. Bohr emphasizes that they link the quantities E, $\boldsymbol{p}$ of the classical particle picture and the quantities ν, $\boldsymbol{k}$ of the classical wave picture:

> In these formulae the two notions of light and also of matter enter in sharp contrast. While energy and momentum are associated with the concept of particles, and hence may be characterised according to the classical point of view by definite spacetime co-ordinates, the period of vibration and wavelength refer to a plane harmonic wave train of unlimited extent in space and time.[20]

These remarks indicate that Bohr regards the relations $E = h\nu$ and $\boldsymbol{p} = \hbar\boldsymbol{k}$ as the origin of wave–particle duality, as suggested in Sect. 7.1. In his view, wave–particle duality is based on the existence of the wavelike and particle-like quantum phenomena which underlie the respective relations. Immediately after introducing his concept of complementarity with the claim about spacetime coordinates and causality which was quoted above, Bohr explains his view of wave–particle duality. He summarizes the phenomena supporting the classical theory of electromagnetic waves, Einstein's light quantum hypothesis, the existence of electric corpuscles, and de Broglie's hypothesis of matter waves as follows:

> In fact, here again we are not dealing with contradictory but with complementary pictures of the phenomena, which only together offer a natural generalization of the classical mode of description.[21]

---

[18]Bohr 1937 and later writings; see Scheibe 1971, 31–32; or Redhead 1987, 50: "Complementarity is a relationship that exists between mutually exclusive QM phenomena. Although complementary phenomena cannot occur simultaneously, their mutual possibility is necessary for the complete description of quantum-mechanical reality." However, at least implicitly this is already assumed in the Como lecture.

[19]Bohr 1928, 90.

[20]Bohr 1928, 92.

[21]Bohr 1928, 91. Here, Bohr considers the wave and particle descriptions of free subatomic particles to be complementary. This is on the lines of the Planck–Einstein and Einstein–de Broglie relations.

Bohr's explanations indicate that he regards the relation between the wave and particle pictures as the *primary complementarity*. It should be noted here that Bohr's view of the complementary wave and particle pictures is more comprehensive than a mere wave–particle *duality*. Complementarity means that these pictures *mutually exclude and complement each other*. However, the complementarity of causality and spacetime coordination derives from this. Following de Broglie, Bohr considers wave packets and estimates their dispersion. In a heuristic consideration, he derives the following dispersion relations for time t and frequency ν respectively for the spatial coordinates x, y, z and the components $k_{x,y,z}$ of the wave vector:

$$\Delta t \Delta \nu = \Delta x \Delta k_x = \Delta y \Delta k_y = \Delta z \Delta k_z = 1 . \tag{7.1}$$

Together with $E = h\nu$ and $\boldsymbol{p} = \hbar\boldsymbol{k}$, he obtains an energy–time unsharpness and a momentum–position unsharpness that looks like the lower bound of Heisenberg's uncertainty relation:[22]

$$\Delta t \Delta E = \Delta x \Delta p_x = \Delta y \Delta p_y = \Delta z \Delta p_z = h . \tag{7.2}$$

This is the origin of the so-called energy–time uncertainty relation. But obviously the above derivation is only heuristic. The relations (7.1) belong to a classical wave model. To them, Bohr adds the quantum mechanical Planck–Einstein and Einstein–de Broglie relations in the spirit of his correspondence principle. The energy–time unsharpness in (7.2) does not have any obvious formal correlate in quantum mechanics. In contradistinction to the analogous dispersion relations for the momentum and position components, it does *not* correspond to an uncertainty relation of non-commuting quantum mechanical observables.[23] Bohr summarizes the physical content of (7.2) as follows:

> [...] according to quantum theory a general reciprocal relation exists between the maximum sharpness of the spacetime and energy–momentum vectors associated with the individuals. This circumstance may be regarded as a simple symbolic expression for the complementary nature of the spacetime description and the claims of causality.[24]

Together with his other considerations, this indicates that he associates energy–momentum conservation with causality. As mentioned before, in his view the energy–momentum representation of quantum systems is associated

---

[22] Bohr 1928, 92–94; equations (1) on p. 92 and (2) on p. 94.

[23] There is no time operator in quantum mechanics. The only formal relation $\Delta E \Delta \nu \geq h$ which can be derived in a quantum theory holds for the line width and the lifetime of unstable quantum states. The derivation requires quantum field theory. For a more detailed discussion, see Messiah 1969, Sect. 4.2.4.

[24] Bohr 1928, 95. The sense of 'symbolic' in Bohr's views is clarified in Pringe 2006.

with the classical particle description. Indeed, in the Compton effect a photon has a local collision with an electron, giving rise to an observable recoil of the electron and an observable energy loss of the photon. Bohr also emphasizes in the Como lecture that the experimental evidence of this phenomenon refuted the BKS theory of 1924 which attempted to save causality at the price of the assumption that energy conservation is violated for the individual scattering events and only valid at the probabilistic level.[25] And therefore, he assumes that Heisenberg's uncertainty relation $\Delta p \Delta q \geq \hbar/2$ and the dispersion relation (7.2) applies to individual scattering events:

> Heisenberg has given the relation (2) as an expression for the maximum precision with which the spacetime coordinates and momentum–energy components of a particle can be measured simultaneously. On one hand, the coordinates of a particle can be measured with any desired degree of accuracy by using, for example, an optical instrument, provided radiation of sufficiently short wavelength is used for illumination. According to the quantum theory, however, the scattering of radiation from the object is always connected with a finite change in momentum which is the larger the smaller the wavelength of the radiation used. The momentum of a particle, on the other hand, can be determined with any desired degree of accuracy by measuring, [...] provided that the wavelength of the radiation is so large that the effect of recoil can be neglected, but then the determination of the spacetime coordinates of the particles becomes correspondingly less accurate.[26]

Hence, Bohr interprets wave–particle duality as dual properties of individual particles. In his view, these dual properties are as well expressed by the relations $E = h\nu$ and $p = \hbar k$ as by Heisenberg's uncertainty relation. Due to them a position measurement changes the momentum to an unknown extent and vice versa. In this way, they give rise to the complementarity of spacetime coordination and causality (or momentum–energy conservation).

Hence, the systematic relationship between wave–particle duality in the sense of the Planck–Einstein and Einstein–de Broglie relations, Heisenberg's uncertainty relation, and Bohr's account(s) of complementarity may be summed up as follows. Complementary quantum phenomena are described in terms of complementary physical quantities. These quantities belong to the complementary classical descriptions of particles and waves which are brought together in the wave–particle duality relations $E = h\nu$ and $p = \hbar k$. In quantum theory, these quantities are related by Heisenberg's uncertainty relation and the heuristic time–energy uncertainty relation. And these uncertainties give rise to complementary descriptions of physical reality in terms of either spatio-temporal coordinates or the momentum–energy representation, which

---

[25] Bohr 1928, 90.
[26] Bohr 1928, 94.

in turn indicate a complementarity of spacetime coordination and causality. The latter is associated with momentum–energy conservation in the Compton effect for individual scattering processes. Bohr considered this account of complementarity to be a 'natural generalization' of the classical description of physical objects.[27]

After all, complementarity is a complex cluster concept. According to Bohr, it applies at several conceptual levels, expressing relations between:[28]

(0) the wave and particle properties of subatomic particles, as expressed in:
   (i)   the relations $E = h\nu$ and $p = \hbar k$,
   (ii) the dispersion relations (7.2), in the wave picture, and
   (iii) Heisenberg's uncertainty relation, in the particle picture;
(1) wave and particle features of the experimental phenomena:
   (i)   wavelike and particle-like quantum phenomena such as interference fringes and particle tracks,
   (ii) the respective experimental devices such as the double slit and the Wilson chamber,
   (iii) the corresponding classical wave and particle pictures, which are expressed in terms of
(2) spatio-temporal and dynamic quantities:
   (i)   spacetime coordinates and momentum–energy,
   (ii) spatio-temporal processes and dynamic conservation laws,
   (iii) 'spacetime coordination' and 'causality'.

Bohr is far from interpreting the uncertainty relations, the many facets of complementarity, and the underlying wave–particle duality in probabilistic terms, as Born's operational interpretation of quantum mechanics would require.[29] In particular, Bohr's understanding of Heisenberg's uncertainty relations as the reason for uncontrolled momentum changes due to position measurements and vice versa is based on applying the classical pictures of particle and wave in heuristic considerations to individual subatomic processes. Heisenberg himself presented this kind of heuristic reasoning in his

---

[27]Bohr 1928, 91. In addition, it can be shown in detail that he considered the complementarity between spatio-temporal and causal description as a rational generalization of Kant's account of physical reality. See Pringe 2006.

[28]Other complementarities which do not directly concern wave–particle duality have to be added (see Bohr 1928 and Pringe 2006): the possibilities of defining physical objects and measuring their properties; Schrödinger's wave mechanics and Heisenberg's matrix theory: "Indeed, the two formulations of the interaction problem might be said to be complementary in the same sense as the wave and particle idea in the description of the free individuals" (Bohr 1928, 111); and more.

[29]In the Como lecture, Bohr only *mentions* Born's probabilistic interpretation but he claims that "wave mechanics just as the matrix theory [...] represents a symbolic transcription of the problem of motion of classical mechanics adapted to the requirements of quantum theory and *only to be interpreted by an explicit use of the quantum postulate*" (Bohr 1928, 586; my emphasis).

paper on the uncertainty relations when he discussed the position measurement of an electron by means of a gamma microscope.[30] It was taken up in the Bohr–Einstein debate and in Feynman's suggestion of the light microscope.[31]

7.2.3 Heisenberg's Analogies

Heisenberg's 1930 book *The Physical Principles of Quantum Theory* made Bohr's ideas more precise. Heisenberg emphasized that wave–particle duality is based on interpreting the quantum phenomena in terms of classical analogies. He showed that the classical pictures or models of particles and waves still apply to quantum phenomena. However, Heisenberg also notes that when the term 'particle' is applied to the cause of a particle track, it no longer has a literal meaning. His 1930 textbook explains the particle concept of subatomic physics as follows:

> A track of water droplets appearing in the Wilson chamber directly represents the trajectory of a single particle. Here, a particle or corpuscle always has to be understood as something which moves like a mass point of classical mechanics.[32]

Here, the expression 'particle' is introduced in terms of *analogy*. For Heisenberg, a particle is something which moves *like* a classical mass point. This definition is operational. The particle trajectory is the phenomenon which is observed in the Wilson chamber. The phenomenon approximately underlies the law of the classical Lorentz force. The subatomic particle or quantum state which gives rise to the phenomenon is understood in terms of the track and the law of point mechanics that describes it. In this way, the classical particle concept still applies to the quantum phenomenon of a track in the Wilson chamber. But the application only has the character of an analogy. The track *looks as though* it is due to a classical particle insofar as it corresponds to a model of classical point mechanics. The same is true of the complementary phenomenon which appears under other experimental conditions, namely in diffraction experiments. It compares to a classical wave. But as Heisenberg also points out, the comparison of a quantum object with a classical particle or wave is always deficient. The classical analogues do not give exact theoretical descriptions:

---

[30]Heisenberg 1927, 174–175 and 198. See also Heisenberg 1930b, 21–23. In contradistinction to Bohr's Como lecture, Heisenberg's paper makes use of the probabilistic interpretation in the statistical calculation of a particle track (Heisenberg 1927, 186–188).

[31]Bohr 1949; Feynman et al. 1965, 1-4-1-9. From these thought experiments emerged the recent which-way experiments of quantum optics; see Scully et al. 1991 and Sects. 7.4–7.5.

[32]Heisenberg 1930a, 4; my translation. The English version is much shorter here. It does not contain any explanation of the particle concept; see Heisenberg 1930b, 4.

Now it is obvious that a thing cannot be a form of wave motions and be composed of particles at the same time – the two concepts are too different. [...] As a matter of fact, it is experimentally certain only that light sometimes behaves as if it possessed some of the attributes of a particle, but there is no experiment which proves that it possesses all the properties of a particle; similar statements hold for matter and wave motion. The solution of the difficulty is that the two mental pictures which experiments lead us to form – the one of particles, the other of waves – are both incomplete and have only the validity of analogies which are accurate only in limiting cases.[33]

Neither the classical particle picture nor the classical wave picture can capture all experimental effects of electrons or other subatomic particles. They only capture part of them. The complementary pictures of waves and particles are needed in order to explain the experiments. Nevertheless, both pictures are incomplete in two regards. First, the experiments do not prove that light or subatomic matter constituents have all of the classical particle or wave properties. Second, the complementary classical pictures of particles and waves are incomplete, too. They do not give a full understanding of light or matter. They rest on analogies. Therefore, they are only approximately valid, as far as the quantum phenomena correspond to classical phenomena and their models. Here, Heisenberg is far from claiming that the classical analogues give a complete account of the quantum domain. Quite on the contrary, he emphasizes that they are both incomplete.

Today it is obvious that the quantum phenomena which correspond to the classical wave or particle picture do not exhaust the quantum domain. There are many quantum phenomena *without* any classical correspondence or analogue. Heisenberg's views about wave–particle duality are perfectly compatible with this fact.

7.3 Prepare Waves, Detect Particles

The 'orthodox' or Born–von Neumann interpretation of quantum mechanics is probabilistic.[34] It underlies physical practice. It is taught in the universities and shared by all physicists as an uncontroversial basis of quantum physics. Obviously, the probabilistic interpretation is *not* identical with the Copenhagen interpretation of quantum mechanics. The latter is based on Bohr's Como lecture, Heisenberg's 1930 book on quantum mechanics, and the Bohr–Einstein debate.[35] In the preceding section it was shown that the two differ mainly in two points. The *Copenhagen interpretation* accepts the

---

[33]Heisenberg 1930b, 10; similarly, Heisenberg 1930a, 7.

[34]Born 1926a,b; von Neumann 1932.

[35]Bohr 1928; Heisenberg 1930; Einstein et al. 1935; Bohr 1935; Bohr 1949. According to Howard 2002, in addition Heisenberg's post-war views about the mea-

probabilistic interpretation but it neglects some of its crucial consequences. Above all, it neglects the idea that wave–particle duality and Heisenberg's uncertainty relations primarily have probabilistic grounds. Bohr and Heisenberg focus rather on the possibilities of applying these concepts to individual quantum processes in terms of analogies, correspondence, and complementarity. Hence, the Copenhagen interpretation employs the unifying bridge principles discussed in Sect. 5.4 rather than Born's probabilistic concepts. The *probabilistic interpretation* in turn does not care about such bridge principles, which indeed express inter-theoretical relations rather than aspects of quantum theory itself. Born's probabilistic interpretation of $|\Psi|^2$ and von Neumann's projection postulate explain how axiomatic quantum mechanics is related to the empirical phenomena. Wave functions belong to the axiomatic structure, and particle detections to the operational content of the theory. The gap between them is only filled at the ensemble level of many particle detections.

In the preceding chapters I argued that without unifying principles quantum mechanics and quantum field theory have a quite meager empirical content. In particular, the interpretation of subatomic structure and the construction of the scales of physical quantities turned out to be based on tacit use of a generalized correspondence principle in Bohr's sense and other unifying principles.[36] I claimed that there is a lot of tacit use of something like Heisenberg's generalized correspondence principle, in particular whenever the quantum mechanical wave function is applied to individual systems. In this regard, the Copenhagen interpretation has more empirical import than the strict operational Born–von Neumann interpretation. Indeed it enters into many heuristic considerations of quantum physics. Above all, the heuristic application of Heisenberg's uncertainty principle to individual subatomic scattering processes is quite usual; even though physicists know that strictly speaking it has probabilistic meaning. Neither the textbooks nor the articles make this explicit. With the exception of the Ehrenfest theorem, the bridge principles employed in the application of quantum mechanical wave functions to individual systems or the construction of semi-classical models are never explained. It seems that physicists do not need such explanations. But why? What do *they* think about the quantum mechanical wave function and the question of how it applies to the individual quantum processes which they investigate in their experiments?

Many experimenters have found a sloppy way to explain wave–particle duality as follows. They claim that in an experiment of quantum physics, they *prepare waves* and *measure particles*. Hence, they talk about the spurious probability waves of quantum mechanics *as if they were real, individual physical objects which can be manipulated in experiments.* They even talk in

---

surement process were decisive for what was finally called 'the' Copenhagen interpretation.

[36] See Sects. 4.2–4.5, and 5.4–5.5.

the same way about the more spurious field modes of quantum fields. Indeed, this is reminiscent of Hacking's reality criterion which was discussed in the contexts of the realism debate and the quasi-particle concept (QP), in Sects. 2.3 and 6.4. Are the quantum waves real because the physicists can 'spray' them? I will come back to this crucial question later.[37] For the moment let me just state that the sloppy way in which physicists talk about quantum waves is interesting. It is the implicit philosophy of physical practice, and it is therefore worth making it explicit.

The Nobel prize winner Wolfgang Ketterle once stressed this pragmatic philosophy of quantum physics in a popular talk about the Bose–Einstein condensate. He told the public that it is very hard to understand quantum mechanics but after several years of physical practice one gets used to *preparing waves and detecting particles*.[38] In this nice statement, the experimental procedures of preparation and measurement are both understood as empirical operations, but they are considered to be different.

7.3.1 What Makes the Difference

A closer look at this account of wave–particle duality reveals that it is based on the following axiomatic and operational aspects of quantum theory: (i) the usual probabilistic interpretation of quantum theory, (ii) the asymmetry of spacetime coordinates and momentum states, and (iii) a resulting asymmetry of preparation and detection:

(i) **The Usual Probabilistic Interpretation of Quantum Theory.** In order to have well-defined particle detections which correspond to a particle of given mass, energy, spin, and eventually charge(s), one has to *prepare* a well-defined quantum state $|\Psi\rangle$ and *measure* an observable O in this state. Preparation and measurement are usually performed by means of *different* experimental devices. Quantum theory gives the expectation value $\langle\Psi|O|\Psi\rangle$ and the probabilities $|c_k|^2$ of measuring the eigenvalues O_k of the observable O. These probabilities correspond to the relative frequencies of the single particle detections of value O_k, taken in the limit of infinitely many measurements or particle detections. Hence operationally, there are a certain number of particle detections of a given type, relative to the normalization of the prepared quantum state. In a high energy scattering experiment this is the number of detections relative to the intensity of the particle beam (the so-called particle flux).[39] This is a relative frequency. It corresponds to the transition probability between quantum states. Axiomatically, this is described in terms of

---

[37] See Sect. 7.6 and the general conclusions concerning the realism debate in Sect. 8.1.

[38] Talk given at the annual meeting of the German Physical Society in Hannover, March 2003.

[39] See the definition of the total cross-section of a particle reaction in Appendix C.

the expectation values of operators which act on wave functions, field modes, or the S-matrix.

(ii) **The Asymmetry of Spacetime Coordinates and Momentum States.** In Hilbert space, the representations of a quantum state in spacetime coordinates or in momentum space are equivalent. They just correspond to distinct choices of the basis. A sharp momentum state is related by a Fourier transform to the superposition of infinitely many states in the position representation, and vice versa. In this regard, spacetime coordinates and the momentum representation are thoroughly symmetric. However, the laws of a quantum dynamics *prefer momentum states*. According to the Schrödinger equation of non-relativistic quantum mechanics, any wave packet with a finite spatial distribution and finite momentum width spreads, whereas a plane wave of sharp momentum p or wave number $k = p/\hbar$ with an infinite spatial distribution is stable. The S-matrix of interacting quantum fields is also preferably calculated in terms of incoming and outgoing momentum states. In contradistinction to the occupation numbers, phases, and field strengths in given spacetime coordinates, the *momentum* is a well-defined observable of the quantum state of an *asymptotically free particle*. *For all practical purposes*, position states may be considered to exist immediately after position measurements. But they are not stable. According to the principles of relativity sharply or unsharply localized quantum states are even *impossible*.[40] Hence, the *natural* representation of quantum states is in terms of *momentum* states. Even though particles are approximately localized by position measurements, quantum theory describes them in terms of waves and wave packets which propagate in *momentum states* before and after the detections.

(iii) **A Resulting Asymmetry of Preparation and Detection.** As an immediate consequence of the asymmetry of spacetime coordinates and the preferred momentum states, the preparation and the detection of quantum states are asymmetric, too. For the purposes of the experimental method, *stable, reproducible quantum states have to be prepared*. But this is only possible for *momentum states*. This insight belongs to the safe theoretical background knowledge of all experiments of quantum physics. After a preparation procedure, waves or stable states of (approximately) sharp momentum states are expected to be there. And at most for a short time period after detection, (approximately) sharp localized states are assumed to be there. Therefore, the preparation procedures of subatomic physics predominantly aim at preparing particles or particle beams of a given dynamic type in well-defined *momentum* and *energy* states.[41] Only then are particles or approximately localized quantum

---

[40] Clifton and Halvorson 2002.

[41] Even for an atom which is caught in a trap and hence localized, the position does not matter. What matters is rather its energy levels.

states measured. This may happen by means of a photographic plate, a scintillator screen, an electron microscope which scans the charge structure of a surface, or a particle detector. It is surely true that preparation and measurement are on a par in being *experimental procedures*. In accordance with the projection postulate of quantum mechanics,[42] both reduce the wave function to some eigenstate of some observable. Wave preparation reduces it to a well-defined momentum state of a very small momentum width. Particle detection reduces it to an approximately localized wave packet. However, the two procedures have *different* goals. Preparation aims to produce momentum states of the electrons, photons, etc., of the beam of a particle accelerator or a laser which are as sharp as possible. Particle detection or measurement aims to localize particles of well-defined mass, charge, and spin within a well-defined spacetime region.[43]

Indeed the asymmetry of preparation and detection is typical of the macroscopic quantum phenomena which have been studied over the last few decades. These experiments lose all paradoxical features as soon as the probabilistic character of wave–particle duality is taken into account. In particular, this holds for the famous EPR correlations.[44] In the Aspect experiment, the wave function of an entangled two-photon state is prepared, whereas after a while two particles are detected. That is, two particle detectors measure the spin components or the polarization states of the photons at a spacelike distance. The distant branches of the experiment are connected by a correlation counter. Quantum theory predicts that Bell's inequality will be violated in certain kinematical regions. In order to prove this, however, it is necessary to detect a *large number* of correlated photon pairs for two different polarizations. All this is well known. In the present context only one point matters. The prediction of quantum theory is made in terms of a tensor product wave function which describes an entangled 2-particle state. The squared amplitude of the wave has the usual probabilistic meaning. That is, it predicts the relative frequency of measurement outcomes. In addition, it predicts that the corresponding conservation law is valid at the level of the individual processes or pair detections. (This is what Born already stressed in 1926 when he noted that the likewise conserved quantities energy and momentum "are transferred as if corpuscles really were flying around."[45]) The crucial measurement results which prove that Bell's inequality is violated are only obtained at the probabilistic level. They are given in terms of correlation functions of the relative frequencies of the correlated particle detections.

---

[42]Von Neumann 1932.

[43]If the momentum or energy is also measured, this is only approximately possible. However, the experimental devices usually have measurement errors which are 10 orders larger than the uncertainty relations.

[44]See Einstein et al. 1935.; Aspect 1982.

[45]Born 1926b, 803.

Another example is the measurement of the Bose–Einstein condensate.[46] Ketterle who claimed that in quantum physics one prepares waves and detects particles got the Nobel prize for the experiment. More examples will be discussed below.

After all, the resulting account of wave–particle duality only *seems* to apply at the level of individual subatomic processes. A closer look reveals that it comes in thoroughly *operational* terms. It is based on the experimental possibilities of generating particle-like and wavelike quantum states and phenomena. In addition, it is supported by the usual probabilistic interpretation of quantum theory. The particle-like properties of quantum phenomena are measured by a particle detector or screen which detects single photons or electrons. The wavelike properties become obvious from the diffraction of an electron or photon beam at a crystal or a double slit which gives rise to an observable interference pattern. To perform the double slit experiment, one needs coherent monochromatic laser light or an electron beam with a very small momentum distribution which comes close to a plane wave. In this way, a wave is prepared in a well-defined, sharp momentum state.

Then the screen which is placed behind the double slit detects single, uncorrelated photons or electrons. In this way a wave is prepared and particles are detected in the operational sense (OP) of Sect. 6.3. According to the probabilistic interpretation of quantum theory, in the limit of infinitely many events the relative frequency of the particle detections corresponds to the squared amplitude of the diffracted wave. If such an experiment runs for hours and hours, it is observed that the particle detections make up the typical double-slit interference pattern. Most strikingly, the interference fringes only show up at the probabilistic level, from many individual particle detections.[47] And it is well known that they differ substantially from the sum of two one-slit diffraction patterns.

In such an experiment, the electrons or photons are prepared as waves by means of the double slit, whereas at the screen they are localized or detected as particles. In the interference pattern which appears after *many* particle detections, the particle-like and the wavelike properties both appear at the probabilistic level. Indeed, in all experiments which demonstrate macroscopic quantum effects, the quantum system under investigation is prepared in a well-defined wavelike quantum state which propagates in a well-defined way, in accordance with the laws of some quantum dynamics. But the measurement results can only be obtained by particle detections, at least in a high precision measurement. And the squared amplitude of the wave function predicts the probability for detecting the particles in given states. The wavelike or non-local features of quantum phenomena only become manifest in the relative frequencies of particle detections. Hence, the dictum *prepare waves*

---

[46] See Dürr 2006.

[47] The double-slit experiment with single electrons is, e.g., reported in Greenstein and Zajonc 1997, 1–7.

and detect particles expresses the non-local features of quantum theory on the safe grounds of the probabilistic interpretation. The double-slit experiment with single electrons or photons is the paradigm case of such experiments.

However, the waves or sharp momentum states prepared by an experimental device are *not* directly measured. From a strictly operational point of view, they only give rise to *conditional probabilities*. But several preparation procedures seems to produce *more* than conditional probabilities, namely some kind of individual, unobservable, spurious subatomic entity. This will now be illustrated for two spectacular experiments with single photons. They rule out the idea of a particle trajectory or path.

7.3.2 Two Lasers, One Photon

In a certain sense, the spectacular experiments of recent quantum optics *prepare single particles*, namely a low intensity beam of electrons or photons. In order to get single particle detections, the beam intensity is strongly damped. It is so much damped that only one electron or photon remains in the matter wave or radiation field. One may also use an electron gun or a short pulsed laser which prepares wave packets in coherent states. It is even possible to prepare a photon field in a sharp number state. At first, the preparation of single photons in a sharp number state was only possible by generating EPR-correlated photon pairs and detecting the *other* photon.[48] This preparation was obviously associated with a measurement. Today, it is even possible to generate single photons without any entanglement and subsequent measurement.[49] One may argue that in this case particles rather than waves are prepared. The wave packet of a single electron after a position measurement spreads towards a plane wave. A sharp photon number state, however, is defined to belong to the field mode of a plane wave and it *remains* in a sharp number state as long as neither the preparation is changed, nor a measurement is made, nor any other absorption process occurs. Is this a case against the above claim that *waves* are prepared? To put it in different terms: Is it possible that a single photon experiment prepares a particle proper with a well-defined path and an unambiguous causal story that connects the preparation and the detection? (This is what hidden variable theories assume.)

For the Pfleegor–Mandel experiment, however, this assumption is untenable. It was performed in 1967.[50] Two independent lasers were made to interfere for a coherence time of 20 μs during which it was possible to keep their phases coherent. Repeated measurements with laser pulses of this duration from both lasers were made, recorded, and analyzed. The results showed

---

[48] Grangier et al. 1986. For a discussion of different single photon light sources, see Greenstein and Zajonc 1997, 32–35.

[49] Keller et al. 2004.

[50] Pfleegor and Mandel 1967. See Greenstein and Zajonc 1997, 43 (the nice title "Two Lasers, One Photon" of this paragraph is taken from there) and the discussion there, 43–53.

interference. Then, the lasers were attenuated so much that for each measurement with a very high probability *only one photon* remained in the radiation field of the experiment. The results still showed weak, but significant interference fringes. This means that the waves or field modes generated by the two lasers interfered, giving rise to an entangled state $\Psi_1 \otimes \Psi_2$, even though the occupation number of the interfering photon field(s) was only 1. Quantum field theory explains this result in terms of a superposition of interfering field modes of occupation numbers 0 and 1. The superposition is composed of one photon coming from laser L_1 and no photon coming from laser L_2, and vice versa:

$$\Psi_1 \otimes \Psi_2 = \frac{1}{\sqrt{2}}(|1_1\rangle|0_2\rangle + |0_1\rangle|1_2\rangle) \,. \tag{7.3}$$

In the wave picture, this is obvious. In the particle picture, however, it is absolutely hopeless to explain the result. None of the single photons which are finally detected at the screen and which in the long run make up an interference pattern actually has an unambiguous causal story. In a classical particle picture, one would assume that each photon either comes from laser L_1 or from laser L_2. But then, no interference would be possible. The quantum phenomenon shows that *all* photons must somehow have been prepared in *both* lasers. They are not in a mixture but in the superposition (7.3). A mixture of photons that stem either from laser one or from laser two would not give rise to interference. But this means that the single photons which are detected neither have a definite path nor an unambiguous causal story. The experimental results can only be interpreted on the lines of the probabilistic interpretation of quantum mechanics. Hence, the Pfleegor–Mandel experiment *prepares waves and detects particles* in the sense explained above.

7.3.3 Polarized Photons

The wave nature of photons does not only give rise to interference phenomena. It also shows up in polarization experiments. In quantum field theory, the polarization is an internal degree of freedom in the description of single photons. In the classical theory of radiation, the polarization is a wave property which is associated with the relative phase of the components of the electric field strength. The following simple but striking polarization experiment with single photons is analogous to the polarization of classical electromagnetic waves with crossed polarizers. Performed with ordinary classical white light, it shows that light consists of waves rather than particles. Performed with single photons, it reveals the non-local features of a quantum field.

Let white light from an ordinary thermal light source pass through three subsequent polarizers $P_|$, $P_/$, and P_- which are crossed with respect to one another at angles of $45°$ and $90°$. On a screen behind P_-, one observes whether light passes through or not. Behind the polarizer $P_|$ the light is polarized in the vertical direction, behind $P_/$ it is polarized at 45° relative to $P_|$, and behind P_- it is polarized horizontally. If $P_/$ is removed, the remaining polarizers are perpendicular to each other and no light passes P_-. Then nothing is seen at the screen. Even in the classical case it is amazing to see the light on the screen appear and disappear when the second polarizer is put in and out.

The quantum analogue is performed with a low intensity light beam coming from a short pulsed laser which emits single photons. The corresponding quantum phenomenon was first discussed in Dirac's textbook on quantum mechanics.[51] Instead of the light screen of the classical polarization experiment, a photon counter is now installed behind the sequence of polarizers. The beam intensity (or the pulse time of the short pulsed laser) should be so low that on average at most one photon at a time is in the radiation field. However, this does *not* necessarily mean that the quantized Maxwell field is in a well-defined number state with occupation number 1. This depends on the preparation. Even if by appropriate devices a field mode with occupation number 1 is prepared, any polarizer destroys this preparation if it does not agree exactly with the polarization state of the photon. In general, the polarizer $P_|$ prepares a superposition rather than a well-defined number state of the photon field.

In correspondence to the classical wave picture, quantum theory predicts that single photons are detected behind P_- if and only if the second polarizer $P_/$ is put in. In the quantum case this is striking. The polarizers obviously have the physical effect of selecting photons of a given polarization. They absorb all photons with perpendicular polarization. This seems to be a measurement. But, how can *three* absorbers let the single photons pass given that they cannot pass *two* of them? The answer has already been given above. The polarizers prepare wavelike modes in which the occupation number of the quantum field is *not* well-defined. They prepare completely *different* field modes with or without the second polarizer $P_/$. The photon counter, however, measures the average occupation number by detecting single photons.

The quantum light consists of photons of wave number k. Now, let the photon state $\Psi_{k,1}$ represent a field mode of this wave number and occupation number 1. Before it arrives at the system of polarizers, its polarization is unknown or undefined. For a quantum field, the occupation number and phase cannot both be sharp or well-defined at the same time. However, the highly non-classical properties of low-intensity light do not affect its description in terms of wavelike field modes which are occupied by photons of well-defined polarization. A state of well-defined polarization, however, is a state of well-defined phase difference. Hence, after a new polarization of the field mode has been prepared, the occupation number of the field mode can no longer be sharp or well-defined. Quantum field theory describes the effect of the

---

[51]Dirac 1958, 4–7. A nice computer simulation can be found at
http://www.physik.uni-muenchen.de/didaktik/Computer/interfer/.

polarizers in terms of field operators for the annihilation and creation of polarized field quanta, $a_|, a_|^+$; $a_/, a_/^+$; a_-, a_-^+. The polarizers $P_|$, $P_/$ or P_- prepare the polarized states $\Psi_|, \Psi_/, \Psi_-$. Let the field operators corresponding to these effects be $\mathcal{P}_|, \mathcal{P}_/, \mathcal{P}_-$.

Behind the first polarizer $P_|$ the light is polarized in the vertical direction. It is in the quantum state $\Psi_|$. $P_|$ reduces the wave function Ψ_ϵ of the short pulsed laser beam to a state of well-defined, vertical polarization, which no longer corresponds to a well-defined occupation number. In terms of the second polarizer $P_/$, this state is a superposition of a photon wave or field mode $\Psi_/$ which can pass $P_/$ (polarization of $-45°$, relative to $P_|$) and a photon wave or field mode $\Psi_\backslash$ which *cannot* pass it (polarization orthogonal to $P_/$). Hence, the effect of $P_|$ on Ψ_ϵ is:

$$\mathcal{P}_|\Psi_k = \Psi_| = \frac{1}{\sqrt{2}}(\Psi_/ + \Psi_\backslash) . \qquad (7.4)$$

Behind the second polarizer $P_/$ the photon field is in the state $\sqrt{2}\Psi_//2$, due to reduction of the wave function. In the basis of eigenstates of the polarizers $P_|$ and P_-, this state is a superposition $(\Psi_| + \Psi_-)/2$, with the state Ψ_- of horizontal polarization being orthogonal to $\Psi_|$. The quantum state $\Psi_/$ corresponds to a superposition of photons that *could* pass the first polarizer $P_|$ and photons that could *not* pass it but could only pass the perpendicular polarizer P_-. The effect of $P_/$ is then

$$\mathcal{P}_/\Psi_| = \frac{1}{\sqrt{2}}\Psi_/ = \frac{1}{2}(\Psi_| + \Psi_-) . \qquad (7.5)$$

Finally, let the photon wave pass through the third polarizer P_- which is perpendicular to the first one. Behind P_- single photons can be detected if and only if $P_/$ is between $P_|$ and P_-. The observations correspond exactly to the classical case described above. With $P_|$ and P_- alone, without the second polarizer $P_/$, no light passes. Not a single photon is detected. In this case, the effect of P_- on $\Psi_/$ is the vacuum state:

$$\mathcal{P}_-\Psi_| = 0\Psi_- = \Psi_{k,0} . \qquad (7.6)$$

But if $P_/$ is put in between them, some photons are again detected at the screen. In this case, inserting the polarizer results in a superposition of a photon state of occupation number 1 and the vacuum state:

$$\mathcal{P}_-\Psi_- = \frac{1}{\sqrt{2}}\Psi_- = \frac{1}{2}(c_0\Psi_{k,0} + c_1\Psi_{k,1}) . \qquad (7.7)$$

Obviously, each of the polarizers reduces the wave functions to a well-defined polarization state, as in a measurement. This gives rise to the absorption of some photons in the polarizers. Each polarizer damps the amplitude of the wave function or field mode by a factor $1/\sqrt{2}$. This has the result of damping the photon intensity (or the average photon number, which is given by the expectation value of the occupation number operator) by a factor $1/2$. The resulting photon intensity or average photon number of the effect (7.7) is $1/4$ of the effect (7.4). In terms of the transition probabilities or counting rates, this means that *on average*, 3 out of 4 photons finally get lost at the photon counter. However, the photons are detected with a completely *irregular* counting rate, due to the ubiquitous vacuum state of the quantum field which predicts ubiquitous probabilistic quantum field fluctuations. For a single field mode and for a non-ideal particle detector with efficiency lower than 1, the probability of the photon counts is a binomial distribution.[52]

With the classical particle concept in mind, one *might* be inclined to ask the following: But what happened to the single photons, given that each of them must *either be absorbed* at one or the other of the polarizers *or detected* behind P_-? Quantum electrodynamics tells us that this is the wrong question. Photons are indistinguishable. And due to the effect of the polarizers, the occupation number of the photon field is not sharp. Therefore, one has to be agnostic about photon absorption at any of the polarizers as long as it is not measured there. From a realistic point of view, the absorption in the polarizers objectively happens due to decoherence. But this again only explains what happens at the probabilistic level.

In the single photon case, the preparation of the photon state by any of the polarizers before the screen must differ substantially from a photon measurement. Otherwise, it would not be possible to undo it by inserting the second polarizer.[53] In the experimental setup described above no absorption of 'lost' photons is observed before the single photons have been detected and their average number has been counted behind P_-. However, to measure the absorption of single photons at the polarizers would require additional experimental devices, and this would constitute a different experimental setup. For the experimental setup discussed here, only the following statement holds. As long as only one photon at a time is in the quantum field, whenever a photon is finally detected no photon was absorbed at any of the polarizers. This is due to the fact that energy conservation holds at the level of the individual quantum processes.[54]

The transition probabilities of quantum field theory tell us that on average 3 out of 4 photons are absorbed at one of the polarizers. If the probability of photon detection is $1/2$ given that $P_/$ is there, it is $1/4$ given that P_- is also there. However, this statement only makes sense at the probabilistic level and with an *ensemble interpretation* of quantum field theory. It is impossible to apply this view to the individual photon detections. Obviously, no polarizer can damp the intensity of a single photon by a factor $1/2$. This precludes any

---

[52]See Walls and Milburn 1994, 51–52; Busch et al. 1995, 9–10 and 177–180.

[53]This point is investigated in more detail below for the recent which-way experiments with quantum erasers. See Sects. 7.4–7.5.

[54]Remember Born's remark in Born 1926a, quoted in Sect. 7.2.

interpretation of the effects discussed above in terms of individual particles, localized photon wave packets, or whatever individual causes of the single photon counts.

But the effects are real. According to the principle of causality, their causes should be too. Indeed, the way in which the photon polarization is prepared has causal relevance for the transition probability which predicts the relative frequency of photon counts. Axiomatically, it is described in terms of field modes, polarization states, occupation numbers, and superpositions. Operationally, the causal relevance of this preparation can only be expressed in terms of conditional probabilities. The gap between the axiomatic and the operational descriptions of photons remains. Operationally, there are particles. Axiomatically, there are fields or waves. There is no uncontroversial link between them beyond the probabilistic interpretation. Nevertheless, physicists consider the waves they prepare to be as real as the particles they detect.

7.4 The Double Slit Reconsidered

For several decades, physicists dreamt of diffraction experiments with single electrons or photons but their dreams remained thought experiments in the spirit of Heisenberg's idea of the gamma microscope.[55] The thought experiments of the famous Bohr–Einstein debate were most influential. It is crucial for their understanding that this debate was based on applying Heisenberg's uncertainty relations to individual subatomic particles. The debate arose after Bohr's Como lecture.[56] In order to show that Heisenberg's uncertainty relation is not unavoidable, Einstein invented one thought experiment after another. All of them were countered by Bohr with more or less convincing physical arguments. As is well known, the debate culminated in the famous EPR thought experiment which was *not* convincingly countered by Bohr.[57] The which-way experiments of recent quantum optics trace back to another thought experiment of the debate, Einstein's recoiling slit. One of Einstein's ideas was to measure the path of the photons which pass through the double slit by means of the slit recoil. Bohr argued that the recoil of the slit would wash out the interference fringes because it makes the momentum measurement uncertain in such a way that Heisenberg's uncertainty relation is respected.[58] Later, Feynman developed another version of this thought experiment. He suggested including a light source in the experimental device

---

[55]Heisenberg 1927, 174–175 and 1930b, 21–23.

[56]See Bohr 1949.

[57]Einstein et al. 1935. Bohr's reply in 1935 was primarily philosophical. For the simple reason that it was too hard to understand, it did not convince any physicist opposed to Bohr's views.

[58]Bohr 1949, 215–216. By the way, this is one of the few probabilistic arguments in Bohr's reasoning.

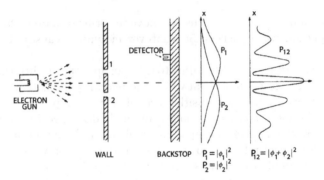

Fig. 7.1. Feynman's double slit without light scattering (Feynman et al. 1965, 1-4)

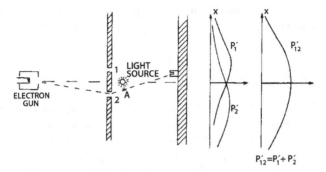

Fig. 7.2. Feynman's double slit with light scattering (Feynman et al. 1965, 1-9)

behind the double slit at which the electron scatters on one side or the other, depending on the slit it comes through.[59]

Only in recent quantum optics has it become possible to realize experiments with single electrons or photons. Finally, some of the thought experiments of the famous Bohr–Einstein debate could also be realized. The EPR correlations were measured in the famous Aspect experiment when the technology for the production of single photons was available.[60] The which-way experiments of recent quantum optics trace back to Einstein's recoiling slit and Feynman's light microscope. In 1991, Scully, Englert, and Walther proposed a refined version of these thought experiments.[61] It is based on the methods of atomic optics, on the passage and diffraction of single atoms through a double slit and other optical devices. (One consequence of wave–particle duality is that in quantum optics atoms and photons, or matter and light particles or waves are interchangeable.)

---

[59]Feynman et al. 1965, 1-4–1-9.

[60]Aspect et al. 1982. See also Greenstein and Zajonc 1997, Chap. 2.

[61]Scully et al. 1991.

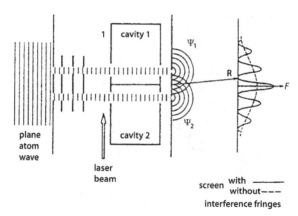

Fig. 7.3. Scully's thought experiment. Reprinted by permission from Macmillan Publishers Ltd: Scully, M.O., Englert, B.-G., Walther, H.: Nature **351**, 111–116 (copyright 1991)

7.4.1 How to Store and Erase Path Information

The proposal is to excite an atom by means of a laser into a well-defined excited state and to mark its path by means of two micromaser cavities in which the excited state decays into the ground state. The experimental arrangement is such that inside the cavity region the excited state decays with probability very close to 1 to the ground state by photon emission. The emission of a photon in one of the cavities marks the path of the atom because the cavity with a photon is in a different field state to the other cavity without a photon. In this sense, the field states of the cavities store the path information.

By additional devices this which-way information can be erased. The quantum erasure is achieved by means of a photon detector in the wall between the cavities. It detects all photons in both cavities. In this way, it destroys the which-way information of the previous experimental arrangement. In order to re-establish the experimental situation with path information and without photon detection, two shutters are placed in the cavities which cover the photon detector and which may be opened. When the shutters are closed, the photon of the decayed atom is left in one of the two cavities and which-way information is obtained. When the shutters are opened, the photon is detected and the which-way information is lost.

Scully, Englert and Walther predict that the interference pattern of the double slit will be destroyed as soon as the path information is stored inside one of the cavities, even if this information is *not* read out. In addition, they predict that the interference pattern will reappear if the stored information is erased, even if the photon from the cavities is detected long after the passage of the atom.

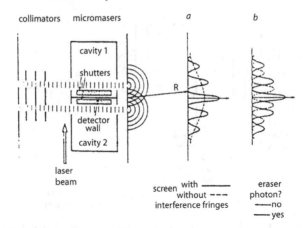

collimators   micromasers   a   b

cavity 1

shutters

R

detector
wall

cavity 2

laser
beam

screen with ——   eraser
        without ---   photon?
interference fringes   ——no
                       ——yes

Fig. 7.4. Scully's quantum eraser. Reprinted by permission from Macmillan Publishers Ltd: Scully, M.O., Englert, B.-G., Walther, H.: Nature **351**, 111–116 (copyright 1991)

The proposal is a thought experiment. Scully, Englert, and Walther describe it in order to make the basic idea clear, namely to mark the path of an atom by means of internal degrees of freedom and then let it pass a double slit. They explain the passage of the atom through the cavities and the double slit, the storage of the path information, and the way in which this information is erased in a quantum mechanical model. In order to demonstrate that the respective quantum process is free of any random scattering or other stochastic perturbations, they refer to separate calculations which prove that in this model the momentum transfer is negligible and the wave function of the atom is almost unaffected by the emission of the photon.[62] The prediction for the outcome of the experiment without quantum eraser is as follows:

> To be sure, we find that the interference fringes disappear once we have which-path information, but we conclude that this disappearance originates in correlations between the measuring apparatus and the systems being observed.[63]

Here, they express their result in terms of quantum information. They claim that the interference fringes disappear once the path information (namely the emitted photon) is stored in the cavity. To be more precise, the information is only stored in the field states of the cavities if they differ 'in a detectable manner' with or without the photon emitted by the atom; this requires that

---

[62]Scully et al. 1991, 113, referring to Englert et al. 1990. The claim that Heisenberg's uncertainty relation is not therefore employed in wave–particle duality gave rise here to a debate within quantum optics; see below.

[63]Scully et al. 1991, 111.

the fields are in a state of well-defined low occupation number.[64] The storage
of the path information makes the interference fringes disappear. This loss
of interference is explained in terms of 'correlations' between the measuring
device and the atom.

The quantum mechanical calculation shows that these 'correlations' are
due to an entanglement of the quantum state $|\Psi\rangle$ of the atom and the un-
known field states $|n_1\rangle$, $|n_2\rangle$ of the cavities 1 and 2 before the two slits. It is
assumed that these field states differ in a detectable way by one photon in
their occupation numbers (i.e., $|n_1 - n_2| = 1$). The respective field states are
orthogonal. This is sufficient to make the interference fringes of the double
slit at the screen disappear. Finally, the photodetector and the shutters are
installed. The arrangement with the opened shutters prepares a wave function
in which the emitted photon is detected *without* any path information. This
kind of detection erases the path information and makes the fringes at the
screen reappear. The different experimental setups and the effects to which
they give rise are described in the following simplified quantum schemes:[65]

Double Slit [DS]

The atom is prepared in a sharp momentum state or plane wave $|\Psi\rangle$ and sent
through a double slit. The double slit prepares the wave function $|\Psi^{DS}\rangle$ as
a superposition of the waves $|\Psi_1\rangle$ and $|\Psi_2\rangle$ which stem from the two slits;
this gives rise to superposition terms and hence to interference fringes at the
screen:

$$|\Psi^{DS}\rangle = \frac{1}{\sqrt{2}}\left(|\Psi_1\rangle + |\Psi_2\rangle\right), \qquad (7.8)$$

$$|\Psi^{DS}|^2 = \frac{1}{2}\left(|\Psi_1|^2 + |\Psi_2|^2 + \langle\Psi_1|\Psi_2\rangle + \langle\Psi_2|\Psi_1\rangle\right). \qquad (7.9)$$

Path Marker [PM]

Now the cavities 1 and 2 are installed before the double slit, the atom is excited, and the paths through the two cavity-plus-slit systems are marked by

---

[64]Scully et al. 1991, 114. They emphasize that the micromasers will *not* serve
as which-way detectors if they contain classical microwave radiation with a large
average photon number N and the associated fluctuations of $\sqrt{N} \gg 1$; ibid.

[65]In comparison to Scully et al. 1991, 113–115, the following technical details
are omitted: the wave function of the internal state of the atom which does not
add any new terms to the sums below in (7.8–7.11); the assignment of the field
states to the cavities 1 and 2 which must be symmetrized in the end (here, the
field states are simply given in terms of occupation numbers); and the probability
density of the entangled detector-screen states which describes the correlations of
photon detection in the cavity wall and atom detection on the screen.

the orthogonal quantum field states $|n_1\rangle$, $|n_2\rangle$. The photon emission couples the cavity field states to the passage through the respective slit. The whole arrangement prepares the wave function $|\Psi^{\mathrm{PM}}\rangle$ in which the components of the double-slit wave function $|\Psi^{\mathrm{DS}}\rangle$ are entangled with the detector states $|n_1\rangle$, $|n_2\rangle$. Since these states are orthogonal, any superpositions of them disappear, and the entangled wave function predicts the loss of interference fringes:

$$|\Psi^{\mathrm{PM}}\rangle = \frac{1}{\sqrt{2}}\left(|\Psi_1\rangle|n_1\rangle + |\Psi_2\rangle|n_2\rangle\right), \qquad (7.10)$$

$$|\Psi^{\mathrm{PM}}|^2 = \frac{1}{2}\left(|\Psi_1|^2 + |\Psi_2|^2\right). \qquad (7.11)$$

Quantum Eraser [QE]

Finally, the photodetector and the shutters are installed. The arrangement with closed shutters is identical with [PM]. But when the shutters are opened, the photon detector will detect photons from both cavities. Hence, the arrangement gives rise to the preparation of some superposition of a photon in one cavity and no photon in the other one. This simplified view of the experimental situation still neglects the detection of photons. To take it into account, the detector atom is described in terms of a sum of symmetric and antisymmetric quantum states. The photon field only couples to the symmetric part of the wave function of the detector atom. This reduces the detector efficency to $1/2$.[66] Due to the coupling, up to this efficiency factor, $|\Psi^{\mathrm{PM}}\rangle$ is entangled with the superposition $(|n_1\rangle + |n_2\rangle)$, the path infomation is erased, and the original wave function of the double slit is re-established:[67]

$$|\Psi^{\mathrm{QE}}\rangle = \frac{1}{\sqrt{2}}|\Psi^{\mathrm{DS}}\rangle = \frac{1}{2}(|\Psi_1\rangle + |\Psi_2\rangle). \qquad (7.12)$$

The language of quantum information suggests that the photon stored in the cavity is as objective as if it were read out by a measuring device. This is associated with the idea that the photon emission is as real (say, due to decoherence) as the death of Schrödinger's cat, even if no one observes these

---

[66]Scully et al. 1991, 115.

[67]The antisymmetric part of the wave function of the detector atom (which does not couple to the photon field states of the cavities) is entangled with $|\Psi^{\mathrm{PM}}\rangle$ in such a way that anticorrelations between the screen and the photon detector and corresponding antifringes are measured; see Scully et al. 1991, 115, and the discussion in Sect. 7.5. The original double slit wave function would be re-established by simply removing the wall between the cavities. This requires another experimental arrangement, namely an opening between the cavities and a shutter which can be closed. Such an arrangement would give rise to another kind of quantum eraser which is similar to the polarization experiment discussed in Sect. 7.3 rather than to the EPR correlations.

events. This tacit quantum realism *seems* to be at odds with the claim that quantum erasure of the path information is possible. The quantum eraser only works if the storage of the which-way information is a reversible process. This requires that neither a measurement reduces the quantum mechanical wave function nor any incoherent scattering destroys the interference of its components. Hence the importance of the claim that marking the path is recoil-free and lacks any random scattering or other stochastic processes.[68]

In informal terms, the sequence of quantum phenomena described above is perhaps best described in terms of the distinction of *preparing waves* and *detecting particles* which was suggested above in Sect. 7.3. The double slit prepares two coherent, interfering waves. The cavities prepare two waves which are entangled with well-distinguished cavity field states, contain path information, and no longer interfere. Finally, the photodetector measures field states which are entangled with the propagation of the atom behind the double slit but which neglect this path information.

However, the quantum eraser gives rise to the following puzzle. In the experimental arrangement [QE] with the photodetector and the two shutters the shutters may be opened a long time after the atom passed the cavities. This experimental situation is a quantum eraser with delayed choice. How can the interference pattern reappear if the time delay between the passage of the atom and the opening of the shutters is so long that the atom can no longer be influenced by what goes on in the cavities? Once the atom is at a spacelike distance from the cavities, no signal can be transferred between them. This is required by Einstein causality.[69] Scully, Englert, and Walther adapt some remarks of Edwin Jaynes to their own experimental arrangements:[70]

> We have, then, the full EPR [Einstein–Podolsky–Rosen] paradox – and more. By applying the eraser mechanism before measuring the state of the microcavities we can, at will, *force* the atomic beam into either: (1) a state with a known path, and no possibility of interference in any subsequent measurement; (2) a state with both Ψ_1 and Ψ_2 present with a known relative phase. Interference effects are then not only observable but predictable. And we can decide which to do after the interaction is over and the atom is far from the cavities, so there can be no thought of any physical influence on the atom's centre-of-mass wavefunction.

This apparently paradoxical situation is indeed easily resolved on the basis of the probabilistic interpretation of quantum mechanics. Entanglement finally

---

[68]This point is emphasized several times in the paper; see Scully et al. 1991, 111; 113, where the calculations concerning the negligible momentum transfer are reported; and 114, where they "emphasize once more that the micromaser *welcher weg* detectors are recoil-free; there is no significant change in the spatial wave functions of the atoms," before they come to explain the quantum eraser.

[69]Einstein et al. 1935.

[70]Scully et al. 1991, 114, after Jaynes 1980.

gives rise to correlations proper, that is, to correlated particle detections. The physical effect of the preparation of an entangled wave function only shows up at the probabilistic level, in the detection of *many* pairs of correlated particles. This is so in the case of EPR correlations, and this is also the crucial point in the thought experiment under investigation.

Before this puzzle is resolved and before the recent realizations of the proposed thought experiment are discussed, another puzzle must be explained and resolved. It is related to the conceptual background of the proposal. Its basic idea is to mark the path of an atom by entanglement with the field states of the cavities in such a way that the path information can afterwards be erased. Therefore, Scully, Englert and Walther claim that in this way the atom's path is marked without any substantial momentum transfer. The question of whether this claim is tenable gave rise to a debate about the question of whether Heisenberg's uncertainty relation or complementarity is more fundamental.

7.4.2 Complementarity Without Uncertainty?

The conceptual background of the proposed thought experiment and its realizations is twofold: the Bohr–Einstein debate on the one hand and the quantum mechanics of scattering on the other hand. The former is based on a particle picture. It proposes in a heuristic way to apply Heisenberg's uncertainty relation and Bohr's complementarity view to individual subatomic particles. The latter tells that this only makes sense for a probabilistic ensemble and it is based on a wave picture.

The Bohr–Einstein debate neglects the probabilistic interpretation of quantum mechanics in favor of the correspondence of individual subatomic interactions to their classical analogue, the particle picture. Bohr and Einstein discuss the possibilities of measuring the position and momentum of individual subatomic particles. In doing so, they apply the classical pictures of quantum systems and their interactions. They discuss the tenability of Heisenberg's uncertainty relations and Bohr's complementarity principle on the lines of Bohr's account of wave–particle duality discussed in Sect. 7.2. Einstein's thought experiments and Bohr's arguments against them conceive of photons or electrons *as if they were particles in a classical sense*. But one should not forget Heisenberg's point stressed in Sect. 7.2. The view of an electron or photon as a particle is only based on analogies. To explain its interactions in a classical model presupposes that Bohr's correspondence principle holds. In any case, the classical analogies only give an incomplete account of the quantum domain.

The very name of the which-way experiments is based on the classical idea of a particle trajectory. In order to interpret them it is tempting to employ Bohr's concept of complementarity and Heisenberg's uncertainty relation at the level of individual subatomic interactions and to think of particle propagation. This idea, however, is at odds with the probabilistic interpretation of quantum mechanics, non-local quantum field fluctuations, and the wavelike quantum phenomena discussed in Sect. 7.3. The only way to escape conceptual confusion is to take wave–particle duality and its probabilistic interpretation seriously. This means that the *particle path* can only have *operational meaning* in these experiments.

In their seminal paper, Scully, Englert and Walther argue that complementarity is more fundamental than Heisenberg's uncertainty relation. They introduce a generalized concept of complementarity which brings the probabilistic aspects of complementary observables into play:[71]

> Complementarity distinguishes the world of quantum phenomena from the realm of classical physics. [...] the credit for teaching us that we have to accept complementarity as a fact [...] belongs to Niels Bohr. In 1927, [...] at Como [...] all examples used to illustrate complementarity referred to the position (particle-like) and momentum (wave-like) attributes of a quantum mechanical object, be it a photon or a massive particle. [...] Complementarity, however, is a more general concept. We say that two observables are 'complementary' if precise knowledge of one of them implies that all possible outcomes of measuring the other one are equally probable. [...] Here, then, is the 'Principle of Complementarity': For each degree of freedom the dynamical variables are a pair of complementary observables. A less formal, less precise version in practical terms is: No matter how the system is prepared, there is always a measurement whose outcome is utterly unpredictable.

Here, complementarity in Bohr's sense is understood as the wave–particle duality of an individual quantum object to which either momentum or position may be attributed. This is opposed to a *more general* account of complementarity which is associated with non-commuting quantum mechanical observables and comes in *probabilistic* terms. This concept of complementarity is then applied to Einstein's thought experiment, Feynman's version of it in which an electron passes a double slit and its path is measured by means of a light source, and Heisenberg's microscope.[72] Scully, Englert, and Walther claim that these thought experiments

> emphasize this complementarity in quantum mechanics. [...] In these examples [...] Heisenberg's position–momentum uncertainty relation $\delta x \delta p \geq \hbar/2$ makes it impossible to determine which hole the electron (or photon) passes through without at the same time disturbing the electrons (photons) enough to destroy the interference pattern.[73]

---

[71]Scully et al. 1991, 111. For a related discussion of the following topics, see also Busch and Lahti 2005; Shillady and Busch 2006.

[72]Bohr 1949, 225–226; Feynman et al. 1965, 1-4–1-9; Heisenberg 1927; 1930a, 15; 1930b, 21.

[73]Scully et al. 1991, 111.

The interference pattern is only obtained and washed out at the probabilistic level. We have seen that according to Bohr's views, the uncertainty relation underlies complementarity. Scully et al. take this point up and claim:

> In the present work we have found a way around this position–momentum obstacle. That is, we have found a way, based on matter interferometry, and recent advances in quantum optics, [...] to obtain which-path or particle-like information without scattering or otherwise introducing large uncontrolled phase factors into the interfering beams.[74]

Obviously, here the scattering of subatomic particles is modelled *in the wave picture*, that is, in the quantum mechanics of scattering. According to this, the scattering results are given by a scattering amplitude which is described in terms of "phase factors" of scattered spherical waves. In particular, the scattering of a particle beam at a double slit gives rise to two "interfering beams" which have a well-defined phase difference depending on the scattering angle. Here, Scully, Englert, and Walther claim that in the proposed experimental setup *no* additional "large uncontrolled phase factors" come into play. Hence, they interpret Heisenberg's uncertainty relation in the wave picture which they presumably interpret probabilistically, like their generalized "principle of complementarity".

The authors go on to discuss Einstein's recoiling-slit arrangement, Feynman's version of the double-slit experiment in which a light source measures the electron's path, and Heisenberg's microscope.[75] Feynman claimed that the scattering of light at the electron will destroy the interference pattern and that there never was a way around Heisenberg's uncertainty principle:

> If an apparatus is capable of determining which hole the experiment goes through, it *cannot* be so delicate that it does not disturb the pattern in an essential way. No one has ever found (or even thought of) a way around the uncertainty principle.[76]

Against this, Scully, Englert, and Walther claim:

> Is this mechanism always at work? No! We have recently found a way around it.[77]

It should be noted that two different models are employed here. In the mentioned examples, Bohr, Heisenberg, and Feynman explained the scattering results in a *heuristic classical particle picture*. For the Einstein–Bohr recoiling slit arrangement, Scully, Englert, and Walther explain the validity of the

---

[74]Ibid.

[75]Scully et al. 1991, 112.

[76]Feynman et al. 1965, 1-9.

[77]Scully et al. 1991, 112., referring to Scully et al. 1989, Scully and Walther 1989, and Englert et al. 1990.

uncertainty relation in *this* model.[78] The two models explain the momentum change to which a position measurement gives rise (and hence the meaning of Heisenberg's uncertainty relation) in completely different terms. In the wave model of quantum mechanical scattering theory, a position measurement gives rise to uncontrolled phase factors. In the particle model, it gives rise to a momentum kick.

In the two models the above claim of finding a way round the uncertainty relation has different meanings. In the heuristic particle model it means making a double-slit experiment with a *recoil-free* path measurement without momentum kick. In quantum mechanics it means obtaining "which-path or particle-like information without scattering or otherwise introducing large uncontrolled phase factors into the interfering beams."[79] The two models have in common that the crucial question is whether or not a substantial momentum transfer is associated with the path measurement. But the description of this momentum transfer differs substantially.

In any case, Scully, Englert and Walther propose a which-way measurement *without momentum transfer* of the measuring devices to the atoms which pass the double slit. Compared with Einstein's and Feynman's thought experiments, their basic new idea is that the path measurement is *recoil-free* (in the particle picture) and *does not give rise to uncontrolled scattering amplitudes* (in the wave picture). The proposal is to achieve this goal by marking an internal degree of freedom of the atoms, as explained in Sect. 7.4.

In 1994, however, a debate arose about the question of whether Heisenberg's uncertainty relation or Bohr's complementarity principle is more fundamental. Storey and her coworkers claimed that the thought experiment of Scully, Englert, and Walther *must* indeed give rise to substantial momentum transfer, or uncontrolled phase factors and momentum kicks, because

> in any path detection scheme involving a fixed double slit, the amount of momentum transferred to the particle by a perfectly efficient detector (one capable of resolving the path unambiguously) is related to the slit separation in accordance with the uncertainty principle.[80]

The authors of the paper employ the particle picture in their informal expressions even though they make the correct quantum mechanical calculation in the wave picture. They take the path measurements into account in terms of a convolution of the incoming wave with phase factors which give rise to a "momentum kick amplitude distribution".[81] This *very* informal expression nicely *combines* ideas about particle kicks and wave amplitudes. The crucial assumption is that the quantum mechanical momentum distribution is associated with a maximum momentum transfer p_m. The calculation gives

[78]Scully et al. 1991, 114.
[79]Scully et al. 1991, 112; see above.
[80]Storey et al. 1994, 626.
[81]Storey et al. 1994, 627.

the result that this maximum momentum transfer is related to the slit separation d by $p_m d \geq \hbar$, in accordance with (Bohr's version of) Heisenberg's uncertainty relation.[82]

Englert, Scully and Walther countered that these objections are not substantial. On the one hand they propose a different experimental arrangement to the one discussed by Storey and her group. On the other hand they claim that their detailed calculations prove quite on the contrary that there is *no* substantial momentum transfer.[83] The defense was printed together with a reply[84] and the debate went on.

There was even an attempt to reconcile the positions by distinguishing a classical random momentum kick distribution from the phase factors in a quantum mechanical scattering amplitude. At the probabilistic level, the classical particle and wave model both predict that many position measurements make the interference pattern disappear. This is the basis of the heuristic considerations of the Einstein–Bohr debate and their followers. In a classical particle model of the scattering, this is due to uncontrolled individual random momentum kicks. In the wave model, this is due to the phase factors which enter the scattering amplitude. Due to the interference terms in a quantum mechanical scattering amplitude the two models generally give different results. Wiseman and Harrison claimed that the correspondence or non-correspondence to the classical particle model of scattering makes the difference for the experimental arrangements under debate.[85] On the one hand, they claim that Einstein's recoiling slit and Feynman's light microscope may be interpreted in terms of a *classical* random momentum kick distribution,[86] whereas this is not possible for the experimental scheme proposed by Scully, Englert, and Walther. On the other hand, however, they claim that the proposal does not escape Storey's results which are based on quantum mechanical random phases. They seem to think that complementarity may only be regarded as more fundamental than the uncertainty relation if the latter is understood in terms of uncontrolled *classical* momentum kicks:

> There is thus room for Scully et al. to claim that complementarity is more fundamental than the uncertainty principle; but there is also room for claims by Storey et al. that one can always consider complementarity as being enforced by the uncertainty principle, if

---

[82]Storey et al. 1994, 627. See also the above discussion of Bohr 1928 in Sect. 7.2.2.

[83]Englert et al. 1995, 367–368. Indeed, Scully et al. 1994 do not suggest any position measurement in the sense of Storey et al. 1994. As they emphasize in Englert et al. 1995, they propose to place the micromaser cavities *before the double slit*, in order to *avoid* precise position measurements.

[84]Storey et al. 1995, 368.

[85]See Wiseman and Harrison 1995.

[86]In terms of Bohr's correspondence principle, this means that at the probabilistic level the scattering at a double slit corresponds to the classical particle model, too (like the Rutherford scattering explained in Chap. 2).

instead the latter is interpreted in terms of the more subtle idea of momentum-kick amplitudes.[87]

Hence, the alleged "reconciliation" was indeed unsuccessful. Instead, it only increased the conceptual confusion about complementarity and the uncertainty relation. The "subtle idea of momentum-kick amplitudes" relies on the same quantum mechanical grounds as Scully, Englert and Walther's claim that their arrangement makes it possible to obtain path information *without* introducing large random phases.[88] Accordingly, there are no indications that this attempt to reconcile the positions was successful. When a *which-way* experiment finally succeeded in realizing the proposal, thus making use of internal degrees of freedom in order to mark the path of an atom without momentum transfer, Englert expressed the hope that the debate would now come to an end.[89]

7.4.3 Duality Relations

Indeed there was another way out of the debate, namely to *generalize both* complementarity and Heisenberg's uncertainty principle and to see how they relate *then*. The seminal paper of Scully, Englert, and Walther introduced a generalized concept of complementarity, as mentioned above. According to this, complementarity is explained in terms of non-commuting quantum observables and the probabilities of their measurement outcomes. Heisenberg's uncertainty relation only concerns one pair of such quantum observables, namely position q and momentum p. Other such pairs of non-commuting observables concern generalized complementarities in the sense of Scully et al. However, at this point the distinction between Bohr's complementarity and Heisenberg's uncertainty principle finally begins to vanish. The generalized complementarities of Scully et al. are formally described by operators in Hilbert space, giving rise to what was called Heisenberg–Robertson uncertainty relations.[90] For each pair of Hermitian operators A and B, the following formal uncertainty relation holds:[91]

$$\Delta A \Delta B \geq \frac{1}{2} |\langle [A, B] \rangle| . \tag{7.13}$$

Here, $\langle A \rangle = \mathrm{Tr} \rho A$ is the expectation value of a Hermitian operator A, and $\Delta A = \sqrt{\langle A^2 \rangle - \langle A \rangle^2}$ is its standard deviation. For the operators p and q of

---

[87]Wiseman and Harrison 1995.

[88]Scully et al. 1991; see above.

[89]See Englert 1998. The experiment is Dürr et al. 1998a,b, Dürr and Rempe 2000a, discussed in Sect. 7.4. Englert 1998 gives a short report of the experiment and the debate. Obviously, he includes Wiseman and Harrison 1995 amongst the opponents of the claim that the proposed experimental scheme aims at path information without substantial momentum transfer.

[90]Robertson 1929.

[91]Dürr and Rempe 2000b, 1022.

momentum and position, (7.13) expresses just the correct probabilistic meaning of Heisenberg's uncertainty relation. Hence, according to Scully, Englert, and Walther's definition, at a generalized level there is no formal distinction between complementarity and Heisenberg's uncertainty relation. This already shows that the above debate about complementarity and Heisenberg's uncertainty relation cannot be substantial. The only point which remains is the question of momentum transfer in a double-slit experiment, which will be discussed below. However, it was unnecessarily merged with the question of whether Heisenberg's *specific* uncertainty principle or Bohr's *specific* complementarity view is more fundamental.

For the recent which-way experiments, such generalized Heisenberg–Robertson uncertainty relations were defined in order to express the information concerning the wavelike or particle-like quantum phenomena measured by a given experimental device. This information is basically *probabilistic*, as indicated by (7.13). And it concerns which-way experiments which are *more general* than the either–or decision which Scully et al. proposed. The experimental setup which they suggest *either* measures path information *or* erases it. It is not designed for preparing and measuring *intermediate* quantum states. But this is also possible.

Intermediate States [IM]

In order to do so, one has to prepare a superposition of two wave functions $|\Psi_1\rangle$, $|\Psi_2\rangle$ propagating along different paths 1, 2 with a phase difference which makes them interfere but with different amplitudes c_1, c_2:

$$|\Psi^{\mathrm{WP}}\rangle = c_1|\Psi_1\rangle + c_2|\Psi_2\rangle \qquad (|c_1| \neq |c_2|) . \tag{7.14}$$

Already one of the earliest which-way experiments succeeded in doing so.[92] It used a Mach–Zehnder interferometer in order to split a low intensity laser beam into two beams and to make them interfere. The low intensity beam was sent through a system of two beam splitters and two mirrors in such a way that it could take two possible paths until single photons are detected by means of the two final photon counters. Both paths are provided with such a phase shift that when they are united by means of the second beam splitter, an interference pattern is obtained. Now in one of the two branches of the experiment the beam is attenuated in such a way that significantly fewer particles are detected in this branch than in the other branch of the experiment. What is observed is the following. The interference pattern does not disappear immediately when the intensity of the beam is increased in one branch and decreased in the other branch. The interference fringes are gradually washed out and only when the path information becomes overwhelming

---

[92] Mittelstaedt et al. 1987. A which-way experiment with path information and incomplete quantum erasure has the same effect; see Dürr et al. 2000a, 60–65.

are they no longer observed. Hence, the quantum phenomena do not discontinuously switch between particle-like and wavelike behaviour. They exhibit smooth transitions between the two kinds of phenomena. This corresponds to a joint unsharp measurement of the corresponding quantum observables.[93]

In order to describe such smooth transitions between wavelike and particle-like phenomena, the visibility V of interference fringes and the path predictability P or distinguishability D were defined as quantum observables.[94] The visibility V is simply the relative contrast of the interference fringes, which is given in terms of the maximum and minimum intensities I_{max}, I_{min}:

$$V = \frac{I_{max} - I_{min}}{I_{max} + I_{min}} \ . \tag{7.15}$$

The path information may be expressed in terms of the relative particle flux which the wave function predicts for the two possible paths. This gave rise to the following definition of the predictability P of the particle path:[95]

$$P = |w_1 - w_2| \ . \tag{7.16}$$

Here, w_1 and w_2 are the probabilities that the particle is finally detected at the end of path 1 or path 2, respectively. These probabilities were understood in the sense of *a priori* knowledge about the probable particle path,[96] as if the quantum probability admitted of an ignorance interpretation. However, this is misleading. According to quantum mechanics and quantum field theory, there are no particle trajectories. Hence strictly speaking, there is neither a particle path nor its probability but just the probability of *detecting particles* for a given way of *preparing waves*. These probabilities are nothing but the squared probability amplitudes $|c_1|^2$, $|c_2|^2$ of the components of the wave function $|\Psi\rangle = c_1|\Psi_1\rangle + c_2|\Psi_2\rangle$. The predictability concerns the measurement results for the two possible outcomes rather than a particle path before the measurement.

Alternatively, the path information may be expressed in terms of the likelihood L_W of guessing the particle path correctly from a measurement of the internal degree of freedom in which the path information is stored, according to the experimental scheme proposed by Scully et al.:[97]

$$L_W = \Sigma_i \max \left\{ p(w_i, 1), p(w_i, 2) \right\} \ . \tag{7.17}$$

Here, W is an observable which measures the path information stored by the internal state, w_i are the eigenvalues of these observables, $1, 2$ are the

---

[93] Busch et al. 1995.

[94] Mittelstaedt et al. 1987. See also Dürr et al. 1998b, Dürr and Rempe 2000b.

[95] Englert 1996.

[96] Dürr et al. 1998b, 5705, claim that $|w_1 - w_2|$ gives rise to "an a priori WW [=which-way, B.F.] knowledge due to the difference between the probabilities, w_+ and w_-, that the particle takes one way or the other."

[97] Englert 1996; Dürr et al. 1998a,b; Dürr and Rempe 2000b; Scully et al. 1991.

two possible paths, and $p(w_i, k)$ $(k = 1, 2)$ are the probabilities of measuring the eigenvalue w_i in path k. Then, for the best choice of the observable W (which makes it possible to read out the maximum path information), the path distinguishability D is defined as follows:[98]

$$D = -1 + 2 \max_W \{L_W\} \,. \tag{7.18}$$

The experimenters emphasize that all these observables have *probabilistic meaning*.[99] What they do *not* discuss is the meaning of the probabilities involved in the definitions. However, it was shown that for visibility and predictability or distinguishability, the following duality relations hold,[100] which may alternatively be expressed in the form of a Heisenberg–Robertson uncertainty relation of type (7.13):[101]

$$P^2 + V^2 \le 1 \,, \tag{7.19}$$

$$D^2 + V^2 \le 1 \,. \tag{7.20}$$

To put it another way, these duality relations are equivalent to generalized uncertainty relations. The latter are appropriately expressed in terms of non-commuting operators with the formal properties of an abstract pseudo-spin.[102]

Now, the question of whether Bohr's complementarity or Heisenberg's uncertainty principle is primary can be resolved as follows. From a formal point of view, complementarity or wave–particle duality can be stated in terms of the above duality relations. According to (7.19) and (7.20), the dual observables V and D may describe maximum fringe visibility $V = 1$ and minimum path distingishability $D = 0$, or vice versa, or some intermediate quantum behaviour. (The same holds for visibility V and path predictability P.) The relations (7.19) and (7.20) just express *another* kind of uncertainty, namely some Heisenberg–Robertson uncertainty relations which can be stated in terms of some suitably chosen abstract operators in Hilbert space. The way in which the respective quantum states are prepared and measured depends on the experimental setup. The path information may be obtained due to Heisenberg's position–momentum uncertainty (as in Einstein's recoiling slit or in Feynman's light microscope), or only due to quantum entanglement (as in the scheme proposed by Scully et al.), or both.[103] Hence, whether Heisenberg's momentum–position uncertainty is employed in a which-way experiment or

---

[98] Englert 1996, Jaeger et al. 1995.

[99] Dürr et al. 2000b, 64: "Note that the distinguishability (as well as the visibility) is an ensemble property."

[100] Englert 1996, Jaeger et al. 1995.

[101] Dürr and Rempe 2000b.

[102] Ibid.

[103] Dürr and Rempe 2000b, 1023–1024.

not, depends on the experimental arrangement. The only relevant question is whether any substantial momentum transfer is associated with the storage of path information or not. In the recent which-way experiments it is *not*.

7.5 Recent Which-Way Experiments

The recoil-free which-way experiment proposed in 1991 by Scully, Englert, and Walther consists of the three steps of (1) preparing interference, (2) marking the path, and (3) erasing the path information. The general quantum mechanical scheme is as follows:

(1) Preparing Interference

The quantum system is prepared in a sharp momentum state or plane wave. It propagates through a double slit, or a Mach–Zehnder interferometer, or a similar experimental device. In this way, a superposition of two quantum states $|\Psi_1\rangle$ and $|\Psi_2\rangle$ with different paths and a phase difference is prepared, giving rise to maximum visible interference fringes:

$$|\Psi^V\rangle = \frac{1}{\sqrt{2}}\left(|\Psi_1\rangle + |\Psi_2\rangle\right) . \tag{7.21}$$

(2) Marking the Path

Now some internal degree of freedom of the propagating quantum system is entangled with orthogonal detector states $|1\rangle$, $|2\rangle$. (At this point, Scully's scheme deviates from the experiment sketched above. There, the path was simply *measured* by detecting the split-off photon. This makes a *mixture* out of the quantum state (2), whereas here we are still dealing with a *superposition*.) Due to orthogonality, the detector states do not interfere. Therefore, the interference terms are cancelled and entanglement of the propagating quantum system and the detector states makes the paths maximally distinguishable:

$$|\Psi^D\rangle = \frac{1}{\sqrt{2}}\left(|\Psi_1\rangle|1\rangle + |\Psi_2\rangle|2\rangle\right) . \tag{7.22}$$

(3) Erasing the Path Information

Finally, an additional device is added in order to prepare a superposition of $|1\rangle$ and $|2\rangle$ again. In the proposal of Scully et al., this was the photodetector behind the opened shutters. In general, what is needed is simply a preparation procedure which multiplies the wave function $|\Psi^{PM}\rangle$ with some suitable

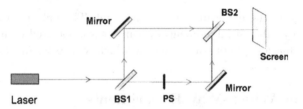

Fig. 7.5. Mach–Zehnder interferometer

superposition of the detector states $|1\rangle$ and $|2\rangle$, again giving rise to visible interference fringes:

$$|\Psi^{V'}\rangle = \frac{1}{\sqrt{2}}\Big[|\Psi_1\rangle\big(a|1\rangle + b|2\rangle\big) + |\Psi_2\rangle\big(c|1\rangle + d|2\rangle\big)\Big] . \qquad (7.23)$$

A first which-way experiment with path information and quantum eraser preceded the proposal of Scully and coworkers discussed in Sect. 7.4. It was based on an earlier proposal of Scully's group.[104] It made use of a Mach–Zehnder interferometer, realizing another experimental scheme. But in many points it was very similar. The idea was very simple, but it was tricky to perform the experiment because it requires optical devices of high precision.

1. In a Mach–Zehnder interferometer, low intensity light is sent through a system of two beam splitters BS_1, BS_2 and two mirrors in such a way that it may propagate along two possible paths and interfere, in accordance with the laws of classical electrodynamics and geometrical optics. In one of the arms of the experiment, a phase shifter PS is inserted in order to vary the phase shift of the two waves. The second beam splitter BS_2 has the effect of superposing them, thus *preparing an interfering quantum state*. Finally, single photons are detected by means of a photodetector D behind the second beam splitter. In the long run, they make up an interference pattern.

2. Next, in each arm of the interferometer behind the first beam splitter a *which-way* detector was installed. Two non-linear down conversion crystals X_1, X_2 are used, making an entangled lower frequency 2-photon state out of any photon passing through it. Two additional photocounters d_1, d_2 detect the lower frequency photons split off by the down converter and *mark the path*. The path measurement reduces the wave function of the left-over photon to one arm of the interferometer and makes the interference pattern disappear. The result is a mixture that gives rise to the same squared amplitude as the state $|\Psi^D\rangle$ of the above general scheme, in (7.22).

---

[104]Zou et al. 1990; based on the proposals Scully and Drühl 1982, Zajonc 1983. The experiment is discussed in Greenstein and Zajonc 1997, 206–209. Another realization of Scully and Drühl 1982 is Kim et al. 1999.

3. However, it is possible to *erase the path information* taken by the two which-way detectors. In order to do so, a third beam splitter BS_3 is installed between the two detectors d_1, d_2 of the split-off photons in such a way, that *none* of the photons detected in the experiment had an *unambiguous path*. Now the two down converters and the three beam splitters prepare interfering quantum states in all arms of the experiment. The resulting wave function has the shape $|\Psi^{V'}\rangle$ of (7.23). A coincidence counter which measures the correlations between the split-off photons detected by d_2 and the left-over photons detected by D will record an interference pattern.

The last point is crucial: it has to be noted

that the interference pattern will only be observed if we look at coincidence counts between two detectors. Interference, of course, is the characteristic sign of quantum behavior: *and this quantum behavior will not be seen if we do the wrong experiment*. Looking only at the single detector D is the wrong experiment: it reveals what appears to be classical behavior – the absence of interference.[105]

In this experimental arrangement, measuring the coincidences between d_1 and D will give rise to an interference pattern, too. But it should be noted that the counts of d_1 and d_2 are anticorrelated. The split-off photon can only be detected by one of the detectors d_1 or d_2. Therefore, the interference patterns obtained from the coincidences of d_1 or d_2 and D will sum up like fringes and antifringes. Together, they give rise to the disappeared interference pattern of arrangement (2). Exactly the same ensemble behaviour was predicted in 1991 by Scully et al. and observed in the quantum eraser experiments reported below.

Scully, Englert, and Walther proposed the analogous experimental arrangement for the double slit. The main difference is the idea of not immediately reading out the path information by photon detection, but only *storing* it. According to their experimental scheme, this is achieved by entangling the quantum waves that propagate through the two slits and the quantum states of the which-way detectors. This requires the quantum system under investigation to have an *internal degree of freedom*. When the system is entangled with the detector states, its internal state marks the path. This entanglement replaces the entangled 2-photon state of the above experimental arrangement. Now, it is no longer necessary to make any photon measurement or particle detection in order to obtain path information. It is sufficient to prepare the waves.

Scully's scheme was recently realized in an atom interferometer.[106] A beam of ^{85}Rb atoms is generated by means of a magneto-optical trap, split,

[105] Greenstein and Zajonc 1997, 208–209.
[106] Dürr et al. 1998a,b, 2000a. In the following, all technical details are omitted.

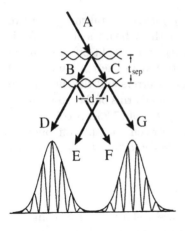

Fig. 7.6. Atom interferometer. Reprinted by permission from Macmillan Publishers Ltd: Dürr, S., Nonn, T., and Rempe, G.: Nature **395**, 33–37 (© 1998)

and made to interfere with itself. The atomic beam is not sent through a double slit but through two pulsed standing light waves. They have the effect of splitting the atom beam twice into a transmitted and a Bragg-reflected beam. The distance between the two standing light waves is such that the transmitted and the Bragg-reflected beam of the first wave split again at the second standing light wave at a distance d, which is the analogue of the distance between the two slits in the double-slit experiment. The spatial distribution of the atoms in the far field is observed by exciting them with a laser beam and detecting fluorescence photons. In stage (0), the atomic beam is split once by one standing light wave only and the atoms are detected in the far field. The atoms from the transmitted and the Bragg-reflected beams arrive at a well-defined position. Trivially, no interference is observed.

In stage (1), the second standing light wave is added, generating two interfering transmitted beams and two interfering Bragg-reflected beams. Now two spatial interference patterns are observed in the far field (see Fig. 7.7).

In stage (2), two internal states or energy levels of ^{85}Rb are used as a path marker. The ground state of ^{85}Rb splits into two hyperfine levels $|2\rangle$ and $|3\rangle$ of different angular momentum and slightly different energy. In order to use them as path markers, the first standing light wave is tuned in such a way that its frequency is exactly in the middle of the frequencies needed to excite the states $|2\rangle$ or $|3\rangle$ to the first excited state $|e\rangle$. In order to generate superpositions of these states, the first standing light wave is put in a microwave resonator, i.e., it is 'sandwiched' between microwave pulses which are resonant with the energy difference between the states $|2\rangle$ and $|3\rangle$. The effect of the first microwave pulse is that the Bragg-reflected beam has a phase shift of π. This phase shift may be used as a path marker since after passing the second microwave pulse, the reflected beam is in state $|2\rangle$ whereas the transmitted beam is in state $|3\rangle$. Hence, the ground state of the ^{85}Rb atoms in the far field stores the information of whether they stem from the transmitted or the reflected beam. However, the states $|2\rangle$ or $|3\rangle$ which

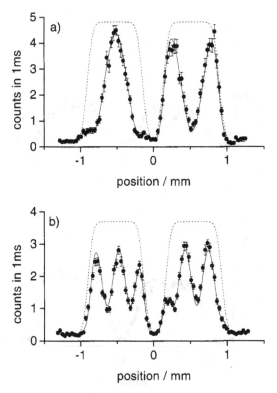

Fig. 7.7. Spatial fringe pattern in the far field of the interferometer. The *solid line* is a fit in the experimental data (*circles*). The *dashed line* shows the independently measured envelope. Reprinted by permission from Macmillan Publishers Ltd: Dürr, S., Nonn, T., and Rempe, G.: Nature **395**, 33–37 (© 1998)

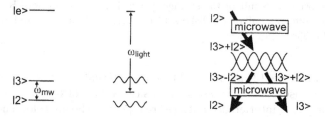

Fig. 7.8. Path marker. Reprinted by permission from Macmillan Publishers Ltd: Dürr, S., Nonn, T., and Rempe, G.: Nature **395**, 33–37 (© 1998)

store the path information are orthogonal. Therefore, the interference terms cancel in accordance with (7.22) of the general scheme sketched above. And indeed, no interference is observed in the far field (see Fig. 7.9).

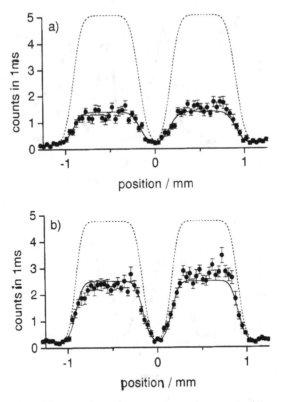

Fig. 7.9. Interference lost. Atomic far-field pattern obtained in the interferometer with which-way information stored in the internal atomic state. In part (**a**), only atoms in state $|3\rangle$ are detected so that the which-way information is read out, in contrast to part (**b**), where all atoms are detected. In both cases, the interference fringes are lost as a result of the storage of which-way information. Reprinted by permission from Macmillan Publishers Ltd: Dürr, S., Nonn, T., and Rempe, G.: Nature **395**, 33–37 (© 1998)

Finally, in stage (3) an observable of the internal states with eigenvectors $|3\rangle \pm |2\rangle$ is measured.[107] In these superpositions the which-way information is lost. Accordingly, interference patterns reappear in the far field. The interference patterns belong to sub-ensembles of the ensemble without interference of stage two. They form fringes and antifringes which add up to the total ensemble of *all* atoms which pass through the first standing light wave, are afterwards prepared in different ways, and are finally detected in the far field.

One crucial remark needs to be made concerning stage (2) of the experiment. Dürr et al. emphasize that in order to destroy the interference pattern it is sufficient to *store* the which-way information. The internal states of the

---

[107]Dürr et al. 2000a, 57.

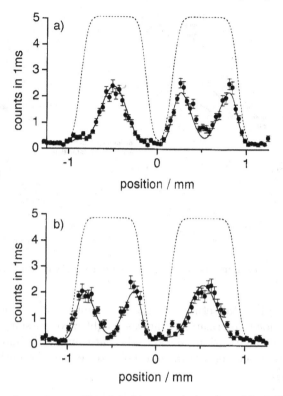

Fig. 7.10. Quantum erasure. Reprinted by permission from Macmillan Publishers Ltd: Dürr, S., Nonn, T., and Rempe, G.: Nature **395**, 33–37 (© 1998)

atoms arriving on the screen need not be read out.[108] The interference fringes disappear by the mere fact that the atoms in the far field store the information of whether they stem from the transmitted or the Bragg-reflected atomic beam. The waves are simply prepared in such a way that the interfering terms cancel due to the orthogonality of the internal states. Hence, the term *path information* must not be taken literally. In all stages of the experiment, the atomic beam is prepared as a wave. The experimental setups of stages (1) and (3) prepare interfering waves, whereas the arrangements (0) and (2) prepare non-interfering waves.

In 2003, Walborn et al. realized a similar experimental scheme with photons and a double slit. In this experiment, the polarization state of the photons is used as the internal degree of freedom which is needed in order to store the path information. The double slit is equipped with quarter-wave plates. They have the effect of generating a phase shift of $\lambda/4$ between the components of the electric field strength and preparing circularly polarized

---

[108]Dürr and Rempe 2000a, 55.

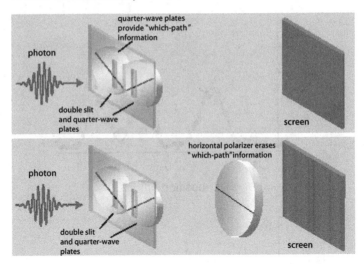

Fig. 7.11. Path marking and its erasure for photons (Walborn et al. 2003, 341)

light. The experimental setup is such that a light wave passing through slit 1 has right-handed polarization, whereas a light wave passing through slit 2 has left-handed polarization. Marking the path makes the interference pattern disappear; inserting a horizontal polarizer erases the path information (see Fig. 7.11):

> [...] quarter-wave plates are placed behind the slits, converting the photons to left-circular and right-circular polarizations respectively. This makes "which-path" information available to the experimenter, and the interference fringes disappear [Fig. 7.11 (top)]. However, a horizontal polarizer [Fig. 7.11 (bottom)] converts either of the circular polarizations to horizontal polarization so there is now no longer any way to distinguish between photons that went through the top and bottom slits. "Which-path" information is erased, and [...] the interference fringes return.[109]

In order to study a *quantum eraser with delayed choice*, an entangled photon pair is generated by a non-linear crystal such as those used in the Mach–Zehnder interferometer experiment discussed above. The crucial point is that the quantum erasure of the path information can now be prepared by means of a polarizer which is put in the *other* branch of the experiment. Now, the experimental arrangement is as follows. The double slit is equipped with quarter-wave plates with the axes at ±45 degrees. They have the effect of generating a phase shift of $\lambda/4$ between the *diagonal* components of the electric field strength and preparing circularly polarized light. Hence, whenever the light is horizontally or vertically polarized, the quarter-wave plates behind

---

[109] Walborn et al. 2003, 341.

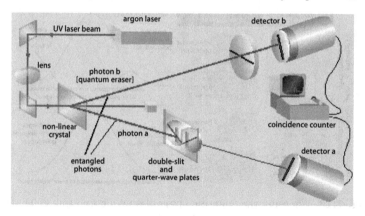

Fig. 7.12. Quantum eraser with delayed choice: "In the delayed-choice experiment, we create a pair of entangled photons, which we will call a and b, in such a way that whenever photon a is observed to have horizontal polarization, photon b will necessarily be vertically polarized, and vice versa." (Walborn et al. 2003, 342)

the two slits prepare circularly polarized light waves of opposite orientation. But on light with diagonal polarization, they have no effect. In this arrangement, the three stages of (1) preparing interference, (2) marking the path, and (3) erasing the path information are as follows:

(1) The double slit prepares interfering waves. The polarizer is not put before the photodetector which detects interference or non-interference (like in Fig. 7.11), but in the other branch of the experiment, in order to polarize the *other* photon of the entangled photon pair (see Fig. 7.12).

(2) A horizontal or vertical polarizer acting on photon b prepares photon a in a quantum state that is a linear combination of two orthogonal diagonal polarizations. In this way, the quarter-wave plates behind the double slit prepare photon a in a state with path information.

(3) However, when a diagonal polarizer acts on photon b, photon a is also diagonally polarized and the quarter-wave plates have no effect on it. Hence, a diagonal polarizer erases the path information stored in the photon polarization and re-establishes interference.

(4) Now, turning the polarizer between horizontal, diagonal, and vertical polarization, makes the wave that passes the double slit and the quarter-wave plates switch between interference fringes, no interference, and anti-fringes. Again, the coincidence counter is crucial. It sorts the total ensemble *without* interference pattern into two sub-ensembles with fringes and antifringes, which add up to the total ensemble (see Fig. 7.13). The delayed choice is just the choice of looking at one sub-ensemble rather than the other. It "amounts to a change in bookkeeping, not a change in history." [110]

---

[110] Walborn et al. 1993, 343.

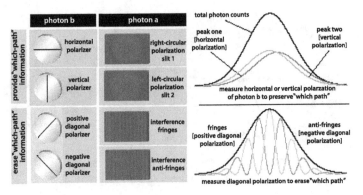

Fig. 7.13. Choosing the quantum ensembles (Walborn et al. 2003, 343)

At this point, Afshar's recent double-slit experiment[111] should be mentioned. It also comes under the title of a which-way experiment, although it is *not* one. Afshar added the following devices to the usual double-slit experiment with single photons. Behind the double slit, he put a lens which images the two openings of the double slit to the screen according to the laws of geometrical optics. The lens has the effect that the interference pattern of the double slit disappears and two maxima of light intensity appear, which are the images of the double slit. Next, he put a grid before the lens which exactly occupies the minima of the double slit interference pattern and lets the maxima pass through. The well-separated images of the double slit still appear at the screen, even though a weak, fuzzy interference pattern is superposed on each of them.

Afshar claims that the imaging of the lens changes the diffraction at the double slit into a path measurement, giving rise to a *which-way* experiment. This is not true, however. The imaging procedure only prepares waves which obey the laws of geometrical optics. The double slit, the lens, and the grid before the lens nevertheless prepare waves. Only in the particle detections on the screen are photons finally localized. Indeed, with a high intensity laser beam, or with coherent classical light, the experiment should give exactly the same results; the only difference being that the photons of a low intensity beam are detected as particles in the sense of the operational particle concept (OP) of Sect. 6.3. But no measurement of whether the photons pass through one or the other slit is made, as a *which-way* experiment would indeed require. Obviously, the lens *neither* detects single photons *nor* stores path information for the passage of the photons. Afshar's experimental device does not give rise to any information about the photon path. Only the idea of a particle trajectory makes him think that a photon which is detected in one of the images *must* have passed through the respective slit.[112] But

---

[111] Afshar 2003.

[112] This point is also criticized in Kastner 2005, 653.

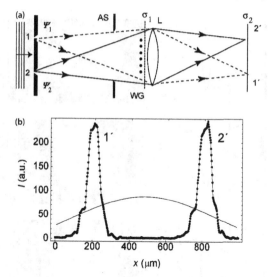

Fig. 7.14. Afshar's double slit experiment (Afshar 2003)

this idea does not give rise to any measurement. In a which-way experiment, a path measurement makes the interference fringes disappear. But here the interference fringes disappear due to optical imaging. This does not give rise to a path measurement. The disappearance of the interference fringes in favor of the images of the two slits can be completely explained in terms of geometrical optics.

Hence, the Afshar experiment is no more or less spectacular than the original double-slit experiment with single photons or electrons. It does not reveal any new features of wave–particle duality at all. Only due to the idea that optical imaging is a path measurement[113] does the experiment *seem* to be striking. However, a *real* path measurement would require the storage of path information *before the lens*. In contradistinction to the which-way experiments discussed above, Afshar illegitimately claims that his experiment is of the same type. In his experiment, a sequence of preparation procedures prepares waves in different superpositions. Without the lens, the light waves superpose to the interference pattern of the double slit. With the lens, they superpose to the images of the two slits. Only at the end are particles detected. If the particle picture is applied *before* the final measurement to the waves that are prepared, conceptual confusion arises.

---

[113] In his note 23, Afshar attributes this idea to Wheeler, but this seems hardly compatible with Wheeler's views, e.g., expressed in Wheeler 1979–1981.

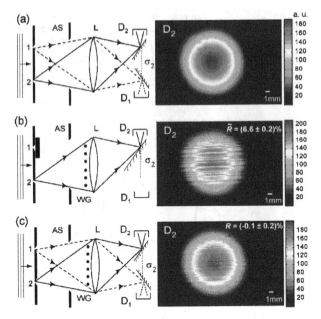

Fig. 7.15. Afshar's results (Afshar 2003)

7.6 The Causes of the Phenomena

Wave–particle dualism and the phenomena of quantum optics shed strikingly new light on the causal particle concept. The results of the experiments with single photons or atoms discussed in this chapter do not have unambiguous causes. This holds in a two-fold sense. On the one hand, it is no longer possible to reconstruct unambiguous causal stories of individual particle detections. On the other hand, the informal concept of causality dissolves into *several* theoretical concepts of causality. They are all employed in the explanation of a quantum phenomenon such as interference, its disappearance due to the storage of path information, and its reappearance due to quantum erasure.

For all the experiments discussed in this chapter, it is no longer possible to tell unambiguous causal stories about the detection of the particles under investigation. There are events that make up an ensemble with interference, as if each event of the ensemble had *two* causes. This is most strikingly demonstrated in case of the Pfleegor–Mandel experiment. Here, one photon comes from two coherent lasers, and the interference fringes observed in an ensemble of *many* individual measurements can only be explained by assuming that each photon interferes with itself (as Dirac emphasized[114]). They come in the superposition (7.3) and not in a mixture of non-interfering photons from lasers L_1 and L_2. The case of subsequently polarized photons is by no means

---

[114]Dirac 1958, 9.

better from a causal point of view. Polarizers are absorbers. They absorb all radiation polarized orthogonally to them. Therefore, two polarizers crossed at 90° do not let any light pass through. If a third polarizer is put between them, however, a light signal of weak intensity reappears as soon as it is no longer orthogonal to any of the other two polarizers. In the particle picture, no causal story can explain why a third absorber makes a signal reappear that was previously cancelled by two absorbers. In the which-way experiments too, the particles detected on the screen do not have an unambiguous causal story, as long as there is no path information stored or as soon as the stored path information is erased. A causal story in terms of path information is only available at the price of vanishing interference.

The reconstruction of an unambiguous causal story only employs *one* concept of causality, namely the requirement of an *unambiguous relation and complete connection between cause and effect*. According to the traditional principle of causality, *any* effect must have a cause that satisfies this requirement. Due to the notorious quantum measurement problem, the traditional principle of causality fails at the level of individual events, giving rise to causal gaps in the explanation of subatomic processes. The missing unambiguous causal stories of quantum optics are most striking examples for this. (The case of particle tracks is completely different. Here, an unambiguous *retro*diction of what happened along a given track is usually possible, even though the quantum mechanics of scattering in general precludes a *pre*diction.) Beside these causal gaps, however, stand several strict physical laws. On the one hand, they *connect* the individual events or particle detections at the ensemble level by probabilistic laws. On the other hand, they put *constraints* on these connections, for example concerning the possibilities of signalling.

Indeed, in any of the experiments discussed above *several* accounts of causality are involved. For subatomic particles, their entanglement, propagation, and detection, the relation of cause and effect undergoes a similar conceptual dissociation to the particle concept. In quantum physics, causality becomes a cluster concept. The different physical theories involved in the models of quantum physics make causality precise in different axiomatic terms, but these do not come together. In the experiments of quantum optics, the causal concepts introduced by quantum theory and special relativity are employed in parallel with the traditional principle of causality but these accounts of causality do not all come together. In particular, the following causal concepts are relevant for a theoretical explanation of the phenomena of quantum optics:

(C0) **The Traditional Principle of Causality.** This is the principle that every effect has a cause to which it is unambiguously related and completely connected. Whenever experimenters prepare a single atom, electron, or photon and send it through an experimental device such as an atom interferometer, a Mach–Zehnder interferometer, or a double slit, use a tricky method to obtain path information, and detect individual particles on a screen, they employ the traditional concept of causality.

(C1) **The Unitary Evolution of the Wave Function.** This is causality in the sense of determinism. The unitary evolution of the wave function is determined by a wave equation such as the Schrödinger equation, the Dirac equation, the Maxwell equations, or some coupled quantized field equations. According to this kind of causality, the wave function propagates. This is what the preparation procedures of quantum experiments take into account.

(C2) **Conservation Laws.** This is causality in the sense of Bohr's account of complementarity, conceptual level (3) (see Sect. 7.2.2). Conservation laws such as the principles of momentum and energy conservation are constraints for all subatomic scattering processes, including the emission, propagation, and absorption of light quanta.[115] Causality in the sense of conservation laws is closely related to the symmetries and superselection rules of physics, as discussed in Sects. 5.4 and 6.4.

(C3) **Einstein Causality.** This is the principle that no signalling is possible over spacelike distances. It follows from the principles of special relativity, i.e., from Lorentz invariance, and is closely related to relativistic energy–momentum conservation. In quantum field theory, this principle is also called *micro-causality*. According to this, all field operators commute at spacelike distances.

(C4) **The Non-Unitary Reduction of the Wave Function.** Measurement reduces the wave function. This gives rise to probabilistic causality. In addition, quantum measurements are irreversible. They are associated with decoherence and some kind of energy dissipation due to the spontaneous emission of field quanta, particle absorption by the detector, ionization of detector atoms, etc., and the amplification of such effects to some signal in a macroscopic measurement device.

The causal concept (C0) is informal and pre-theoretical. As far as physics aims at causal explanations, it should be made precise in terms of the causal concepts (C1)–(C4). However, (C4) is obviously at odds with (C0). Quantum theory does not assign to the outcome of a measurement an individual cause to which it is unambiguously related and completely connected. This is the quantum measurement problem. In order to circumvent it, physicists regard the quantum probability waves they prepare in accordance with (C1) *as if these waves were individual physical objects*. For entangled quantum systems, the unitary evolution (C1) of the wave function and the conservation laws (C2) predict non-local correlations of the measurement results (C4). Einstein thought that his signalling causality (C3) precludes such non-local quantum correlations. The EPR paper is about this apparent conflict.[116] Today it is

---

[115] However, this holds only for real field quanta and scattering processes, not for the virtual processes discussed in Sect. 6.4.3.

[116] Einstein et al. 1935.

known whether the non-local quantum correlations violate Einstein causality. They do *not*. Signalling over spacelike distances by means of quantum entanglement is impossible. Hence, the four kinds of causality (C1)–(C4) involved in quantum entanglement coexist peacefully. Nevertheless, their relations remain a challenge for any variant of *scientific realism* that is based on Newton's and Planck's belief in the unity of Nature, or unified laws of physics. Hence, Bohmian hidden variable theories, the many-worlds interpretation, and other forms of quantum metaphysics arose, bringing with them their own causal problems.

8 Subatomic Reality

Quantum phenomena show that subatomic reality does not have a classical structure. The world is *not* made up of particles in the sense of the traditional mereological and causal particle concept. Nature is *not* as "conformable to herself and simple" as Newton thought.[1] In the course of the quantum revolution, the traditional metaphysical assumptions about the constituent parts of matter were refuted at the same time as the light flashes and particle tracks caused by α-rays showed that *there must be* subatomic particles. The quantum revolution showed that subatomic reality is *not* as classical physics and traditional metaphysics wanted it to be. Nevertheless, the philosophical debate about the interpretation of quantum theory focused mainly on the following options: *either* be agnostic about or even deny the existence of subatomic reality; *or* try to re-establish the metaphysical picture of a reality that comes as close as possible to the lost world of classical physics.

In this book, the concepts, methods, and experimental results of subatomic physics have been analyzed in order to shed new light on the old debates about subatomic reality. In particular, some crucial metaphysical assumptions behind the experimental method of modern physics and the traditional particle concept have been made explicit; the empirical basis, models, language, and unifying principles of particle physics have been investigated; the family of current particle concepts has been analyzed; various accounts of wave–particle duality have been compared; and some crucial experiments of quantum optics have been discussed in order to demonstrate what wave–particle duality means in physical practice.

The only remaining issue is to summarize the philosophical conclusions. They come in the following order. I summarise my position regarding *scientific realism* (Sect. 8.1), draw my conclusions concerning the meaning of quantum concepts (Sect. 8.2), discuss the fate of the *traditional particle concept* (Sects. 8.3–8.4) and *wave–particle duality* (Sect. 8.5), and sketch my *critical view* of subatomic reality (Sect. 8.6).

---

[1] Newton 1730, *Query 31*.

8.1 Scientific Realism Reconsidered

In contradistinction to the empiricist and constructivist positions discussed in Chaps. 1–2, I defend a mild epistemic version of scientific realism. According to this, the well-confirmed measurement results of subatomic physics give rise to knowledge of empirical reality. This knowledge supports three versions of scientific realism: property realism, structural realism, and causal realism. However, the measurement results of particle physics and quantum optics support each of them only to a certain degree.

1. Property Realism

The contingent events observed in a particle detector and the measured values of physical quantities attributed to them certainly belong to empirical reality. Hence, particles exist in the sense of the *operational particle concept* (OP) of Sect. 6.3:

(OP) Operationally, particles are

> (MESQ)   collections of *mass m, energy E, spin s, charge q,*
>
> (LOCPD)   *localizable* by a particle detector,
>
> (INDEP)   *independent* of each other.

The local events observed in a particle detector are independent particle detections that obey conservation laws and sum rules for mass, energy, spin, and charge. The operational particle concept supports a realism of the collections of these dynamic properties. These collections of dynamic properties are *natural kinds* detected under well-defined experimental conditions in a macroscopic environment. But this property realism requires *neither* a metaphysical realism of carriers of these properties in a classical sense, *nor* a platonism of properties. It admits that neither the properties nor their collections exist on their own. They may only exist in a macroscopic measuring device.[2]

2. Structural Realism

In the experiments of particle physics and quantum optics, the physical properties of subatomic particles exhibit *quantum structures*. The algebraic structure of quantum observables is a true description of subatomic phenomena, whereas the classical construal of physical reality fails in view of the unexpected quantum phenomena. This is a case for structural realism. The empirical basis of subatomic scattering events (Sect. 3.3) and the conservation

---

[2]See Falkenburg 1993b.

laws according to which they are classified give further support to structural realism. This justifies the physicist's belief in the dynamic or 'internal' *symmetries* associated with these conservation laws. Full-fledged belief in the corresponding quantum field theories and the underlying concept of gauge invariance,[3] however, is quite another topic. The reasons for such a belief and the ontological claims to which it commits one are subject to the philosophical discussion about the philosophical interpretation of quantum theory *beyond* measurement, which has *not* been discussed in this book.

3. Causal Realism

Finally, a moderate version of causal realism is in accordance with the phenomena and measurement results of particle physics. The causes of particle detections, tracks, scattering events, resonances, pointlike scattering cross sections, jet events, etc., belong to empirical reality. (Here, the famous miracle argument in favor of scientific realism may be employed.) Even the causes of the hydrogen Lamb shift or the scaling violations in deep-inelastic lepton–nucleon scattering are real. The causes of all these phenomena are quantum processes that obey certain conservation laws and symmetries. However, in the quantum domain these causes are non-local and context-dependent. In general, they can be isolated neither by experimental methods nor in terms of the Feynman diagrams of quantum field theory. The non-local correlations of particle detections observed in quantum optics support this, too.

The quantum structure of the subatomic domain refutes the classical particle concept as well as the underlying metaphysical assumptions. Subatomic particles are neither substances on their own, nor well-distinguished parts of matter or light in a classical sense, nor are they local causes of particle tracks and scattering events. Classical metaphysical realism is untenable. Similarly, a global realism that interprets the quantum mechanical wave function literally (instead of probabilistically) is untenable. It is not only *without* any empirical support. It is *at odds* with the empirical basis of the operational particle concept itself, for the following reason: all quantum processes obey *probabilistic laws*. Empirically, they only appear as certain kinds of *events* that happen with certain *relative frequencies*. Here, the very nature of an event is that something happens locally, in a macroscopic measuring device or environment. According to a quantum theory on its own, i.e., the unitary development of the quantum mechanical wave function, *nothing* actually happens. In this sense, the wave function Ψ is only about possible measurement results but not about the actual world. In addition, any realism that comes in terms of many worlds or hidden variables must cope with the non-local correlations of *relativistic* particles or field quanta, which is a difficult, if not impossible task.

---

[3] See Kuhlmann et al. 2002 and Lyre 2004.

8.2 The Meaning of Quantum Concepts

One of the obstacles for classical realism is that the meaning of quantum concepts on their own is quite meager. The meaning of a physical concept has operational, axiomatic, and referential aspects (see Appendix A). In Chap. 5 it was shown in detail how the operational, axiomatic, and referential aspects of quantum concepts fall apart. The attribution of physical properties to subatomic particles is not based on one unified theory but on *several incommensurable theories*. The current concepts of subatomic particles have axiomatic aspects which stem from quantum field theory, operational aspects which rest on classical measurement laws, and referential aspects which are unclear.

In particular, the operational aspects of the mass, charge, spin, etc., attributed to subatomic particles depend on classical measurement methods and laws; the axiomatic aspects of these dynamic quantities are associated with the conservation laws and symmetries of some quantum field theory; and the way in which these quantities refer to subatomic particles or quantum processes remains unclear. The group theoretical particle concept (GT) of Sect. 6.4 only refers to particle *types*, but not to individual particles. The gap between it and the operational particle concept (OP) is only closed by belief, by a trust in the unity of physics that remains unjustified in view of the unresolved quantum measurement problem.

Indeed, the gap between the operational particle concept and its quantum counterparts is the strongest case for incommensurability in Kuhn's sense. The operational particle concept is based on classical measurement methods. It is given in terms of the familiar quantities mass and charge *plus* the quantum concept of spin which, however, is measured in half-integer units of angular momentum. Classical physical quantities such as mass, charge, and angular momentum are represented by *real-valued functions*, whereas the observables of a quantum dynamics correspond to *operator-valued measures*. The former are attributed to *individual phenomena* and the latter to *probabilistic ensembles*, that is, to *many* measurement results obtained under the same experimental conditions.

Therefore, the semantic gap between classical concepts and quantum concepts can only be closed at the probabilistic level. The decoherence approach, too, is only a probabilistic solution of the quantum measurement problem. At the individual level, quantum theory is semantically inconsistent in the sense that the axiomatic and the operational aspects of quantum concepts fall apart. The heterogeneous measurement theory of particle physics closes the gap between them only partially. It employs classical and average quantum laws for the data analysis of individual particle tracks and combines them with statistical methods in order to correct the results at the probabilistic level.

This measurement method is based on constructing the *scales of physical quantities* down to atoms, nucleons, and quarks. Axiomatically, the construction of the scales relies on the Archimedean axiom of the theory of real numbers (see Appendix A). In addition, it is based on several unifying meta-theoretical principles of physics. The most important of them are the tacit assumption that physical quantities are dimensionally invariant (see Appendix B), the symmetries and conservation laws of physics, the superselection rules of quantum mechanics, and Bohr's (generalized) correspondence principle. All these principles together make it possible to construct the length, time, and mass scales. These scales have a quasi-transcendental status. Since they underlie all measurements of physics, they are conditions of the possibility of physical experience. However, the extension of the scales down to the quantum domain is not arbitrary. It is empirically supported by the use of redundant measurement methods, that is, by the coherent parallel use of semi-empirical and quantum theoretical measurement methods. Hence, even though there is no axiomatic unity of classical and quantum physics, there is a certain hidden semantic unity of physics. The coherent construction of the length, time, and mass scales gives additional credit to the *property realism* defended above.

In the scales of physical quantities, the *axiomatic* and *operational* concepts of particle physics are brought together as far as possible. In addition, Bohr's generalized correspondence principle is employed to establish *reference to subatomic scattering centers*. In the chain of models discussed in Chap. 4, the correspondence principle links classical Rutherford scattering to the measurement of form factors in nuclear physics, to the concept of 'scaling behavior' in deep-inelastic lepton–nucleon scattering, and to the quark–parton model. It was shown how, along this chain of models, the classical construal of subatomic reality does not break down at once but stepwise. Subatomic scattering centers may still be described in terms of classical charge distributions, as far as correspondence to the classical model of scattering holds. Thus, 'looking into the atom' by means of scattered probe particles gives an unambiguous, undistorted picture of subatomic scattering centers, as long as specific quantum mechanical scattering effects (exchange and spin effects) may be neglected and the scattering is non-relativistic.

Hence, quantum physics presupposes the language of classical physics. Without the familiar scales of physical quantities, any quantum theory remains an abstract, uninterpreted formalism. In this regard, Bohr's long neglected views about the indispensability of the classical language are absolutely right. Without it, even the symmetries, conservation laws, and superselection rules of quantum physics remain pure mathematics. In addition, the application of any quantum theory to individual systems makes tacit use of one or the other version of a generalized correspondence principle. It has to be emphasized again that, according to the unitary development of a quantum dynamics alone, *nothing* does happen in the world: no *click* in a particle detector, no definite measurement outcome, no particle track in a Wilson chamber, no interference pattern at a scintillation screen, and no observable effects of an atom in a Paul trap.

8.3 The Mereological Particle Concept

Traditionally, particles were considered to be the microscopic constituent parts of matter and light. When the electron, the α-particle, and the atomic nucleus were discovered and the light quantum hypothesis was established, these particles were considered to be matter and light constituents in this sense. After the quantum revolution, only the operational particle concept (OP) and the family of quantum particle concepts discussed in Chap. 6 remained.

The mereological particle concept spells out the constitution of matter or light in terms of a part–whole relation (see Appendix E). It is based on the idea that matter may be decomposed into constituent parts and recomposed from them. This idea is closely related to the methods which are constitutive for modern physics, namely the traditional method of analysis and synthesis or Galileo's *resolutive–compositive method* applied in experiments of particle physics. As shown in Chap. 2, the experimental and mathematical methods of modern physics are intimately related to the traditional metaphysical presuppositions of physical theories sketched in Chap. 1. They aim at the decomposition and composition of natural phenomena, and they rely on the separability of causal agents in nature. Traditional metaphysics conceived of these causal agents in terms of substance. Modern physics conceives of these causal agents in terms of atoms, forces, and subatomic particles.

Hence, the mereological and the causal particle concepts are closely related, even though in the quantum domain the latter finally vanishes. Insofar as the experiments of modern physics aim at causal analysis, they aim at the isolation of the constituent parts of matter and light, too. Traditional metaphysics considered these causal agents as independent entities in their own right. Modern causal realism is indebted to this metaphysical assumption up to the present day. However, 20th century physics taught us that the subatomic causal agents are *not* like this. In the quantum domain, experimental analysis underlies certain limitations. In particular, according to quantum mechanics and quantum field theory the causes of non-local quantum correlations are neither local nor isolable. Therefore, only a weakened causal realism is tenable, as suggested in Sect. 8.1.

But for atoms and their constituent parts, the mereological particle concept has to be weakened, too. According to classical mechanics, the constituent parts of matter are well-distinguished in spatio-temporal terms, i.e., in terms of a well-defined extensional mereology (see Appendix E). The bodies that make up a mechanical system such as the solar system do not overlap and they are individuated by unambiguous trajectories.

This is different in quantum mechanics. Here, the constituent parts of a compound system may be entangled and the constituent parts of matter obey weaker criteria. In mereological terms, they overlap spatio-temporarily as well as dynamically. Here, spatio-temporal overlap means that not all subatomic particles are individuated by trajectories, whereas dynamic overlap only concerns the quark–antiquark and gluon field contents which contribute more to the nucleon momentum as the energy of a scattering experiment increases (see Sect. 6.5). In general, the constituent parts of molecules and atoms are described by a totally (anti-) symmetrized N-particle wave function:

$$\Phi^N = \sum_{k_1, k_2, \ldots, k_N} c_{k_1, k_2, \ldots, k_N} \phi_{k_1}^{(1)} \otimes \phi_{k_2}^{(2)} \otimes \ldots \otimes \phi_{k_N}^{(N)} \pm \text{permutations}. \quad (8.1)$$

Hence, they do not have well-distinguished parts in the sense of separable dynamical states in Hilbert space. What remains are scattering experiments that *localize pointlike structures* at a certain energy scale on the one hand (see Chap. 4 and Appendix D), and the *sum rules* for the dynamic quantities of a whole and its parts on the other hand. In particle physics, the sum rules for *mass–energy*, the *electric charge*, the other (generalized) charges *flavor* and *color*, *spin*, and *momentum* are crucial.[4]

With regard to the mereological particle concept, too, the classical picture of subatomic reality does not break down at once, but stepwise. This holds in several regards:

1. **Isolability of Matter Constituents.** Even though electrons, protons, and neutrons do not have separable wave functions inside the atom, they may be split off an atom and localized in a particle detector. They are empirical parts of matter in the sense of the operational particle concept and certain conservation laws and sum rules. For quarks, this is different. No isolated fractional charges have been measured. That there are no isolated quarks is explained in terms of quark confinement. The quark constituent parts of matter can only be localized *inside* the nucleon, in deep-inelastic lepton–nucleon scattering with scaling behavior, in terms of pointlike structures which obey certain sum rules for mass, charge, spin, and momentum.

2. **Distinctness of Field Quanta.** Fermions have half-integer spin. They obey Fermi statistics and Pauli's principle, according to which each particle of a many-particle compound system is in another dynamic state, where the states have distinct quantum numbers. Hence, even though the fermion parts of matter are not spatio-temporally individuated like classical particles, they are at least dynamically distinct, i.e., well-distinguished in terms of their quantum numbers. In contradistinction to them, bosons have integer spin and obey Bose–Einstein statistics. Many bosons may occupy the same quantum state or field mode. Hence, the photon and the

---

[4]See, e.g., Perkins 2000, 166–168.

field quanta of the other interactions cannot be distinguished in terms of their quantum states. Here, *any* criterion for mereological distinctness fails. In terms of formal mereology, they overlap (see Appendix E). Therefore, one may say that matter is made up of electrons, protons, neutrons, and quarks as constituent parts. But one should *not* say that light consists of photons. Light is *not* constituted by well-distinguished light atoms or quanta in the way Newton conceived of them in 1704 or Einstein still conceived of them in his famous 1905 paper on the photoelectric effect.

3. **Energy-Dependent Field Content of a Compound System.** A fermion compound system may turn out to contain additional boson fields which have no well-distinguished parts. The experimental test of the momentum sum rules for the quarks in the nucleon shows that a substantial part of the nucleon momentum is *not* carried by the three 'valence' quarks that make up the proton or neutron. The amount of this additional field content depends on the scattering energy of the probe particles. It is explained in terms of the field quanta of the strong interaction, the gluons, and a so-called 'sea' of virtual quark–antiquark pairs. The amount of both increases with the energy of the probe particles of a scattering experiment. Similar effects are observed in measurements of the sum rules for spin and angular momentum.[5] Hence, with increasing scattering energy, the dynamic content of the proton or neutron increases, too. In this sense, subatomic structure is *relational*. It changes with the scattering of the probe particles. It has even been argued that with increasing energy (or spatial resolution) deep-inelastic scattering reveals fractal structures:

> With increasing resolution, quarks and gluons turn out to be composed of quarks and gluons; which themselves, at even higher resolutions, turn out to be composite as well [...]. The quantum numbers (spin, flavour, colour, ...) of these particles remain the same; only the mass, size, and effective coupling α_s change. Hence, there appears to be in some sense a self similarity in the internal structure of strongly interacting particles.[6]

It has to be noted that this interpretation is based on quite a literal interpretation of virtual processes described in terms of Feynman diagrams. In order to support such a claim, the same considerations may be made as in the case of the hydrogen Lamb shift or the electron $(g-2)/2$ measurement and calculation.[7] However, on the basis of *jet events*, physicists hope to single out gluonium states and quark–gluon plasma as matter constituents.[8] Hence, the search for better experimental methods for disentangling the subatomic dynamic content of matter continues even in a

---

[5] See Perkins 2000, 162–192, in particular 166–168 and 186–190; Povh et al. 1999, 107–111.

[6] Povh et al. 1999, 111.

[7] See Sect. 6.4, in the part on virtual particles.

[8] Perkins 2000, 190–192.

domain where the classical picture of well-distinguished parts of matter completely breaks down.

8.4 The Causal Particle Concept

The causal particle concept is not just *weakened* in the subatomic domain. It simply *fails*. There are particles and there are causes, but the particles are *effects* and their causes are *not particles* but quantum waves and fields.

The causal particle concept is closely related to the traditional metaphysical assumptions about the causal agents of nature (see Chaps. 1–2 and Sect. 8.3). According to this, particles are the local causes of local effects in particle detectors, in particular of event sequences or particle tracks. The lesson from quantum theory is that there are no such local causes. This has two aspects. *First*, one event does not cause the next event (as the traditional concept of causality would require), since the relation between the events is not deterministic.[9] *Second*, in general the causes do not act locally, as the identification of the causes with particles would require. The causal agents behind the repeated position measurements which make up a particle track are *not* local. The same causal agents may cause the particle detections which make up the interference pattern behind a double slit.

For all the reasons explained in Chaps. 5–7, particles are observable local effects rather than the unobservable causes of such effects. What remains is the operational particle concept (OP). Current quantum theories no longer support a causal particle concept. They are at odds with any attempt to explain the experimental results of particle physics in terms of individual causes of particle tracks, scattering events, resonances, etc. It is an irony of the history of science and philosophy that the most serious objections against the causal particle concept do not result from an empiricist or instrumentalist view of physics. They come from physics itself.

In particular, the results of the recent experiments of quantum optics *cannot* be due to local causal agents. There are no single photons taking this or that path through the branches of a Mach–Zehnder interferometer, even though the experiments are still misleadingly called *which-way* experiments. There are no single photons *created* by a polarizer (i.e., *absorber*) when the latter is put in-between two crossed polarizers together absorbing *all* photons, in the polarizer experiment discussed in Sect. 7.3. And also, most strikingly, there are no single photons coming from this or that laser in the Pfleegor–Mandel experiment. The idea that the photon field *consists* of photons is simply misleading (see above Sect. 8.3). According to quantum field theory, there is only a fluctuating quantum field that gives rise to a certain statistical

---

[9]The relation is probabilistic. But probabilistic causality does not help either, since the probabilistic cause is not an *event* but the quantum state that stems from a preparation procedure or a measurement.

distribution of photon detections, in accordance with the principle of energy conservation.

So what is the cause of a particle track? And what is the cause of the cor- related photon detections in the interference experiments of quantum optics explained in Chap. 7? In view of relativistic quantum field theory, neither the idea of a physical pilot wave nor hidden variables in Bohm's sense will do. In addition, a *variety of causal concepts* is employed in the physical and philosophical interpretation of such quantum phenomena (see Sect. 7.6). In the Pfleegor–Mandel experiment, it is no longer possible to tell *unambiguous* causal stories about the single particle detections. In the polarizer experiment mentioned above, *no* causal story in a traditional sense may explain the reap- pearance of *some* photons after inserting the third polarizer. In both cases, the pre-theoretical (i.e., traditional) concept of causality is at odds with the unitary evolution *plus* the non-unitary reduction of the wave function. And in the case of EPR correlations, the unitary evolution of the product wave function of an entangled 2-particle state, its non-unitary reduction, and the validity of conservation laws are at odds with Einstein causality – as long as the latter is interpreted in terms of local causal agents and not just in terms of the possibilities of signalling.

Regarding the perhaps hopeless quest for a *pre-theoretical* account of causality, the only viable answer in view of such quantum phenomena is this: The causal agents behind them are non-local quantum processes which obey certain conservation laws, give rise to local particle detections, and exhibit wave–particle duality.

8.5 Wave–Particle Duality

Due to the quantum revolution, there is no longer a unique particle concept today but rather a *family of particle concepts*. At a macroscopic scale, the classical particle model is still in use. (Its best realization is the celestial bodies in the solar system.) In subatomic physics, however, there are several quantum particle concepts. At this point, conceptual confusion may arise between philosophers and physicists. The particles of particle physics are *not* considered to be classical but to show *wave–particle duality*. Indeed, today this view belongs to the tacit theoretical background knowledge of any physicist.

In Chap. 6, the family of particle concepts was analyzed. What all suc- cessors of the classical particle concept (except virtual field quanta) still have in common is the *independence* property (INDEP). It is a crucial feature of the operational particle concept (OP), too (see Sects. 6.3 and 8.1). Like the mereological and causal particle concepts discussed above, the independence property is closely related to the meta-theoretical principles of experimental analysis, or to Galileo's resolutive–compositive method which is constitutive of modern physics. Traditional atomism and classical statistical mechanics postulated separable, isolated parts of matter as the causal agents of physical phenomena. These parts of matter and causal agents were conceived to be independent, i.e., substances in their own right. In the course of the quantum revolution, the degree of independence attributed to the parts of matter and the causes of particle phenomena shrank. This shrinking independence corresponds to the way in which the classical models of subatomic scattering centers and the parts of matter break down in a stepwise manner.

In Chap. 6, I subsumed several different concepts under the independence property (INDEP), and I was not very happy with this. The many particle concepts discussed there roughly employed two sub-meanings of (INDEP) that stem from classical mechanics but had to be further differentiated:

(INDEP) Independence means that particles

(UNCOUP)   *may* be in *non-interacting* or uncoupled states, and

(UNCORR)   their *initial conditions* are *statistically uncorrelated*.

(UNCOUP) is dynamic independence, (UNCORR) is statistical independence. In classical mechanics, (UNCOUP) implies that the carriers of dynamic properties are entities in their own right, that is, substances in the sense of traditional metaphysics. Such substances interact independently of each other. For classical particles, this implies the statistical behavior (UNCORR). Hence, in classical mechanics, (UNCORR) derives from (UNCOUP). Both independence properties agree with the methodological (and metaphysical) principle of decomposing matter into independent parts. Dynamic independence means that particles are independent causal agents on their own. This gives rise to statistical independence and in particular to independent particle detections. Both meanings are closely related to the goals of causal and mereological analysis and to the constructive features of the experimental method explained in Chap. 2. Nevertheless, (UNCOUP) and (UNCORR) are more than mere construals, because otherwise the experiments of physics would not be possible and the concepts of a physical dynamics would not be tenable (see Sect. 6.6).

In particle physics, (UNCORR) is closely related to the statistical properties of single particle detections, whereas (UNCOUP) is a meta-theoretical property of particles and fields alike. In the quantum domain, together with the classical particle concept, the independence property breaks down stepwise and becomes differentiated with the different quantum particle concepts. The group theoretical particle concept (GT) involves (UNCOUP) without (UNCORR); conversely, the operational particle concept (OP) employs (UNCORR) without (UNCOUP). In addition, in the quantum domain, *both* meanings of (INDEP) underlie various restrictions. In particular, (UNCORR) is substantially weakened for fermions and bosons; non-local correlations occur; and in view of the perturbation expansion of a quantum field theory and the need for renormalization, (UNCOUP) turns out to be a first-order approximation. (In this point, the particles of particle physics are exactly like the quasi-particles of solid state physics; see the end of Sect. 6.4.) The differentiation of these meanings of (INDEP) was forced by the limitations quantum theory puts on these assumptions.

When independence breaks down, *quantum entanglement* remains. It is described in terms of quantum mechanical many-particle states and their superpositions. Superpositions of the components of entangled quantum states were also the clue for understanding the experiments of quantum optics and their striking results, including the quantum erasers (see Sects. 7.3–7.5). In terms of particles alone, their results seem paradoxical. But once the quantum waves are implemented in the models, all causal mysteries disappear; admittedly, at the price of identifying a probability wave with a causal agent. However, what do these quantum waves mean from a philosophical point of view, adopting the modest version of *scientific realism* described in Sect. 8.1?

In physical practice, we found roughly two accounts of wave–particle duality. One follows Max Born, claiming pragmatically that in any quantum process *waves propagate* (and should therefore be prepared in approximately pure momentum states, in experiments), whereas *particles are detected* (in agreement with the probabilistic interpretation of the wave function, after Born and von Neumann). The other follows Bohr's views about quantum phenomena, claiming that waves interfere and are associated with a sharp momentum, whereas particles are independent, local, do *not* interfere, and are associated with a sharp position.

The two accounts of wave–particle duality have in common that they consider quantum waves to be real causal agents, *without* identifying them with substances or physical objects in a classical sense or local carriers of physical properties. But the experiments of quantum optics show that both views finally deal with the *same relation between particles and waves*: The probabilistic interpretation of the wave function predicts the relative frequency of empirical particle detections. Hence, wave–particle duality is closely related to the gap between theoretical probabilities and empirical relative frequencies. It is a duality of the probabilities (which are usually calculated from momentum states) and the empirical event frequencies (which are counted from the outcomes of position measurements).

What does this mean, given that there is no tenable physical interpretation of *relativistic* wave functions (or field modes) beyond the usual probabilistic interpretation? Even though the Born–von Neumann interpretation is *operational* rather than *referential*, the wave function must have *some* referential import. Indeed, the which-way experiments with quantum eraser discussed in Sect. 7.5 deal with information in an objective sense. They show that in order to destroy the interference pattern it is sufficient to *store* the path information. It is *not* necessary to read it out.[10] Hence, the stored information has a certain objectivity, even though it is *not* the objectivity of a full-blown physical object.

But in which sense is the stored path information *objective*, given that it may be erased afterwards by adding an additional experimental device? Here, it has to be emphasized again that the term 'path information' is as misleading as the term 'which-way experiment'. Neither expression must be taken literally. In all stages of such an experiment, the atomic beam is *prepared as a wave*. Therefore, the crucial objectivity claim is: In quantum physics, *the results of preparation are as objective as the measurement outcomes.*

The difference between the preparation and detection of quantum states is a neglected topic in the philosophical discussion. The main distinction between preparation and detection is the kind of quantum state obtained. Preparation usually gives rise to a non-local wave, whereas detection results in an approximately localized particle state. Operationally, and in contradistinction to measurement, preparation means *not to read out* the information contained in the wave function.[11] Recent which-way experiments indicate that the storage of information (which may or may not be read out later) is sufficient for making the difference between measurement and preparation. This point needs further conceptual analysis, but its investigation is beyond the scope of the present book.

From an empiricist point of view, the only reality of the wave function consists in conditional probabilities. Successful preparation procedures and the predictions to which they give rise remain a miracle. No empiricist can explain why the preparation of a particle beam of well-defined energy is useful for a scattering experiment. Here, Hacking's reality criterion should be employed, according to which unobservable causal agents must exist if it is possible to use them as a regular tool in a successful experiment.[12]

Indeed, Hacking's own example referred to the beam of an electron gun.[13] Hacking only neglected that the beam described by a quantum wave is *not* as real as a hammer used to knock the nail into the wall, because it is no *object* in a classical sense. From a traditional metaphysical point of view, the reality of the wave function beyond the probabilities remains a mystery. A *critical view* of quantum reality should find a middle way between empiricism and metaphysical realism.

---

[10] This was predicted by Scully et al. 1991 (see my discussion of the paper in Sect. 7.4.1) and tested in the experiment of Dürr et al. (see Dürr and Rempe 2000a, 55, and my discussion in Sect. 7.5).

[11] See the remarks in Peres 1993, 12, and my discussion in Sect. 7.3.

[12] See Hacking 1983, 22–25 and Sect. 2.4.

[13] Hacking 1983, 266–273.

8.6 Subatomic Reality: A Critical View

From a Kantian point of view, the options of metaphysical realism here, empiricism or instrumentalism there, and the current debates around them are reminiscent of the pre-critical struggles between 18th rationalism and empiricism. Is there any way out, in terms of a *critical* account of subatomic reality? For such an account, the following Kantian features of a critical view of physical reality should be reconsidered:[14]

(I)   Physical reality is *empirical*, not metaphysical. That is, the subject of physical knowledge is empirical reality in Kant's sense and not a world of things-in-themselves.

(II)   Physical reality is *structured by our cognitive tools*. To them belong the concepts and principles of:
(i)   extensive and intensive magnitudes,
(ii) substance, and
(iii) causality.
Here, it has to be noted that Kant's category of substance is *not* identical with the traditional metaphysical concepts of substances as things-in-themselves (i.e., independent entities or things-on-their-own that carry primary qualities). It is more modest. For Kant, substance is only *our* concept a priori of an entity with some stable, reidentifiable property. According to Kant's *Metaphysical Principles of Natural Science*, the substance of classical mechanics is the quantity of momentum, which is conserved. A substance of physics is subject to a conservation law.[15] Similarly, for him causality is only *our* concept a priori of a necessary link between subsequent events that helps to establish an objective time order.[16]

(III) Physical reality cannot be known completely. The concepts of *ultimate* substances and causes are only *regulative ideas*, giving rise to the methodological principles of analyzing matter into smaller and smaller constituent parts and searching for the fundamental forces of physics.

(IV)   Physical reality is thoroughly *relational*.[17] There are no substances on their own, only objects of physical knowledge related to other objects of physical knowledge. For Kant, this is the ontological consequence of rejecting things-in-themselves as knowledge-independent entities.

What remains of these Kantian topics and how do they help to develop a critical view of subatomic reality? Items (I) and (II) are in a certain tension.

---

[14]For obvious reasons, most of Kant's theory of knowledge has to be neglected here.

[15]See Kant 1786, 537–542, and von Weizsäcker 1971, 383–404.

[16]See Kant 1781/87, A 189–211/B 232–256.

[17]See Kant 1781/87, A 265/B 321: "The inner determinations of a *substantia phaenomenon* in space [...] are nothing but relations, and it is itself a sum total of mere relations."

According to (I), physical reality is empirically given; according to (II), it depends on our construals. The former claim *seems* to support empiricism, the latter constructivism. But in Kant's view, they complement each other. For Kant, the fundamental construals of our understanding *belong* to empirical reality. He thought that these construals (in particular the concepts of substance and causality) are *a priori*, not open for revision, and express the structure of the phenomena.

Quantum physics shows that subatomic reality is structured in a different way. The empirical results of atomic, nuclear, and particle physics put limitations on the use of the concepts of substance and causality and on the mathematical and experimental analysis of the phenomena. Niels Bohr thought that these limitations are indicated by Planck's constant as a minimum action.[18] In this book I have investigated the way in which the mathematical and experimental analysis of the phenomena in particle physics gave rise to the discovery of more and more particles as the microscopic constituent parts of matter and the putative causes of the local events in particle detectors, while the underlying mereological and causal particle concepts broke down stepwise and completely failed, respectively. The dynamic structure of the atom can be measured in scattering experiments and analyzed in quasi-classical terms down to the form factors of the proton and neutron (Chap. 4), but the constituent parts of the latter, namely quarks, cannot be isolated and with increasing scattering energy add field-like constituents to them, namely quark–antiquark pairs and gluons (Sect. 6.5). The quantum description of the energy loss along particle tracks shows how the causal particle concept fails. The cause of a particle track is not an isolated causal agent but a non-local quantum process (Sect. 5.3). Finally, the recent experiments of quantum optics show that neither a traditional concept of substance nor an unambiguous concept of causality applies to genuine quantum phenomena without classical correspondence (Sects. 7.3–7.6). Nevertheless, several crucial aspects of the above Kantian topics (I)–(IV) remain:

(I) Subatomic reality consists of *empirical quantum structures*.

(II) Subatomic reality is structured by the following cognitive tools:
   (i) the *scales of physical quantities* that derive from the *classical* length, time, and mass scale;
   (ii) the *conservation laws* for dynamic quantities such as mass–energy, charge, and spin; and
   (iii) the assumption of *non-local causal agents* that are described in terms of quantum states, can be prepared in a well-defined way, used in experiments, and give rise to non-local quantum correlations.

Here, the above remarks on substance and causality apply. The operational particle concept (OP) assumes that particles are *collections* of conserved quantities. These collections of dynamic properties have to

---

[18]Bohr 1928, 580; see note 90 of Chap. 1.

be understood in the sense of Kant's concept of substance, that is, as stable, reidentifiable bundles of properties (see Sects. 6.3 and 6.6). The (subjective) assumption that there is a unique causal agent behind repeated particle detections is the (only) metaphysical glue that makes the collection of properties stick together.

(III) Subatomic reality is far from being completely understood. The methodological principles of analyzing matter into smaller and smaller constituent parts and searching for the fundamental forces of physics are still employed in never-ending attempts to:

  (i) find new structures and detect new kinds of particles, in scattering experiments of higher and higher energy;

  (ii) resolve the quantum measurement problem in terms of hidden variables, many worlds, and other approaches, decoherence being up to now the most promising; and

  (iii) find a quantum gravity that escapes the current conceptual problems of unifying quantum theory and general relativity.

(IV) Subatomic reality is thoroughly relational, that is:

  (i) It is *context dependent*. The kind of quantum phenomena that is observed (particle detections, interference patterns, or both) depends on the experimental arrangement. In addition, all quantum phenomena occur in a classical world.

  (ii) It is only *defined relative to classical concepts*. All quantum phenomena are eventually described in terms of length, time, and mass. In addition, the scattering experiments of high energy physics are interpreted relative to classical models of scattering centers, charge distributions, etc.

  (iii) It is *energy dependent*. At increasing scattering energies, the quark–antiquark and gluon content of nucleons increases (see Sect. 8.3).

Taken together, (I)–(IV) do indeed make up a critical view of subatomic reality occupying a middle ground between empiricism and metaphysical realism. And the resulting position is closely related to Bohr's complementarity view of quantum mechanics. Let us first see in which sense it is stronger than empiricism but weaker than metaphysical realism, and then sketch its relation to Bohr's views.

All the empirical phenomena discussed in Chap. 3 belong to the empirical quantum structures (I) of subatomic reality. In particular, the position measurements, particle tracks, scattering events, and resonances of particle physics are empirical quantum structures. The phenomena of quantum optics discussed in Chap. 7 also belong to them. Here, the empirical quantum structures are the event distributions obtained in the double slit experiment with single photons or electrons, the Pfleegor–Mandel experiment, the single photon experiments with polarizers, and the which-way experiments with or without quantum eraser. Even an empiricist like van Fraassen would agree with the claim that all these phenomena belong to empirical reality. The

Kantian assumptions (II)–(IV) are stronger. In particular, the assumption (II) (iii) of non-local causal agents behind the quantum phenomena of particle physics and quantum optics is stronger than strict empiricism.

But the resulting position is weaker than metaphysical realism, since, on the lines of Kant's theory of nature, the constructive features of (II) are taken into account. The scales of physical quantities, the conservation laws of physics, and the causal assumptions that underlie the preparation procedures and physical interpretation of quantum experiments belong to physical practice, not to an independent physical reality in its own right. The genuine Kantian point is that such an independent reality does not exist *for us*. Physical reality is conceived to depend on the cognitive tools of physics. In particular, the scales of physical quantities are necessary conditions for the possibility of physical experience. They underlie all measurements of physics. The metaphysical concepts of substance and causality also belong to them. According to Kant, they are necessary conditions for the possibility of *any* experience. Indeed, physical practice makes tacit use of them. In the pragmatic attitude of preparing waves and detecting particles, the probability waves function as substances with causal powers and the particle detections as the effects caused by them. Without the concepts of conserved dynamic quantities, their reidentifiable collections, and the causal agents behind their reappearance, no data analysis of particle physics or quantum optics would be possible. In this sense, the metaphysical glue mentioned above has transcendental status.

However, the transcendental status of this metaphysical glue differs substantially in the classical and the quantum domains. In the classical domain, the principles of substance and causality are constitutive for the knowledge of physical objects with completely determined properties. Any quantum theory is at odds with this kind of physical knowledge. The quantum structure of the spatio-temporal and dynamic properties of subatomic particles precludes the existence of physical objects with completely determined properties. The principles of substance and causality cannot be constitutive for the knowledge of quantum *objects*.

Pringe's recent Kantian approach to quantum theory suggested interpreting them in terms of Kant's ideas of reasons and regulative principles, in particular in terms of the systematic unity of the empirical phenomena of nature analyzed in Kant's *Critique of Judgment*.[19] By drawing a parallel with Kant's account of organisms, Pringe spells out what it means to say that physical practice deals with quantum objects *as if they were real causal agents*: it means to unify complementary quantum phenomena that are generated from the same source, but in different experimental arrangements. But the unity of the quantum domain is limited. As shown above in Chap. 6 and Sect. 7.6, there is neither a unified particle concept nor a unified account

---

[19]Pringe 2006. Here, the systematic relations of this approach to Bohr's complementarity view of quantum phenomena are also investigated in detail.

of causality. The regulative use of the principles of substance and causality in physical practice comes together with a plurality of theoretical particle concepts and meanings of causality.

The most interesting philosophical problem regarding subatomic reality is how to understand wave–particle duality as the legitimate successor of the classical particle concept, *given that neither waves nor particles in a classical sense underlie the wavelike and particle-like phenomena of quantum physics*. The quantum waves and fields are not just fake entities. Physicists prepare them *as if* they were real physical objects, even though they are *not*. They have causal relevance, as the experiments of quantum optics discussed in Chap. 7 show so strikingly. They determine the conditional probabilities of particle detections. Probabilities are more than empirical relative frequencies. Quantum events obey the probabilities calculated from the wave function Ψ and in doing so, they happen as if they were governed by an *invisible hand* or follow a "law without law".[20] The critical position suggested here comes close to this. It considers the idea of an underlying causal agent to be a regulative principle that unifies the quantum events of a probabilistic ensemble.

The Kantian topic (IV) was not previously discussed. It is the claim that physical reality is thoroughly relational. For the quantum domain this has to be qualified, too, resulting in a relational semantics and ontology of quantum physics that comes close to Bohr's views. (i) The *context dependence* of quantum phenomena expresses two insights of Bohr's complementarity philosophy. *First*, the appearance of complementary quantum phenomena such as particle tracks and interference patterns depends on the experimental arrangement. *Second*, all quantum phenomena occur in a macroscopic experimental device or environment. There are no quantum objects in their own right that may be isolated from the measurement device. (ii) According to Bohr, the *language of classical physics is indispensable*. Bohr's complementarity view is based on his correspondence principle. For him, complementary quantum phenomena correspond to mutually exclusive classical phenomena, and in order to interpret them the language of classical physics must be employed. Even though this view cannot cope with genuine quantum phenomena without classical correspondence, Bohr's underlying ideas remain valid up to the present day: all measurement results must be expressed in terms of the classical length, time, and mass scales; and the scattering experiments that investigate dynamic subatomic structures can only be interpreted in terms of models that finally correspond to the classical Rutherford scattering. (iii) In addition to these Bohrian topics, the *energy dependence* of subatomic structure has to be taken into account. The higher the energy of a scattering experiment, the smaller the structures that are investigated. However, the smaller the structures, the harder it becomes to decompose them and to distinguish the constituent parts of matter (see Chap. 4 and Sect. 6.5). Finally, the nucleon constituents that show up with increasing scattering energy are no longer

---

[20]Wheeler 1979–1981.

particles in the sense of the mereological particle concept. According to the quark–parton model, the quark–antiquark and gluon content of the nucleon is field-like. But the quantum fields, too, are described and measured *as if* they were real physical objects, even though they are *not*.

Bohr's complementarity view of quantum mechanics already suggested long ago a way of avoiding the dilemma of either empiricism or metaphysical realism. Due to the difficulties in understanding his writings, no unambiguous philosophical interpretation of quantum theory emerged out of them. Today, most physicists (mis-)identify the Copenhagen interpretation of quantum mechanics with Born's and von Neumann's probabilistic interpretation; whereas most philosophers consider Bohr's views to be obsolete and only of historical interest. In addition, it has to be taken into account that quantum physics has changed a lot since Bohr's day. There are many non-local quantum phenomena that do not fit in with Bohr's view that the quantum domain is exhausted by complementary quantum phenomena with some corresponding classical phenomena behind them. But in Chaps. 4–5 it was shown in detail to what extent the central claims of Bohr's complementarity view hold up to the present day. The measurement theory of particle physics is built on classical laws, even though it also contains quantum corrections applying at the probabilistic level and several new quantities such as spin, parity, flavor, and color. The latter, however, are also defined in terms that build bridges to classical physics, namely mathematical symmetries, conservation laws, and the physical dimensions of classical quantities (such as angular momentum). Finally, all measurement results are expressed in terms of length, time, mass, and energy. The scales of these quantities are extended to the subatomic domain and even down to the Planck scale. Providing these scales, the language of classical physics is indispensable for all quantum physics.

In all these regards, subatomic reality is *relational*. Quantum physics investigates nothing but *phenomena in a classical environment*. The various kinds of non-local quantum correlations show that quantum phenomena may be *very big*, i.e., extended over spacelike distances. And the local particle detections that hint at subatomic structures only emerge as objective events in a classical world.

The relational account of subatomic reality defended here results from a *top–down* approach. The opposite *bottom–up* explanation of the classical macroscopic world in terms of electrons, light quanta, quarks, and some other particles remains an empty promise. Any attempt at constructing a particle or field ontology gives rise to a *non-relational* account of a subatomic reality made up of *independent* substances and causal agents. But any known approach of this type is either at odds with the principles of relativistic quantum theory or with the assumption that quantum measurements give rise to actual events in a classical world. As long as the quantum measurement problem is unresolved, an independent quantum reality is simply not available.

After all, to our present knowledge subatomic reality is not a micro-world on its own but a part of empirical reality that exists relative to the macroscopic world, in given experimental arrangements and well-defined physical contexts outside the laboratory. This sub-reality is investigated with macroscopic measuring devices. The existence of electrons, quarks, and photons is inferred from the highly sophisticated experiments of particle physics and quantum optics. From a realist point of view, there are sufficient reasons to believe that there are entities such as electrons, quarks, and photons in physics laboratories and in the world beyond. But any *scientific realism* about subatomic particles must take into account that they do *not really* behave like particles; and that current physics only supports the belief in the existence of *quantum processes within a classical world*.

Appendices

A Measurement Theory

Measurement theory was elaborated by Helmholtz and Campbell; Nagel, Hempel and Carnap set it in the framework of logical empiricism. The Suppes school reformulated it in model-theoretic terms and developed it as an abstract axiomatic theory. Today, abstract measurement theory is a part of applied mathematics.[1]

A.1 Empirical Relational Structures

Formal measurement theory deals with the axiomatization of relational structures that correspond to the properties of empirical objects or processes. The underlying philosophy is empiricism. According to Carnap or Suppes, measurement stems from ordering empirical objects or processes in a well-defined way by empirical operations. The ordering gives rise to an *empirical relational structure*, i.e., a relational structure that is an approximate model of the empirical properties of these entities. The axiomatic structure of measurement is established in three steps, by defining, axiomatizing, and numerically representing an empirical relational structure $\langle \mathcal{E}, \circ, \preceq \rangle$:

1. For a class $\mathcal{E}$ of empirical objects or processes, two empirical operations are defined:
   (i) a *concatenation* $\circ$ (e.g., combining rods along a straight line), and
   (ii) a *comparison* $\preceq$ (e.g., setting two rods parallel to each other and observing whether one of them is longer).
2. A set of *axioms* is chosen. They fix an algebraic structure with an ordering relation that defines equivalence classes on $\mathcal{E}$:

$$\forall a, b \in \mathcal{E} : a \sim b \iff a \preceq b \wedge b \preceq a \,. \tag{A.1}$$

3. A *representation theorem* postulates the existence of a homomorphism (that is unique up to isomorphism)

$$f : \langle \mathcal{E}, \circ, \preceq \rangle \longmapsto (\mathbb{R}, +, <) \,, \tag{A.2}$$

---

[1] Helmholtz 1887, Campbell 1920, Nagel 1931, Hempel 1952, and Carnap 1966; Krantz 1971; Narens 1985.

from the structure $\langle \mathcal{E}, \circ, \preceq \rangle$ to the real numbers $\mathbb{R}$ with their usual arithmetic structure. In particular:

(i) a *unit* is chosen, i.e., an element $e \in \mathcal{E}$ that maps to 1 (e.g., the standard meter);

(ii) to the other elements of the empirical structure, numbers $r \in \mathbb{R}$ are assigned that express their magnitude in multiples of the unit.

The axioms (2) fix an algebraic structure that corresponds to the empirical relational structure (1). Since measurement is the numerical comparison of any element of an empirical structure to some unit, a relational structure is only good for measurement if it is unambiguously representable by numbers. This is guaranteed by the representation theorem (3). The homomorphism has the formal properties of a measure. In particular, it is additive and it is uniquely determined up to isomorphic transformations of the scale of a quantity. The representation theorem is *not* an axiom. It derives from the postulate that the empirical relational structure $\langle \mathcal{E}, \circ, \preceq \rangle$, the axioms of measurement, and the structure of the real numbers must be compatible.

Abstract measurement theory deals with the formal problems of giving axioms for relational structures and mapping them into the real numbers. However, it describes ideal measurements, that is, the mapping of *sharp* relational structures into the real numbers. For more realistic applications, the theory becomes complicated. Measurement errors (that is, the fuzziness of empirical structures) may be taken into account by adding probabilistic assumptions to the axioms.[2]

A.2 Physical Quantities

A *physical quantity* is a function that assigns real numbers to empirical objects or processes and their physical properties. According to axiomatic measurement theory, the objects or processes are elements of an empirical structure $\langle \mathcal{E}, \circ, \preceq \rangle$, and the function is a homomorphism f in the sense of the above representation theorem. The equivalence classes defined on $\mathcal{E}$ by the ordering relation $\preceq$ are the classes of objects or processes that have the same physical properties, say, the classes of all bodies of the same mass. Let $\mathcal{E}_r$ be the class of all $o \in \mathcal{E}$ to which f assigns the real number r:

$$\mathcal{E}_r = \{o \mid o \in \mathcal{E} \land f(o) = r\} . \tag{A.3}$$

Then f obviously maps the equivalence classes $\mathcal{E}_r$ to the real numbers r. Correspondingly, a physical quantity is the set of ordered pairs $(\mathcal{E}_r, r)$. Since the domain of the function consists of classes of entities with the same properties, physical quantities are *second order concepts*. The *scale* of a physical quantity is the range of the homomorphism f. In general, the scale of a physical quantity ranges from 0 to ∞.

---

[2]Kyburg 1984, 183.

A.3 The Archimedean Axiom

The extension of the length, time, and mass scale from the size of the universe down to the Planck scale is based on the Archimedean axiom.[3] The Archimedean axiom belongs to the theory of real numbers. It guarantees that the unit of a scale may be arbitrarily chosen:

$$\forall a, b \in \mathbb{R} \text{ with } a < b, \quad \exists n \in \mathbb{N} \text{ such that } na > b. \qquad (A.4)$$

According to the Archimedean axiom and the representation theorem, *any* unit $a \in \mathcal{E}$ may serve to measure the magnitude of any element $b \in \mathcal{E}$. Hence, the unit may be arbitrarily chosen: the size of the universe may be expressed in terms of light years, cm, or the Planck length. *Without* the Archimedean axiom, it no longer follows that all elements of a relational structure are representable by real numbers,[4] with the consequence that there are *mathematically* incommensurable elements of a relational structure.

Kuhn introduced his notion of incommensurability as an analogue of the mathematical concept of incommensurability.[5] By the Archimedean axiom of measurement theory, the two concepts are related. In physics, the Archimedean axiom guarantees that the scales of physical quantities may be extended to the relativistic as well as the quantum domain. Hilbert (1918, 149) claimed that the validity of the Archimedean axiom can be empirically tested. But its empirical validity is no more (and no less) than the possibility of constructing the scales and making measurements in all domains without contradiction. Since subatomic, terrestrial, and cosmological distances, times, or masses are subject to incommensurable theories in Kuhn's sense, the Archimedean axiom supports the bridge principles that connect incommensurable theories. If it holds, the *measured* quantities in the subatomic, terrestrial, and celestial domains are *not* incommensurable in Kuhn's sense, due to axiomatic measurement theory. The extension of the physical scales into the relativistic and the quantum domain is closely related to dimensional considerations and the Π-theorem of dimensional analysis.

A.4 The Metaphysics of Measurement

From an empiricist point of view, measurement has three aspects:

1. It has an *operational basis* that gives rise to an empirical ordering.
2. It depends on *axioms* that determine a numerical representation.
3. It implies the *referential claim* that the axioms and their numerical representation express the operational basis.

---

[3] See Hilbert 1918, 149.
[4] Narens 1985, part II.
[5] Kuhn 1960, 1970.

These aspects simply correspond to the above three steps of (1) operationally defining, (2) axiomatizing, and (3) numerically representing an empirical relational structure $\langle \mathcal{E}, \circ, \preceq \rangle$. Measurement theory *per se* is metaphysically neutral. It is just a mathematical theory about the structure of the use of numerical methods in empirical science. In this regard, it is equal to *any* mathematical theory of empirical science itself. [This view is presupposed in Chaps. 3–5. The experiments of particle physics may easily be reconstructed in terms of measurement theory. (1) They are based on *empirical operations*, resulting in the construction and calibration of a beam and a particle detector, a period of data-taking, and the comparison of the observed properties of the particle tracks recorded by the detector. (2) The design, performance, and data analysis of such experiments are based on theoretical laws that function as *axioms of measurement*. These measurement laws depend crucially on classical and quantum models of what goes on in the experiment. (3) These models are based on the scales of physical quantities and on the assumption that the observed particle tracks are *uniquely representable* by numbers that denote the physical magnitudes of particles.]

Philosophers start to disagree about measurement, however, when they overemphasize isolated aspects of measurement. Disregarding the referential aspects of the representation theorem gives rise to operationalism, holism, or conventionalism. Disregarding the operational aspects of measurement gives rise to Platonism. Many philosophers of science emphasize the representational (or referential) aspects of measurement. They either tend to empiricism or to scientific realism. In this way, the following metaphysical views arise:

(I.a) *Holism* focuses on the axiomatic aspects of measurement. Any measurement depends on a theory of the measured phenomena. Sneed and his followers emphasized the theoreticity of dynamic quantities such as mass or force.[6] But the spatio-temporal quantities of non-relativistic mechanics also depend crucially on theory; relativistic space-time theories give rise to distinct axioms of measurement. The measurement of a quantity is always embedded in a framework of assumptions about the laws of physics, *ceteris paribus* clauses, etc. This is taken into account by the Duhem–Quine thesis and by Kuhn's view that the crucial concepts of rival theories are incommensurable, even though they may give rise to approximately equal numerical predictions.

(I.b) *Conventionalism* is a related anti-representational view of measurement. It results from focusing on the arbitrary assumptions necessarily built into any measurement. The unit of the scale of a quantity is always arbitrarily chosen. In addition, many measuring methods depend on theoretical concepts without any operational content. Typical examples are: the basic assumption of ordinary length measurement that rods are rigid; or Einstein's convention concerning the relativistic definition of simultaneity (Einstein 1905, §1).

---

[6] Sneed 1971, Balzer 1987.

(II) *Operationalism* exclusively considers the operational aspects of measurement.[7] Bridgman proposed the extreme view that each measuring method defines another quantity. Ellis defends a more liberal operationalism. For him, physical quantities are cluster concepts that derive from all measuring methods giving approximately the same quantitative results. However, Ellis rejects the assumption of real physical properties that correspond to such cluster concepts, because he considers them to be universals [see position (V) below].

(III) *Empiricism* focuses on the empirical structures underlying any measurement and emphasizes that they are constraints for the axioms of measurement. Even though any measuring method necessarily contains arbitrary elements, the axioms must be empirically adequate; therefore, their choice is more than a matter of convention. As regards the domain of the axioms, empiricism recommends ontological parsimony; there is no empirical justification for extending the axioms of a measurement to an infinite domain of unobservable entities.

(IV) *Scientific realism* claims that physical quantities such as length, mass, or charge express the real properties of natural kinds such as electrons. From a realist point of view, quantities are classes of magnitudes, magnitudes are properties that come in degrees, and these properties adhere to the physical systems and processes that actually exist in nature.

(V) *Platonism* is the strongest version of scientific realism about physical properties. Platonists reify properties and the corresponding magnitudes.[8] Swoyer and Ellis emphasize that physical properties are first order universals, and their quantitative relations second order universals. [This is the reason why Ellis prefers a moderate operationalism; see position (II) above.] Kyburg defends the views that physical magnitudes are abstract objects, and that quantities should not be interpreted as functions from empirical structures into the real numbers but as functions from empirical structures into classes of magnitudes.[9]

The formal structure of measurement theory nicely explains this diversity of metaphysical views. Taken in reverse order, they more or less correspond to the positions in the debate on *scientific realism* (see Sect. 1.2). Holism and conventionalism give rise to constructivist views. Operationalism is a variant of strict empiricism and a more liberal empiricist view of physical quantities gives rise to moderate empiricism. Scientific realism about physical properties comes in weaker or stronger versions. A critical account of physical quantities considers the construction of the scales to be indispensable (see Sect. 1.6). Finally, platonism about physical properties is even stronger than Planck's or Einstein's metaphysical realism (see Sects. 1.5 and 2.1).

---

[7] Bridgman 1927; Ellis 1968, 34–36.

[8] Armstrong 1987a,b.

[9] Swoyer 1987; Ellis 1987; Kyburg 1984, 17.

B The Π-Theorem of Dimensional Analysis

Dimensional analysis is the method behind the heuristic dimensional considerations of physics. It serves to derive physical laws as far as they are constrained by the dimensions of the physical quantities involved in a problem.[1] The method is based on the Π-*theorem* of dimensional analysis, which holds for all dimensional invariant (or 'homogeneous') functions of physical quantities,[2] i.e., for all physical laws that do not depend on the choice of the unit of the quantities involved. Any physical law L may be brought into the following form:

$$L : f(x_1, \dots, x_n) = 0 , \tag{B.1}$$

where f is a function that relates n physical quantities x_i $(i = 1, \dots, n)$. Let k be the maximum number of the algebraic basis of all possible measurements of (L), i.e., the maximum number of independent quantities that enter the dimensions of the x_i. Then, there are $n - k$ physical quantities that depend on this basis of measurement, their dimensions being products of the dimensions of the k basic quantities. For example, in classical mechanics the basic quantities are the *length* L, the *time* T, and the *mass* M $(k = 3)$. The law of force contains in addition to the basic quantity of mass the quantities of velocity and force $(n = 5$ and $n - k = 2)$, the dimensions of which $(LT^{-2}$ and $LT^{-2}M)$ are products of length, time, and mass.

The Π-theorem states that the physical law L may be reformulated into an equivalent law (L') with a function F that only depends on $n - k$ *dimensionless quantities* $\pi_j (1 \leq j \leq n - k)$:

$$L' : F(\pi_1, \dots, \pi_{n-k}) = 0 . \tag{B.2}$$

The Π-theorem may be used to derive from the basic quantities of measurement a system of linear equations which relates the powers of the n quantities x_i that characterize a physical system or process to the $n - k$ independent dimensionless quantities π_j that may be built out of them (for examples, see

---

[1] Its theoretical and meta-theoretical foundations were investigated by Bridgman 1922 and Campbell 1920. Krantz et al. 1971, Chap. 10, present the axiomatic foundations of dimensional analysis and discuss its justification.

[2] Krantz et al. 1971, 464.

Krantz et al. 1971, 472 ff). If the x_i are chosen specifically for the respective system, the calculation of the dimensionless quantities π_j yields information about a physical law which is a function $F(\pi_1, \dots, \pi_j)$ and describes the system under investigation, *without* any need to know an underlying fundamental physical law. Hence, a system is described in terms of n contingent constraints x_i. These constraints are then expressed in terms of k independent familiar physical quantities. Finally, from these constraints and the Π-theorem (which is assumed to hold for physical quantities in general), one derives a specific physical law in order to describe the system. In order to derive the system description, the theory of the system does not have to be known.

The method of dimensional analysis and the related dimensional considerations are abundantly used in physics, for making fast derivations of specific laws from well-known theories as well as for the purposes of theory formation. They make it possible to fix specific cases of an unknown theory, which serve as constraints for theory construction. However, unrestricted use of the method is problematic for the following reasons:

1. The method has to be used carefully in order to avoid errors when the physical quantities that characterize a physical problem are determined.[3] The correct use of dimensional considerations and the underlying physical knowledge belong to the crucial skills of a paradigm in Thomas S. Kuhn's sense.
2. To be more precise, these skills belong to the paradigm of *classical* physics. Beyond the domain of classical physics, the method generalizes the familiar assumptions about physical quantities in such a way that the algebraic properties of length, time, mass, and temperature are maintained.
3. When used within the quantum domain, the method relies on a generalized correspondence principle in the sense of Bohr and Heisenberg (see Sects. 5.4–5.5). It has to be noted that the construction of the Planck scale is based on dimensional analysis, too. Yet no one knows whether the Π-theorem is justified for the domain of quantum gravity.

---

[3]Krantz et al. 1971, 473.

C The Effective Cross-Section

In classical physics there are two very simple models of scattering. The first model is elastic or inelastic *impact*. This purely kinematic model is a specification of the general dynamic model of *potential scattering*; elastic impact corresponds to the scattering at a box potential. The other simple model is scattering at the Coulomb potential (Rutherford scattering). In the general classical model, charged particles (point masses) that move along trajectories are scattered at some arbitrary potential without energy transfer (*elastic* scattering) or with energy transfer (*inelastic* scattering). In the case of elastic scattering, the individual scattering process is described by the *impact parameter* b, which depends on the scattering angle θ and the kinetic energy E of the particle before and after the scattering. (θ and E can be measured. b is the characteristic quantity of the classical model of scattering; it is the minimum distance of the scattered particle from the scattering center.) For the Coulomb potential $V(r) = C/R$, the following relation holds:

$$b = \frac{C}{2E} \cot \frac{\theta}{2} . \tag{C.1}$$

The characteristic quantity of the scattering is the *effective cross-section* or *(scattering) cross-section*. In the classical model, it is calculated from the angle dependence of the impact parameter b. It has dimensions of area and is expressed in units of barn (1 barn = 10^{-24} cm^2). In particle physics, the *differential* and *total* cross-section are distinguished.

The *differential cross-section* $d\sigma/d\Omega$ is proportional to the probability of the scattered particles per scattering angle θ or a corresponding infinitesimal solid angle $d\Omega$. As an empirical quantity, $d\sigma/d\Omega$ is measured from the relative frequency of particles which are scattered into a finite solid angle $\Delta\Omega$. In the theoretical model, $d\sigma/d\Omega$ is defined from the number of particles N^{sc} which are scattered into the differential solid angle $d\Omega$ belonging to the scattering angle θ, per differential surface element dF and per scattering center and taken in the formal limit of infinitely many incoming particles (N_C = number of scattering centers):

$$\frac{d\sigma}{d\Omega} = \lim_{N_{in} \to \infty} \frac{N^{sc}}{N_{in} N_C} \frac{dF}{d\Omega} . \tag{C.2}$$

The formal limit simply expresses the difference between probability and relative frequency, i.e., the unavoidable gap between a probabilistic quantity and its empirical basis. Here, probability is understood as the limit of relative frequency for very big event numbers.

In particle physics, the number N^{in} of incoming particles per differential surface element dF is a theoretical quantity, just like the number of scattering centers N^{ST}. Without any knowledge of these numbers, $d\sigma/d\Omega$ is only known up to some normalization factor. In the classical model, $d\sigma/d\Omega$ depends on the impact parameter b as follows:

$$\frac{d\sigma}{d\Omega} = \frac{b}{\sin\theta}\left|\frac{db}{d\theta}\right| . \tag{C.3}$$

For the Coulomb potential $V(r) = C/r$, *Rutherford's scattering cross-section* [(4.1) in Sect. 4.1] follows immediately from (A.3) and (A.1).

The *total cross-section* σ is simply obtained by integrating $d\sigma/d\Omega$ over all solid angles:

$$\sigma = \int \frac{d\sigma}{d\Omega} d\Omega . \tag{C.4}$$

The total cross-section expresses the probability or relative frequency of a certain kind of scattering event, as compared to other kinds of scattering events. It is a measure for the 'hit ratio' of a kind of reaction. In a very simple mechanical model of the scattering, the total cross-section may be illustrated as the *effective area of the reaction*, that is, as the area of an extended and impenetrable scattering center, off which negligibly small probe particles bounce just like balls from the slats of a garden fence.

To give an example, the differential cross-section of the elastic scattering of pointlike probe particles from a hard sphere of radius R is isotropic and the total scattering cross-section σ is identical to the geometric cross-section of the sphere, i.e., it is $\sigma = \pi R^2$. The expression 'effective cross-section' or 'scattering cross-section' obviously stems from this simple mechanical model.

In classical dynamics, σ or $d\sigma/d\Omega$ in general depends not only on geometric quantities, but also on the kinetic energy of the probe particles and a possible energy transfer. Even in the case of scattering at the Coulomb potential, the simple mechanical analogy fails. Rutherford's cross-section depends on energy and it diverges in the forward direction. However, the following relation holds. According to $V = mv^2/2$ and $C/r = E$, the turning point of a particle of kinetic energy E that is scattered maximally backward is the point of its minimum distance $R_{min} = C/E$ from the scattering center. Hence, the turning point defines an effective radius C/E of the scattering center which depends on the kinetic energy E of the scattered particle: the larger the scattering energy, the closer the probe particle approaches to the scattering center.

In quantum mechanics, there are neither trajectories nor an impact parameter. Therefore, the above illustration fails. Nevertheless, the effective cross-section can be defined as a probabilistic quantity and measured in a scattering experiment from the relative frequency of scattering events of a given type. In a cross-section measurement of particle physics, the number of particle tracks is measured by means of particle detectors. In the effective cross-section of a kind of particle reaction, quantum field theory meets experiment. For a given kind of subatomic scattering process, σ and $d\sigma/d\Omega$ are proportional to the quantum mechanical transition probability and the corresponding element of the S-matrix, respectively.

D Dimensional Analysis of Rutherford Scattering

The dimensional analysis of Rutherford scattering neglects what cannot be measured in the subatomic domain, namely the trajectory of the scattered particle. In this way, one dispenses with the impact parameter b which characterizes the individual scattering in the classical model. The resulting dimensional analysis of the Rutherford scattering is theory-independent. Based on the generalized correspondence principle discussed in Chap. 5 (Sect. 5.4), the following dimensional considerations give rise to a theory-independent model of Rutherford scattering and the definitions of form factors discussed in Chap. 4 (Sects. 4.2–4.4).

Without the classical impact parameter b, Rutherford scattering is described in terms of eight quantities:

1. the differential cross-section $d\sigma/d\Omega$,
2. the kinetic energy E of scattered probe particles,
3. the scattering angle θ,
4. the fine structure constant α, i.e., the coupling constant of the electromagnetic interaction,
5./6. the charge numbers Z and Z', i.e., the charges of the probe particle and the scattering center, in multiples of the electric elementary charge e,
7. Planck's constant $\hbar$, and
8. the speed of light c.

The quantities θ, α, Z and Z' are dimensionless. L (length), T (time), and M (mass) are obviously appropriate basic quantities for the dimensional analysis. According to the Π-theorem of dimensional analysis, there must be $8 - 3 = 5$ dimensionless physical quantities π_i on which the description of the scattering depends. Four of them, namely $\pi_1 = \theta$, $\pi_2 = \alpha$, $\pi_3 = Z$ and $\pi_4 = Z'$, are already known. Expressed in the basic dimensions L, T, M, the differential cross-section $d\sigma/d\Omega$ has dimensions L^2, the kinetic energy $E = mv^2/2$ of the scattered probe particles has dimensions ML^2T^{-2}, and the constants of nature $\hbar$ and c have dimensions ML^2T^{-1} and LT^{-1}, respectively. A simple calculation shows that the missing dimensionless quantity π_5 that may be formed from these four remaining quantities is the expression $(d\sigma/d\Omega)(E/\hbar c)^2$:

$$\pi_5 = \frac{d\sigma}{d\Omega} \left(\frac{E}{\hbar c}\right)^2 . \tag{D.1}$$

This is the expression on the left side of (4.3) in Sect. 4.1. According to the Π-theorem, the only dimensionless quantities of the scattering apart from α, Z, and Z' are the scattering angle θ and the effective cross-section that is made dimensionless by multiplying it by $(E/\hbar c)^2$. Therefore, the specific physical law (L') that describes the scattering according to Appendix B as dependent on the five dimensionless quantities π_i $(1 \le i \le 5)$ can only depend on α, Z, Z', θ, and the dimensionless quantity obtained from the scattering cross-section. It has the following form:

$$\frac{d\sigma}{d\Omega} \left(\frac{E}{\hbar c}\right)^2 = \Phi(\theta) . \tag{D.2}$$

Comparison with (4.3) from Sect. 4.1 shows that $\Phi(\theta)$ has the form

$$\Phi(\theta) = \frac{(ZZ'\alpha)^2}{16 \sin^4 \dfrac{\theta}{2}} . \tag{D.3}$$

The result in which Rutherford was mainly interested, namely the functional dependence of the differential cross-section $d\sigma/d\Omega$ on the scattering angle θ, does not follow from dimensional analysis. However, dimensional analysis shows at least that $d\sigma/d\Omega$ only depends on the electromagnetic quantities α, Z, Z', the scattering angle θ, and the kinetic energy of the probe particles. No other quantities enter the description of the model.

A crucial property of the Rutherford scattering is that $(d\sigma/d\Omega)(E/\hbar c)^2$ is *scale invariant*, i.e., it no longer depends on the scattering energy E. The meaning of this scale invariance becomes obvious if we assume that an additional physical quantity enters the scattering model, namely a quantity R with dimensions of length which describes the spatial extension of the scattered probe particles or the scattering center. In this case, the existence of an additional dimensionless quantity π_6 follows from the Π-theorem, with the value:

$$\pi_6 = \frac{RE}{\hbar c} . \tag{D.4}$$

In this case, the physical law (L') that describes the scattering has the form

$$\frac{d\sigma}{d\Omega} \left(\frac{E}{\hbar c}\right)^2 = \Phi(\theta; RE/\hbar c) . \tag{D.5}$$

In this case, the dimensionless scattering cross-section depends not only on the scattering angle θ, but also on the kinetic energy E of the probe particles and on the length R that describes the scattering center or the probe particles. Now, $(d\sigma/d\Omega)(E/\hbar c)^2$ is no longer scale invariant but depends on the kinetic energy of the scattering experiment.

E Mereology

According to the mereological particle concept, particles are the (microscopic) proper parts of matter. Mereology is the logic of the part–whole relation. In a physical dynamics, the part–whole relation is not spatially but dynamically interpreted. It is primarily based on sum rules for dynamic quantities such as momentum, mass–energy, charge, and angular momentum or spin. In the following it is shown how the constituent models of current physics may be explained in terms of formal mereology.

E.1 Axioms of Mereology

Mereology is a partial ordering based on the following axioms.[1] The primitive relation $\ll$ ('is proper part of') satisfies the principles of *asymmetry* and *transitivity*:[2]

$$x \ll y \quad \Longrightarrow \quad \neg\, y \ll x \,, \tag{E.1}$$

$$x \ll y \,\wedge\, y \ll z \quad \Longrightarrow \quad x \ll z \,. \tag{E.2}$$

x, y and z are variables that run over the logical individuals in a calculus of individuals. (In set theory $\ll$ would define a strict partial ordering. However, set theory is not presupposed here.) With the identity $=$ and the relation $\ll$, the relation $<$ ('is part of') can be defined:

$$x < y \quad \Longleftrightarrow \quad x \ll y \,\vee\, x = y \,. \tag{E.3}$$

The axioms (E.1) and (E.2) are *not sufficient* for distinguishing the part–whole relation from other partial orderings. In particular, an axiom must be added to (E.1) and (E.2) which states that the parts of a whole are in some respect *well distinguished*. In order to do so, the relations of $\diamond$ ('overlaps') and $\wr$ ('is disjunct from') are defined:

$$x \diamond y \quad \Longleftrightarrow \quad \exists\, z\, (z < x \,\wedge\, z < y) \,, \tag{E.4}$$

---

[1] Goodman 1951; Simons 1987.
[2] Simons 1987, 26, and Goodman 1951, 34.

$$x \wr y \quad \Longleftrightarrow \quad \neg\, x \diamond y \,. \tag{E.5}$$

The weakest axiom according to which the parts of a whole are well-distinguished states that a whole only has proper parts if there are at least two of them that do not overlap:

$$x \ll y \quad \Longrightarrow \quad \exists\, z \,(z \ll y \,\wedge\, z \wr x) \,. \tag{E.6}$$

The whole that consists of two and only two distinct proper parts x and y is the *mereological sum* of x and y. The mereological sum may be defined as follows:[3]

$$x \oplus y = \imath z \,\forall\, w \,(w \diamond z \;\Longleftrightarrow\; w \diamond x \,\vee\, w \diamond y) \,. \tag{E.7}$$

According to the definitions (E.7) and (E.4), the mereological sum $x \oplus y$ is the entity z that overlaps only with x and y. For example, a broom is the mereological sum of stick and brush.[4] The axioms (E.1), (E.2), and (E.6) do not imply the existence and unambiguity of the mereological sum. In particular, an entity is not completely determined by its disjunct parts. The same parts may form different mereological sums or wholes.[5] Therefore, Lesniewski's 'classical' mereology is based on stronger assumptions than the axioms (E.1), (E.2), and (E.6) alone. In particular, it assumes a principle of extensionality, according to which two entities may only have the same parts if they too are related by a part–whole relation. The weakest extensionality axiom is:[6]

$$\exists\, z \,(z \ll x) \,\wedge\, \forall\, z \,[(z \ll x \Longrightarrow z \ll y) \Longrightarrow x < y] \,. \tag{E.8}$$

The axiom (E.8) states that x has proper parts z [according to the axiom (E.3), at least two of them are disjunct], and that any proper part of x is only a proper part of y, if x is part of y. A mereology with (E.8) is said to be *extensional*. If (E.8) does not hold, the mereological sum of two entities is not unambiguous and the mereology is called non-extensional. Lesniewski's 'classical' mereology is based on much stronger assumptions than (E.8). In particular, it has the structure of a *Boolean algebra* without zero element and it assumes that, for any two entities x and y, there is a mereological sum $x \oplus y$ (Simons 1987, 25 and 37 ff).

E.2 Mereology and Physics

Mereology may be connected to the usual mathematical formalism of physical theories by taking sets as individuals. Depending on the assumptions about

---

[3] Simons 1997, 32, definition SD7; Goodman 1951, 36, definition D2.047; here, $\imath$ is the denoting operator of formal logic.

[4] Simons 1987, 14.

[5] Simons 1987, 32–37, discusses several axioms that guarantee the existence and uniqueness of the mereological sum.

[6] Simons 1987, 239.

the sets, one obtains a weaker or stronger system of axioms for the part–whole relation. With arbitrary intervals of real numbers as individuals and the set-theoretical relation $\subset$ as the proper part–whole relation $\ll$ (and $\subseteq$ as $<$), an extensional mereology may be constructed in which the relations of overlapping $\diamond$ and being disjunct $\wr$ have the usual set-theoretical meaning.

In the one-dimensional case, this means: two intervals overlap if their intersection is non-empty and they are disjunct if it is the empty set $\emptyset$. Obviously, this mereology satisfies the axioms (E.1), (E.2), and (E.6). The mereological sum with respect to $\subset$ exists for any two intervals and it is unambiguous; it is their union. Hence, for intervals in the set of the real numbers the extensionality principle (E.8) holds (corresponding to the extensionality axiom of axiomatic set theory). However, the existence of the mereological sum is only guaranteed if non-connected subsets of the real numbers are admitted as individuals.

Against Lesniewski's axiomatic system, it was objected that the relation 'consists of' (which underlies the constituent models of molecular, atomic, and subatomic physics) should be explained in terms of a non-extensional mereology.[7] Rescher's objection seems to aim at two facts. *First*, at all constituent levels of matter, the same matter constituents may form different composite systems with very different physical, chemical, and even biological properties. Well-known examples from the current natural sciences are the difference between the clockwise and counterclockwise molecules of milk acid, the different phases of matter (solids, liquids, and gases), the different possibilities of chemical binding, the excitations of an atom, or the compound systems of quarks that give rise to hadrons of different mass and spin. *Second*, the constituent models of atomic and subatomic physics are based on quantum theory. But quantum properties do *not* have the structure of a Boolean algebra that Lesniewski's extensional mereology presupposes.

However, Rescher neglects the fact that any non-extensional interpretation of the part–whole relation in physics *presupposes* an extensional part–whole relation, as far as the underlying mathematics is based on set theory and as far as the dynamics of a bound system presupposes a background spacetime. In addition, the fact that the constituent models of physics have dynamic as well as spatio-temporal aspects must be taken into account. In contrast to the traditional philosophical debates on atomism (e.g., in the Leibniz–Clarke debate or its reflection in Kant's pre-critical and critical work), modern physics focuses on the dynamic aspects of the constituent models of matter. Hence, Rescher's objection disregards the fact that the formal part–whole relation *per se* is much too poor to cope with the constituent models of physics.

---

[7]Rescher 1955, 10; see also Simons 1987, 112–117.

E.3 Matter Constituents

In modern physics, the models of matter constituents are based on a dynamics of bound systems. A constituent model connects an object y and its constituent parts x by considering y to be a bound system $\Sigma(x_1, \dots, x_n)$ of n entities x_i:

$$y = \Sigma(x_1, \dots, x_n) . \tag{E.9}$$

A constituent model of physics has *ontological* and *theoretical* aspects. Ontologically, (E.9) means that y is made up of the x_i; i.e., y is the mereological sum of n well-distinguished proper parts x_i:

$$y = x_1 \oplus \cdots \oplus x_n . \tag{E.10}$$

Since the whole y is a spatio-temporal entity that occupies a certain spacetime region, in addition to the axioms (E.1), (E.2) and (E.6), the extensionality axiom (E.8) is needed. However, according to (E.9), y is more than a mere collection or aggregate of the x_i in spacetime. It is a bound system of the x_i. It is assumed that y as a whole is a spatio-temporal entity which stems from forces that act between the x_i. This means that some crucial physical properties of y are explained by the physical properties and the interactions of the x_i.

Here, physical theory comes in. It is required that a dynamic description T_y of y reduces approximately to a dynamic description T_x of the x_i. In an ideal constituent model the theories T_y and T_x are identical. In this case, a physical dynamics T describes y as a bound system $S(x_1, \dots, x_n)$ of n objects x_i in terms of some physical quantities φ:

$$T\big(\varphi(y)\big) = S\big(\varphi(x_1), \dots, \varphi(x_n)\big) . \tag{E.11}$$

In general, the x_i and y carry dynamic properties such as mass and charge which are given as functions of spacetime coordinates. They interact according to laws of T which are given in terms of differential equations for all these quantities. In this way, the spacetime region occupied by y is explained in terms of the dynamics of the x_i. Hence, the dynamics of a bound system S has two completely different theoretical aspects:

(1) It explains how the motions of the x_i constitute the *spacetime region* occupied by y.
(2) It gives *sum rules* for the way in which the dynamic properties of y add up from the respective properties of the x_i.

According to (1), T explains the extension of y in terms of a *spatio-temporal part–whole relation*. In this regard, the minimum requirement for a whole y having parts x_i is that the parts may be localized within the whole. (2) The sum rules for dynamic quantities such as mass and charge, however, explain the dynamic properties μ_y of y in terms of the respective properties μ_{x_i} of the x_i, giving rise to several *dynamic* part–whole relations for any conserved quantity of T:

$$\mu_y = \sum_{i=1}^{n} \mu_{x_i} \, . \tag{E.12}$$

Whereas the spatio-temporal part–whole relation is extensional, the sum rules of physics *per se* do not support an extensional mereology.

References

Abachi, S., et al. (1995): Observation of the top quark, Phys. Rev. Lett. **74**, 2632–2637

Abe, F., et al. (1995): Observation of top quark production in p$\bar{\text{p}}$ collisions with the collider detector at Fermilab, Phys. Rev. Lett. **74**, 2626–2631

Afshar, S.S. (2003): Sharp complementary wave and particle behaviours in the same 'welcher weg' experiment. Available as an e-print (26 June 2006) at http://irims.bluemirror.net/quant-ph/030503

Aitchison, I.J.R. (1982): *Gauge Theory in Particle Physics*, Hilger, Bristol

Anderson, C.D. (1932): The apparent existence of easily deflectable positives, Science **76**, 238

Anderson, C.D. (1933): The positive electron, Phys. Rev. **43**, 491–494

Anderson, C.D., and Anderson, H.L. (1983): Unraveling the particle content of cosmic rays. In: Brown and Hoddeson (1983) pp. 131–154

Anderson, P.W. (1997): *Concepts in Solids*, World Scientific, Singapore

Andrade, E.N. da C. (1964): *Rutherford and the Nature of the Atom*, New York

Armstrong, D.M. (1987a): Comments on Swoyer and Forge. In: Forge (1987) 311–317

Armstrong, D.M. (1987b): Are quantities relations? A reply to Bigelow and Pargetter, Philosophical Studies **54**, 305–316

Aspect, A., Grangier, P., and Roger, G. (1982): Experimental realization of Einstein–Podolsky–Rosen–Bohm gedankenexperiment. A new violation of Bell's inequalities, Phys. Rev. Lett. **49**, 91–94

Auyang, S.Y. (1995): *How is Quantum Field Theory Possible?* Oxford University Press, New York

Bacon, F. (1620): *Novum Organum*, Apud Joannem Billium, London. New edn. by L. Jardine and M. Silverthorne, Cambridge University Press, Cambridge (2000)

Balzer, W. (1987): Sneeds Zirkel ist nicht wegzukriegen, Conceptus XXI, 103

Balzer, W., Moulines, C.U., and Sneed, J.D. (1987): *An Architectonic for Science. The Structuralist Program*, Reidel, Dordrecht

Balzer, W., and Moulines, C.U. (Eds.) (1996): *Structuralist Theory of Science. Focal Issues, New Results*, de Gruyter, Berlin, New York

Bailer-Jones, D.M. (1997): *Scientific Models. A Cognitive Approach with an Application in Astrophysics*, University of Cambridge, Cambridge

Bailer-Jones, D.M. (2004): Models in philosophy of science, Habil. Thesis, Bonn (to be published)

Bartell, L.S. (1980): Complementarity in the double-slit experiment. On simple realizable systems for observing intermediate particle–wave behaviour, Phys. Rev. D **21**, 1698–1699

Bartels, A. (1994): *Bedeutung und Begriffsgeschichte*, Schöningh, Paderborn

Bell, J.S. (1964): On the Einstein–Podolsky–Rosen paradox, Physics **1**, 195–200

Beller, M. (1999): *Quantum Dialogue. The Making of a Revolution*, University of Chicago Press, Chicago, London

Beltrametti, E.G., and Cassinelli, G. (1981): *The Logic of Quantum Mechanics*, Addison-Wesley, Reading (Mass.), London

Berkeley, G. (1721): *De Motu – sive, de motus principio et natura, et de causa communications mottum*, Impensis Jacobi Tonson, London. New edn. in: A.A. Luce and T.E. Jessop (Eds.): *The Works of George Berkeley, Bishop of Cloyne*, Thomas Nelson, London (1948–1957)

Bernays, P. (1922): Die Bedeutung Hilberts für die Philosophie der Mathematik, Die Naturwissenschaften (David Hilbert zur Feier seines sechzigsten Geburtstages) **10**, 93–99

Bethe, H. (1930): Zur Theorie des Durchgangs schneller Korpuskularstrahlen durch Materie, Annalen der Physik **5**, 325–400

Bethe, H. (1932): Bremsformel für Elektronen relativistischer Geschwindigkeit, Z. Physik **76**, 293–299

Bethe, H., and Heitler, W. (1934): On the stopping of fast particles and on the creation of positive electrons, Proc. Roy. Soc. A **146**, 83–112

Bethge, K., and Schröder, U.E. (1968): *Elementarteilchen*, Wissenschaftliche Buchgesellschaft, Darmstadt

Bjorken, J.D., and Drell, S.D. (1964): Relativistic Quantum Mechanics, McGraw-Hill, New York

Bjorken, J.D. (1965): *Relativistic Quantum Fields*, McGraw-Hill, New York

Bjorken, J.D. (1969): Asymptotic sum rules at infinite momentum, Phys. Rev. **179**, 1547–1553

Bjorken, J.D. (1970): High-energy inelastic neutrino–nucleon interactions, Phys. Rev. D **1**, 3151–3160

Bjorken, J.D., and Paschos, E.A. (1969): Inelastic electron and γ–proton scattering and the structure of the nucleon, Phys. Rev. **185**, 1975–1982

Bloch, F. (1933): Bremsvermögen von Atomen mit mehreren Elektronen, Z. Physik **81**, 363–376

Bloom, E.D., Coward, D.H., DeStaebler, H., Drees, J., Miller, G., Mo, L.W., and Taylor, R.E. (1969): High-energy inelastic e–p scattering at 6° and 10°, Phys. Rev. Lett. **23**, 930–934. Repr. in: Cahn and Goldhaber (1989)

Bogen, J., and Woodward, J. (1988): Saving the phenomena, The Phil. Rev. **97**, 303–352

Bogoliubov, N.N., and Shirkov, D.V. (1959): *Introduction to the Theory of Quantized Fields*, Interscience Publishers, New York

Bohm, D. (1952): A suggested interpretation of the quantum theory in terms of 'hidden' variables, I and II, Phys. Rev. **85**, 166–179, 180. Repr. in: Wheeler and Zurek (1983)

Bohr, N. (1913a): On the theory of the decrease of velocity of moving electrified particles on passing through matter, Phil. Mag. **25**, 10–31. Repr. in: *Bohr Collected Works* (BCW), North-Holland (Elsevier) Amsterdam, Vol. 2, 18–39

Bohr, N. (1913b): On the constitution of atoms and molecules, Phil. Mag. **26**, 1 (Part I); 476 (Part II); 857 (Part III). Repr. in: BCW 2, 161–233

Bohr, N. (1915): On the decrease of velocity of swiftly moving electrified particles on passing through matter, Phil. Mag. **30**, 581–612. Repr. in: BCW 8, 127–160

Bohr, N. (1920): Über die Serienspektren der Elemente, Z. Physik **2**, 423–469. Engl. transl. in: Bohr (1922)

Bohr, N. (1922): *The Theory of Spectra and Atomic Constitution*, Cambridge University Press, Cambridge

Bohr, N. (1923): Über die Anwendung der Quantentheorie auf den Atombau: I. Die Grundpostulate der Quantentheorie, Z. Physik **13**, 117–165. Engl. transl. in: Proc. of the Cambridge Philosophical Society **2** (1924), 1–42. Repr. in: BCW **3**, 458–499

Bohr, N. (1928): The quantum postulate and the recent development of atomic Theory, Nature **121**, 580–590. Repr. in: Wheeler and Zurek (1983) pp. 87–126. Modified version of the Como Lecture (1927). Both versions in: BCW 6, 109–158

Bohr, N. (1935a): Quantum mechanics and physical reality, Nature **136**, 65. Repr. in: Wheeler and Zurek (1983) p. 144

Bohr, N. (1935b): Can quantum-mechanical description of physical reality be considered complete? Phys. Rev. **48**, 695–702. Repr. in: Wheeler and Zurek (1983) pp. 145–151

Bohr, N. (1937): Causality and complementarity, Phil. of Sci. **4** (3), 289–298

Bohr, N. (1948): On the notions of causality and complementarity, Dialectica **2**, 312–318 Repr. in: BCW 7, 330–337

Bohr, N. (1949): Discussion with Einstein on epistemological problems of atomic physics. In: Schilpp (1949) pp. 115–150. Repr. in: Wheeler and Zurek (1983) pp. 9–49

Bohr, N., Kramers, H.A., and Slater, J.C. (1924): The quantum theory of radiation, Phil. Mag. **47**, 785–802. Repr. in: B.L. van der Waerden (Ed.): *Sources of Quantum Mechanics*, Dover, New York (1967) pp. 159–176

Born, M. (1926a): Zur Quantenmechanik der Stoßvorgänge, Z. Physik **37**, 863–867. Engl. transl. in: Wheeler and Zurek (1983) pp. 52–55

Born, M. (1926b): Quantenmechanik der Stoßvorgänge, Z. Physik **38**, 803–827. Repr. in: Herrmann (1962)

Born, M. (1926c): Zur Wellenmechanik der Stoßvorgänge, Nachr. Ges. Wiss. Göttingen **1926**, 146–160. Repr. in: Herrmann (1962)

Bothe, W., and Geiger, H. (1925): Über das Wesen des Comptoneffekts, Z. Physik **32**, 639–663

Breidenbach, M., Friedman, J.I., and Kendall, H.W. (1969): Observed behavior of highly inelastic electron–proton scattering, Phys. Rev. Lett. **23**, 935–939. Repr. in: Cahn and Goldhaber (1989)

Breit, G., and Wheeler, J.A. (1934): Collision of two light quanta, Phys. Rev. **46**, 1087–1091

Bridgman, P.W. (1922): *Dimensional Analysis*, Yale University Press, New Haven

Bridgman, P.W. (1927): *The Logic of Modern Physics*, Macmillan, New York

Brown, H.I. (1987): Naturalizing observation. In: Nersessian (1987) 179–193

Brown, H.R., and Harré, R. (Eds.) (1988): *Philosophical Foundations of Quantum Field Theory*, Clarendon Press, Oxford

Brown, L.M., and Hoddeson, L. (1983): *The Birth of Particle Physics*, Cambridge University Press, Cambridge

Brown, L.M., Dresden, M., and Hoddeson, L. (Eds.) (1989): *Pions to Quarks. Particle Physics in the 1950s*, Cambridge University Press, Cambridge

Busch, P., Lahti, P.J., and Mittelstaedt, P. (1991): *The Quantum Theory of Measurement*, Springer, Berlin

Busch, P., Grabowski, M., and Lahti, P.J. (1995): *Operational Quantum Physics*, Springer, Berlin, Heidelberg

Busch, P., and Mittelstaedt, P. (1991): The problem of objectification in quantum mechanics, Found. Phys. **8**, 889–904

Busch, P., and Lahti, P.J. (1995): The complementarity of quantum observables: Theory and experiments, Rivista del Nuovo Cimento **18**, 1–27. Also http://arxiv.org/pdf/quant-ph/0406132

Butts, R.E., and Brown, J.R. (Eds.) (1989): *Constructivism and Science*, Kluwer, Dordrecht

Cahn, R.N., and Goldhaber, G. (1989): *The Experimental Foundations of Particle Physics*, Cambridge University Press, Cambridge

Callender, C., and Huggett, N. (2001): *Physics Meets Philosophy at the Planck Scale*, Cambridge University Press, Cambridge

Campbell, N.R. (1920): *Foundations of Science*, Dover, New York (1957). A reduced photographic reprint of the edition of *Physics. The Elements*, Cambridge (1920)

Cao, T.Y. (Ed.) (1999): *Conceptual Foundations of Quantum Field Theory*, Cambridge University Press, Cambridge

Carnap, R. (1928): *Der logische Aufbau der Welt*, Weltkreis Verlag, Berlin. Engl. transl. by R.A. George: *The Logical Structure of the World*, Open Court, Chicago, Ill. (2003)

Carnap, R. (1931): Überwindung der Metaphysik durch logische Untersuchung der Sprache, Erkenntnis **2**, 219–241

Carnap, R. (1947): *Meaning and Necessity*, University of Chicago Press, Chicago

Carnap, R. (1956): The methodological character of theoretical concepts. In: H. Feigl and M. Scriven (Eds.): *The Foundations of Science and the Concepts of Psychology and Psychoanalysis*, Minnesota Studies in the Philosophy of Science Vol. 1, University of Minnesota Press, Minneapolis, pp. 38–76

Carnap, R. (1966): *Philosophical Foundations of Physics*, Dover, New York

Carrier, M. (1991): What is wrong with the miracle argument? Stud. in Hist. and Phil. of Sci. **22**, 23–36

Carrier, M. (1993): What is right with the miracle argument. Establishing a taxonomy of natural kinds, Stud. in Hist. and Phil. of Sci. **24**, 391–409

Carson, E., and Huber, R. (Eds.) (2006): *Intuition and the Axiomatic Method*, Springer, Dordrecht

Cartwright, N. (1983): *How the Laws of Physics Lie*, Clarendon Press, Oxford

Cartwright, N. (1989): *Nature's Capacities and Their Measurement*, Clarendon Press, Oxford

Cartwright, N. (1999): *The Dappled World. A Study of the Boundaries of Science*, Cambridge University Press, Cambridge

Cassidy, D. (1981): Cosmic ray showers, high energy physics, and quantum field theories: Programmatic interactions in the thirties, Hist. Stud. Phys. Sci. **12**, 1–39

Cassirer, E. (1937): Determinismus und Indeterminismus in der modernen Physik, Gøteborg. Reprint in: H.-W. Wendt (Ed.): *Zur modernen Physik*, Wissenschaftliche Buchgesellschaft, Darmstadt (1957)

Chevalley, C. (1991): Introduction et glossaire. Analyse et synthèse. In: *Niels Bohr: Physique Atomique et Connaissance Humaine*, Gallimard, Paris, pp. 17–144 and 345–567

Christenson, J.H., Cronin, J.W., Fitch, V.L., and Turlay, R. (1964): Evidence for the 2π decay of the K_2^0 meson, Phys. Rev. Lett. **13**, 138–140

Churchland, P.M., and Hooker, C.A. (Eds.) (1985): *Images of Science*, University of Chicago Press, Chicago

Clauser, J.F., and Horne, M.A. (1974): Experimental consequences of objective local theories, Phys. Rev. D **10**, 526–535

Clifton, R., and Halvorson, H. (2002): No place for particles in relativistic quantum theories? Philosophy of Science **69**, 1–28. Repr. in: *Quantum Entanglements. Selected Papers*, ed. by J. Butterfield and H. Halvorson, Oxford University Press, Oxford (2004)

Cohen, H. (1885): *Kants Theorie der Erfahrung*, Duemmler, Berlin (2nd extended edn.). Reprint in: Werke, Vol. 1.1, Olms, Hildesheim (1977 ff.)

Cohen, H. (1896): Einleitung mit kritischem Nachtrag zur 9. Auflage der Geschichte des Materialismus von Friedrich Albert Lange. Repr. in: Werke, Vol. 5.2, Olms, Hildesheim (1977 ff.)

Compton, A.H. (1923): A quantum theory of the scattering of X-rays by light elements, Phys. Rev. **21**, 483–502

Compton, A.H., and Allison, S.K. (1935): *X-Rays in Theory and Experiment*, Macmillan, Toronto, New York, London

Cowan, G. (1998): *Statistical Data Analysis*, Clarendon Press, Oxford

Cramer, J.G. (1986): The transactional interpretation of quantum mechanics, Rev. Mod. Phys. **58**, 647–687

Crisp, M.D., and Jaynes, E.T. (1969): Radiative effects in semiclassical theory, Phys. Rev. **179**, 1253–1261

Darrigol, O. (1992): *From c-Numbers to q-Numbers*, University of California Press, Berkeley

Daston, L. (Ed.) (2000): *Biographies of Scientific Objects*, University of Chicago Press, Chicago

Davidson, D. (1974): On the very idea of a conceptual scheme. Proc. and Addr. of the Am. Phil. Ass. 47. Repr. in: Davidson, D.: *Inquiries into Truth and Interpretation*, Clarendon Press, Oxford (1984) pp. 183–198

Davisson, C., and Germer, L.H. (1927): Diffraction of electrons by a crystal of nickel, Phys. Rev. **30**, 705–740

de Broglie, L. (1923): Ondes et quanta, Comptes rendus **177**, 507–510

Debye, P. (1923): Zerstreuung von Röntgenstrahlen und Quantentheorie, Phys. Zeitschrift **24**, 161–166. Engl. transl. in: *The Collected Papers of P.J.W. Debye*, Interscience, New York (1954) pp. 80–88

Descartes, R. (1644): *Principia philosophiae*, Apud Ludovicum Elzevirium, Amsterdam. Engl. transl. by Miller, V.: *Principles of Philosophy*, transl. with explanatory notes, Kluwer, Dordrecht (1984)

Dirac, P.A.M. (1927): The quantum theory of the emission and absorption of radiation, Proc. Roy. Soc. A **113**, 243–265

Dirac, P.A.M. (1928a): The quantum theory of the electron, Proc. Roy. Soc. (London) A **117**, 610–624

Dirac, P.A.M. (1928b): The quantum theory of the electron (Part II), Proc. Roy. Soc. (London) A **118**, 351–361

Dirac, P.A.M. (1958): *The Principles of Quantum Mechanics*, Oxford University Press, Oxford, London

Drell, S.D., and Zachariasen, F. (1961): *Electromagnetic Structure of Nucleons*, Oxford University Press, London, Oxford

Duhem, P. (1906): *La theorie physique, son objet et sa structure*, Chevalier & Rivière, Paris. Engl. transl. by P.P. Wiener: *The Aim and Structure of Physical Theory*, Princeton University Press, Princeton (1991)

Dürr, S. (2006): The phase of a Bose–Einstein condensate. To appear in: Physics and Philosophy. Open Access Online Journal:
http://physphil.uni-dortmund.de/

Dürr, S., Nonn, T., and Rempe, G. (1998a): Origin of quantum-mechanical complementarity probed by 'which-way' experiment in an atom interferometer, Nature **395**, 33–37

Dürr, S., Nonn, T., and Rempe, G. (1998b): Fringe visibility and which-way information in an atom interferometer, Phys. Rev. Lett. **81** (26), 5705–5709

Dürr, S., and Rempe, G. (2000a): Wave–particle duality in an atom interferometer, Advances in Atomic, Molecular and Optical Physics **42**, 29–71

Dürr, S., and Rempe, G. (2000b): Can wave–particle duality be based on the uncertainty relation? Am. J. Phys. **68** (11), 1021–1024

Earman, J. (1989): *World Enough and Spacetime*, MIT Press, Cambridge (Mass.)

Eddington, A. (1939): *The Philosophy of Physical Science*, Cambridge University Press, Cambridge. Repr. with corrections: Cambridge University Press (1949)

Einstein, A. (1905): Über einen die Erzeugung und Verwandlung des Lichts betreffenden heuristischen Gesichtspunkt, Annalen der Physik **17**, 132–148

Einstein, A. (1917): Zur Quantentheorie der Strahlung, Physikal. Zeitschrift **18**, 121–128

Einstein, A. (1936): Physik und Realität. Talk in Albany (New York), Oct. 15th, 1936. Reprint in: Einstein (1993) pp. 63–130

Einstein, A. (1949): Autobiographical Notes. In: Schilpp (1949), pp. 1–94

Einstein, A. (1993): *Aus meinen späten Jahren*, 4th edn., Ullstein Verlag, Frankfurt, Berlin

Einstein, A., Podolsky, B., and Rosen, N. (1935): Can quantum mechanical description of reality be considered complete? Phys. Rev. **47**, 777–780

Ellis, B. (1968): *Basic Concepts of Measurement*, 2nd edn., Cambridge University Press, Cambridge

Englert, B.-G. (1996): Fringe visibility and which-way information: An inequality, Phys. Rev. Lett. **77**, 2154–2157

Englert, B.-G. (1998): Von wißbaren und unwißbaren Wegen, Phys. Bl **54**, 999–1000

Englert, B.-G., Schwinger, J., and Scully, M.O. (1990): Center-of-mass motion of masing atoms. In: A.O. Barut (Ed.): *New Frontiers in Quantum Electrodynamics and Quantum Optics*, Plenum Press, New York, pp. 513–519

Englert, B.-G., and Scully, M.O. (1990): Good and bad welcher weg detectors. In: A.O. Barut (Ed.): *New Frontiers in Quantum Electrodynamics and Quantum Optics*, Plenum Press, New York, pp. 507–512

Englert, B.-G., Scully, M.O., and Walther, H. (1991): Quantum optical tests of complementarity, Nature **351**, 111–116

Englert, B.-G., Scully, M.O., and Walther, H. (1995): Complementarity and uncertainty, Nature **375**, 367–368

Ernst, F.J., Sachs, R.G., and Wali, K.C. (1960): Electromagnetic form factors of the nucleon, Phys. Rev. **119**, 1105–1114

Falkenburg, B. (1988): The unifying role of symmetries in particle physics, Ratio (New Series) **1**, 113–134

Falkenburg, B. (1993a): The concept of spatial structure in microphysics, Phil. Nat. **30/2**, 208–228

Falkenburg, B. (1993b): Substanzbegriff und Quantentheorie, Phil. Nat. **30**, 229–246

Falkenburg, B. (1993c): Hegel on mechanistic models of light. In: Petry, M.J. (Ed.): *Hegel and Newtonianism*, Kluwer, Dordrecht, pp. 531–546

Falkenburg, B. (1995): *Teilchenmetaphysik. Zur Realitätsauffassung in Wissenschaftsphilosophie und Mikrophysik*, 2nd edn., Spektrum, Heidelberg

Falkenburg, B. (1996): The analysis of particle tracks. A case for trust in the unity of physics, Stud. Hist. Phil. Mod. Phys. **27**, 337–371

Falkenburg, B. (1998): Bohr's principles of unifying quantum disunities, Phil. Nat. **35/1**, 95–120

Falkenburg, B. (2000): *Kants Kosmologie. Die wissenschaftliche Revolution der Naturphilosophie im 18. Jahrhundert*, Klostermann, Frankfurt

Falkenburg, B. (2000a): How to observe quarks. In: E. Agazzi and M. Pauri (Eds.): *The Reality of the Unobservable*, Kluwer, Dordrecht, pp. 329–341

Falkenburg, B. (2001): Die Maßsetzung im Endlichen. Introduction to: Wind (1934) (new German edn., Suhrkamp Verlag, Frankfurt) pp. 11–59

Falkenburg, B. (2002a): Measurement and ontology: What kind of evidence can we have for quantum fields? In: Kuhlmann et al. (2002) pp. 235–254

Falkenburg, B. (2002b): Correspondence and the non-reductive unity of physics. In: C. Mataix and A. Rivadulla (Eds.): *Física Cuántica y Realidad – Quantum Physics and Reality*, Editorial Complutense, Madrid

Falkenburg, B. (2004): Experience and completeness in physical knowledge: Variations on a Kantian Theme. In: Meixner and Newen (2004) pp. 153–176

Falkenburg, B. (2006a): Functions of intuition in quantum physics. In: Carson and Huber (2006) pp. 267–292

Falkenburg, B. (2006b): Metamorphosen des Teilchenkonzepts. In: H. Fischler and C.S. Reiners (Eds.): *Die Teilchenstruktur der Materie im Physik- und Chemieunterricht*, Logos, Berlin (to appear) pp. 25–45

Falkenburg, B., and Schnepf, R. (1998): Kausalität in Metaphysik und Physik. In: B. Falkenburg and D. Pätzold (Eds.): *Verursachung. Repräsentationen von Kausalität*, Dialektik **1998/2**, 27–48

Falkenburg, B., and Ihmig, K.-N. (2004): "Hypotheses non fingo." Newtons wissenschaftliche Methodenlehre, DFG – Fa 261/5-1

Faye, J. (1991): *Niels Bohr, His Heritage and Legacy. An Anti-Realist View of Quantum Mechanics*, Kluwer, Dordrecht

Feynman, R.P. (1949a): The theory of positrons, Phys. Rev. **76**, 749–759

Feynman, R.P. (1949b): Space-time approach to quantum electrodynamics, Phys. Rev. **76**, 769–789

Feynman, R.P. (1961): *Quantum Electrodynamics*, Benjamin, New York

Feynman, R.P. (1969): Very high-energy collisions of hadrons, Phys. Rev. Lett. **23**, 1415–1417

Feynman, R.P. (1985): *QED – The Strange Theory of Light and Matter*, Princeton University Press, Princeton

Feynman, R.P. (1987): The reasons for antiparticles. In: Feynman, R.P., and Weinberg, S.: *Elementary Particles and the Laws of Physics*, The 1986 Dirac Memorial Lectures, Cambridge University Press, Cambridge, pp. 1–59

Feynman, R.P., Leighton, R.B., and Sands, M. (1965): *The Feynman Lectures on Physics*, Part III, Addison-Wesley, Reading (Mass.)

Fine, A. (1984): The natural ontological attitude. In: Leplin (1984) pp. 83–107

Foerster, H.V. (1981): Das Konstruieren einer Wirklichkeit. In: Watzlawick, P.: *Die erfundene Wirklichkeit*, Piper, München

Folse, H.J. (1985): *The Philosophy of Niels Bohr: The Framework of Complementarity*, North Holland, Amsterdam

Forge, J. (Ed.) (1987): *Measurement, Realism and Objectivity*, Reidel, Dordrecht

Fox, T. (2006): Relativer Atomismus, Dissertation, University of Dortmund

Franklin, A. (1986): *The Neglect of Experiment*, Cambridge University Press, Cambridge

Franklin, A. (1990): *Experiment, Right or Wrong*, Cambridge University Press, Cambridge

Franklin, A. (2001): *Are There Really Neutrinos? An Evidential History*, Perseus Books, Cambridge (Mass.)

Frege, G. (1884): *Die Grundlagen der Arithmetik. Eine logisch mathematische Untersuchung über den Begriff der Zahl*, M. Marcus, Breslau. Repr. by Thiel, Ch. (Ed.): Meiner, Hamburg (1986)

Frege, G. (1892): Über Sinn und Bedeutung, Zeit. f. Phil. u. phil. Kritik **100**, 25–50

Friedman, M. (1974): Explanation and scientific understanding, Journ. of Phil. **71**, 5–19. Repr. in: Joseph C. Pitt (Ed.): *Theories of Explanation*, Oxford University Press, Oxford (1988)

Friedman, M. (1983): *Foundations of Space-Time Theories*, Princeton University Press, Princeton

Friedman, M. (1992): *Kant and the Exact Sciences*, Harvard University Press, Cambridge (Mass.), London

Friedman, M. (1999): *Reconsidering Logical Positivism*, Cambridge University Press, Cambridge

Friedman, M. (2000): *A Parting of the Ways. Carnap, Cassirer, Heidegger*, Open Court, Chicago, La Salle (Illinois)

Galileo, G. (1638): *Discorsi e Dimostrazioni Matematiche, intorno à due nuove scienze Attenti alle Mecanica i Movimenti Locali*, Leyden. Engl. transl.: *(Discourses on the) Two New Sciences*, transl. and ed. by S. Draken, Toronto (2000, 2nd edn.)

Galison, P. (1987): *How Experiments End*, University of Chicago Press, Chicago

Geiger, H., and Marsden, E. (1913): Die Zerstreuungsgesetze der α-Strahlen bei großem Ablenkungswinkel, Phil. Mag. **25**, 604–623

Gelfert, A. (2003): Manipulative success and the unreal, International Studies in the Philosophy of Science **17/3**, 245–263

Giuntini, R. (1991): *Quantum Logic and Hidden Variables*, Bibliographisches Institut (Wiss.-Verl), Mannheim

Giulini, D., Joos, E., Kiefer, C., Kupsch, J., Stamatescu, I.-O. and Zeh, H.D. (1996): *Decoherence and the Appearance of a Classical World in Quantum Theory*, Springer Verlag, Berlin

Goldberger, M.L., and Watson, K.M. (1975): *Collision Theory*, Wiley, New York

Goodman, N. (1951): *The Structure of Appearance*, Kluwer, Harvard, Dordrecht (1977 3rd edn.)

Grangier, P., Rogier, G., and Aspect, A. (1986): Experimental evidence for a photon anti-correlation effect on a beamsplitter, Europhys. Lett. **1**, 173–179

Greenstein, G., and Zajonc, A.G. (1997): *The Quantum Challenge. Modern Research on the Foundations of Quantum Mechanics*, Jones and Bartlett, Boston

Grimsehl, E. (1938): *Grimsehls Lehrbuch der Physik*, ed. by R. Tomaschek, Teubner, Leipzig, Berlin (8th edn.)

Grünbaum, A. (1963): *Philosophical Problems of Space and Time*, Alfred A. Knopf, New York

Hacking, I. (1983): *Representing and Intervening*, Cambridge University Press, Cambridge

Hanson, P.P. (1982): Recension of van Fraassen 1980, Philosophy of Science **49**, 290

Harding, S.G. (1976): *Can Theories be Refuted? Essays on the Duhem–Quine Thesis*, Reidel, Dordrecht

Hedrich, R. (2006): String Theory – From Physics to Metaphysics. To appear in: Physics and Philosophy. Open Access Online Journal: http://physphil.uni-dortmund.de/

Hegel, G.W.F. (1830): *Enzyklopädie der philosophischen Wissenschaften*, Teil 1: *Die Logik* and Teil 2: *Die Naturphilosophie*. In: E. Moldenhauer and K.M. Michel (Eds.): Gesamtwerk, 8. and 9. Bd., Suhrkamp Verlag Frankfurt am Main (1969 ff.)

Heisenberg, W. (1927): Über den anschaulichen Inhalt der quantentheoretischen Kinematik und Mechanik, Z. Physik **43**, 172–198

Heisenberg, W. (1930a): *Physikalische Prinzipien der Quantentheorie*, Hirzel, Leipzig. Repr.: Bibliographisches Institut, Mannheim (1958)

Heisenberg, W. (1930b): *Physical Principles of Quantum Theory*, transl. by C. Eckart and F.C. Hoyt, University of Chicago Press, Chicago

Heisenberg, W. (1967): *Einführung in die einheitliche Feldtheorie der Elementarteilchen*, Hirzel, Stuttgart

Heisenberg, W. (1969): *Der Teil und das Ganze. Gespräche im Umkreis der Atomphysik*, Piper, München. Repr. in: Heisenberg (1985) pp. 3–334

Heisenberg, W. (1971): Der Begriff der kleinsten Teilchen in der Entwicklung der Naturwissenschaft. In: *Meyers Enzyklopädisches Lexikon*, Vol. 2, Bibliographisches Institut, Mannheim, pp. 870–879. Repr. in: Heisenberg (1985) pp. 395–404.

Heisenberg, W. (1976): Was ist ein Elementarteilchen? Die Naturwissenschaften **63**, 1–7. Repr. in: Heisenberg (1985) pp. 507–513

Heisenberg, W. (1985): *Collected Works*, Part C III, ed. by W. Blum, H.-P. Dürr and H. Rechenberg, Piper, München, Zürich

Heitler, W., and Sauter, F. (1933): Stopping of fast particles with emission of radiation and the birth of positive electrons, Nature **132**, 892

Helmholtz, H. von (1887): Zählen und Messen erkenntnistheoretisch betrachtet. Engl. transl. in: R.S. Cohen and Y. Elkana (Eds.): *Epistemological Writings*, Boston Studies 37, Reidel, Dordrecht (1977)

Hempel, C.G. (1952): *Fundamentals of Concept Formation in Empirical Science*, University of Chicago Press, Chicago

Herrmann, A. (Ed.) (1962): *Zur statistischen Deutung der Quantenmechanik, Dokumente der Naturwissenschaft, Abt. Physik*, Bd. 1. Ernst Battenburg Verlag, Stuttgart

Hilbert, D. (1918): Axiomatisches Denken, Mathematische Annalen **78**, 405–415. Repr. in: *Gesammelte Abhandlungen*, Vol. 3, 146–156, Springer, Berlin (1935)

Hofstadter, R. (1956): Electron scattering and nuclear structure, Rev. Mod. Phys. **28**, 214–259

Hofstadter, R. (1989): A personal view of nucleon structure as revealed by electron scattering. In: Brown (1989) pp. 126–143

Hofstadter, R., Bumiller, F., and Yearian, M.R. (1958): Electromagnetic structure of the proton and neutron, Rev. Mod. Phys. **30**, 482–497

Hones, M.J. (1991): Scientific realism and experimental practice in high-energy physics, Synthese **86**, 29–76

Honner, J. (1987): *The Description of Nature: Niels Bohr and the Philosophy of Quantum Physics*, Clarendon Press, Oxford

Hooker, C.A. (1973): Metaphysics and modern physics. In: C.A. Hooker (Ed.): *Contemporary Research in the Foundations and Philosophy of Quantum Theory* Reidel, Dordrecht, pp. 174–304

Howard, D. (2002): Who invented the Copenhagen interpretation?, Philosophy of Science **71**, 669–682

Hoyningen-Huene, P. (1993): *Reconstructing Scientific Revolutions*, University of Chicago Press, Chicago

Huber, R. (2000): *Poincaré und Einstein: Zur philosophischen Beurteilung physikalischer Theorien*, Mentis, Paderborn

Hughes, R.I.G. (1989): *The Structure and Interpretation of Quantum Mechanics*, Harvard University Press, Cambridge (Mass.)

Hume, D. (1739): *A Treatise of Human Nature*, John Noon, London. New edn. by L.A. Selby-Bigge: Clarendon Press, Oxford (1988)

Hüttemann, A. (1997): *Idealisierungen und das Ziel der Physik. Eine Untersuchung zum Realismus, Empirismus und Konstruktivismus in der Wissenschaftstheorie*, de Gruyter, Berlin, New York

Ihmig, K.-N. (2004): Die Bedeutung der Methoden der Analyse und Synthese für Newtons Programm der Mathematisierung der Natur. In: Meixner and Newen (2004) pp. 91–119

Itzykson, C., and Zuber, J.-B. (1985): *Quantum Field Theory*, McGraw-Hill, New York

Jackson, J.D. (1975): *Classical Electrodynamics*, 2nd edn., Wiley, New York, London

Jaeger, G., Shimony, A., and Vaidman, L. (1995): Two interferometric complementarities, Phys. Rev. A **51**, 54–67

Jammer, M. (1966): *The Conceptual Development of Quantum Mechanics*, McGraw-Hill, New York

Jammer, M. (1974): *The Philosophy of Quantum Mechanics*, Wiley Interscience, New York

Janich, P. (1996): *Konstruktivismus und Naturerkenntnis. Auf dem Weg zum Kulturalismus*, Suhrkamp Verlag, Frankfurt

Jaynes, E. (1980): Quantum beats. In: A.O. Barut (Ed.): *Foundations of Radiation Theory and Quantum Electrodynamics*, Plenum Press, New York, pp. 37–43

Jochmann, E., Hermes, O., and Spies, P. (1900): *Grundriß der Experimentalphysik*, 14th edn., Winckelmann, Berlin

Joos, E. (1990): Die Begründung klassischer Eigenschaften aus der Quantentheorie, Phil. Nat. **27**, 31

Kant, I. (1766): *Träume eines Geistersehers*, Kanter, Königsberg. Engl. Transl. by D. Walford and R. Meerbote (Eds): *The Cambridge Edition of the Works of Immanuel Kant. Theoretical Philosophy*, 1755–1770, Cambridge University Press, Cambridge (1992) pp. 923–989

Kant, I. (1781/87): *Kritik der reinen Vernunft*, Hartknoch (ed. A/B), Riga. Engl. transl.: P. Guyer and A.W. Wood (Eds.): *The Cambridge Edition of the Works of Immanuel Kant. Critique of Pure Reason*, Cambridge University Press, Cambridge (1998)

Kant, I. (1786): *Metaphysische Anfangsgründe der Naturwissenschaft*, Riga. Engl. Transl.: M. Friedman: *Metaphysical Foundations of Natural Science*, Cambridge University Press, Cambridge (2004)

Kant, I., Akademie-Ausgabe: Kants gesammelte Schriften. Hrsg. von der königlich preußischen Akademie der Wissenschaften (later: Deutsche Akademie der Wissenschaften zu Berlin). Bd. 1–23 (Werke, Briefe, Handschriftlicher Nachlaß), Berlin (1900–1955). Bd. 24 ff. (Vorlesungen), Berlin (1966 ff)

Kastner, R.E. (2005): Why the Afshar experiment does not refute complementarity, Stud. in Hist. and Phil. of Mod. Phys. **36**, 649–658

Keller, M., Lange, B., Hayasaka, K., Lange, W. and Walther, H. (2004): Continuous generation of single photons with controlled waveform in an ion-trap cavity system, Nature **431**, 1075–1078

Kim, Y.-H., Yu, R., Kulik, S.P., Shih, Y.H. and Scully, M.O. (1999): A delayed choice quantum eraser. Available as an e-print (13 March 1999) at www.americanscientist.org/template/PDFDetail/assetid/35556

Kleinknecht, K., and Burkhard, R. (1987): Z. Phys. C **34**, 209

Knorr-Cetina, K.D. (1981): *The Manufacture of Knowledge. An Essay on the Constructivist and Contextual Nature of Science*, Pergamon Press, Oxford

Knorr-Cetina, K.D. (1984): *Die Fabrikation von Erkenntnis. Zur Anthropologie der Naturwissenschaft*, Suhrkamp Verlag, Frankfurt

Knorr-Cetina, K.D. (1999): *Epistemic Cultures. How the Sciences Make Knowledge*, Harvard University Press, Cambridge (Mass.), London

Krantz, D.H., Duncan, R., Suppes, L.P., and Tversky, A. (1971): *Foundations of Measurement*, Vol. 1, Academic Press, San Diego

Kuhlmann, M., Lyre, H., and Wayne, A. (Eds.) (2002): *Ontological Aspects of Quantum Field Theory*, World Scientific, Singapore

Kuhn, T.S. (1961): The function of measurement in modern physical science, Isis **52**, 161–193

Kuhn, T.S. (1962): *The Structure of Scientific Revolution*, 1st edn., University of Chicago Press, Chicago

Kuhn, T.S. (1970): *The Structure of Scientific Revolution*, 2nd edn., University of Chicago Press, Chicago

Kuhn, T.S. (1977): *The Essential Tension. Selected Studies in Scientific Tradition and Change*, University of Chicago Press, Chicago, London

Kyburg, H.E., Jr. (1984): *Theory and Measurement*, Cambridge University Press, Cambridge

Lamb, W.E., Jr., and Scully, M.O. (1969): The photoelectric effect without photons. In: *Polarisation, Matière et Rayonnement*, Presses Univ. de France, Paris, pp. 363–369

Landau, L.D., and Lifschitz, E.M. (1987): *Lehrbuch der theoretischen Physik*, Vol. I, *Mechanik*, 12th edn., Berlin

Landau, L.D., and Lifschitz, E.M. (1988): *Lehrbuch der theoretischen Physik*, Vol. III, *Quantenmechanik*, 8th edn., Berlin

Latour, B., and Woolgar, St. (1979): *Laboratory Life. The Construction of Scientific Facts*, Sage Publications, Beverly Hills, London. Princeton University Press, Princeton (N.J.) (2nd edn. 1986)

Lattes, C.M.G. (1983): My work in meson physics with nuclear emulsions. In: Brown and Hoddeson (1983) pp. 307–310

Laudan, L. (1981): A confutation of convergent realism, Phil. of Science **48**, 19–49. Repr. in: Leplin (1984) pp. 218–249, or Papineau (1996) pp. 107–138

Leibniz, G.W. (1714): *Monadology*, ed. by N. Rescher, Routledge, London (2002)

Leibniz, G.W., and Clarke, S. (1715/16): *The Leibniz–Clarke Correspondence*, edited with introduction and notes by H.G. Alexander, Manchester University Press, Manchester (1956, repr. 1998)

Leonhard, H.S., and Goodman, N. (1940): The calculus of individuals and its use, Journal of Symbolic Logic **5**, 45–55

Leonard, C.S. (2003): Ideal or real: What is the "nature of science"? Philosophy of Education **2003**, 293–295

Leplin, J. (Ed.) (1984): *Scientific Realism*, University of California Press, Berkeley

Liss, T., and Tipton, P. (1997): The discovery of the top quark, Scientific American **277**, 54–59

Locke, J. (1689): *An Essay Concerning Human Understanding*, London. New edn. by P.H. Nidditch: Clarendon Press, Oxford (1975)

Lorentz, H.A. (1895): *Versuch einer Theorie der electrischen und optischen Erscheinungen in bewegten Körpern*, Section 12, Brill, Leiden. Repr. in: *Collected Papers*, Vol. 5, Martinus Nijhoff, The Hague (1936)

Lorenzen, P. (1974): *Konstruktive Wissenschaftstheorie*, Suhrkamp Verlag, Frankfurt

Losee, J. (1993): *A Historical Introduction to the Philosophy of Science*, 3rd edn., Oxford University Press, Oxford, New York

Ludwig, G. (1990): *Die Grundstrukturen einer physikalischen Theorie*, 2nd edn., Springer, Berlin, Heidelberg, New York

Lyre, H. (2004): *Lokale Symmetrien und Wirklichkeit. Eine naturphilosophische Studie über Eichtheorien und Strukturenrealismus*, Schöningh, Paderborn

Mach, E. (1883): *Die Mechanik in ihrer Entwicklung historisch-kritisch dargestellt*, 1st edn., Brockhaus, Leipzig. Quoted from the repr.: Wissenschaftliche Buchgesellschaft, Darmstadt (1991, repr. of the 9th edn.: Brockhaus, Leipzig 1933). Engl. Transl.: *The Science of Mechanics. A Critical and Historical Account of its Development*, transl. by T.J. McCormack, new introduction by K. Menger, Open Court, La Salle, Ill. (6th edn. 1960)

Mach, E. (1926): *Erkenntnis und Irrtum. Skizzen zur Psychologie der Forschung*, 5th edn., Barth, Leipzig. Quoted from the repr.: Wissenschaftliche Buchgesellschaft, Darmstadt (1991). Engl. Transl.: *Knowledge and Error. Sketches on the Psychology of Enquiry*, with an introduction by E.N. Hiebert. Transl. by T.J. McCormack and P. Foulkes: Reidel, Dordrecht (1976)

Mackie, J.L. (1980): *The Cement of the Universe*, Clarendon Press, Oxford

MacKinnon, E.M. (1982): *Scientific Explanation and Atomic Physics*, University of Chicago Press, Chicago

Malament, D.B. (1996): In defense of dogma. Why there cannot be a relativistic quantum mechanics of (localizable) particles. In: R. Clifton (Ed.): *Perspectives on Quantum Reality*, Kluwer, Dordrecht, pp. 1–10

Mandel, L. (1976): The case for and against semiclassical radiation theory. In: *Progress in Optics*, Vol. 13 (A76-30145), North-Holland, Amsterdam, pp. 13–74

Margenau, H. (1954): Advantages and disadvantages of various interpretations of the quantum theory, Physics Today **7**, 6–13

Matthews, M.R. (2003): Data, phenomena, and theory: How clarifying the concepts can illuminate the nature of science, Philosophy of Education **2003**, 283–292

Maxwell, G. (1962): The ontological status of theoretical entities. In: H. Feigl and G. Maxwell (Eds.): *Scientific Explanation, Space, and Time*, Minnesota Studies in the Philosophy of Science, Vol. 3, University of Minnesota Press, Minneapolis, pp. 3–27

McMullin, E. (1985): Galilean Idealization, Studies in History and Philosophy of Science **16**, 247–273

Mehra, J., and Rechenberg, H. (1982): *The Historical Development of Quantum Theory*, Springer, New York, Heidelberg, Berlin (1982 ff.)

Meixner, U., and Newen, A. (Eds.) (2004): *Schwerpunkt: Geschichte der Naturphilosophie*, Band 7 von: *Philosophiegeschichte und logische Analyse*, Mentis Verlag, Paderborn

Messiah, A. (1964): *Mécanique quantique*, Vol. 2, Dunod, Paris. Engl. transl.: North Holland, Amsterdam (7th edn. 1973)

Messiah, A. (1969): *Mécanique quantique*, Vol. 1, Dunod, Paris. Engl. transl.: North Holland, Amsterdam (7th edn. 1973)

Meyer-Abich, K.M. (1965): *Korrespondenz, Individualität und Komplementarität*, Steiner, Wiesbaden

Mill, J.St. (1872): *A System of Logic, Ratiocinative and Inductive, Being a Connected View of the Principles of Evidence and the Methods of Scientific Investigation*, Parker, London (reprint of 2nd edn.)

Millikan, R.A. (1911): The isolation of an ion, a precision measurement of its charge, and the correction of Stokes's law, Phys. Rev. **32**, 349–397. Partly reprinted in: Shamos (1959) pp. 238–249

Millikan, R.A. (1917): *The Electron*, University of Chicago Press, Chicago

Mittelstaedt, P. (1972): *Philosophische Probleme der modernen Physik*, 4th edn., Bibliographisches Institut, Mannheim

Mittelstaedt, P. (1986): *Sprache und Realität in der modernen Physik*, Bibliographisches Institut, Mannheim

Mittelstaedt, P. (1995): Die wechselseitigen Beziehungen zwischen der Quantentheorie und ihrer Interpretation. In: L. Krüger and B. Falkenburg (Eds.): *Physik, Philosophie und die Einheit der Wissenschaften*, Spektrum, Heidelberg, pp. 97–117

Mittelstaedt, P. (1997a): Is quantum mechanics a probabilistic theory? In: R.S. Cohen et al. (Eds.): *Potentiality, Entanglement and Passion-at-a-Distance*, Kluwer, Dordrecht, pp. 159–175

Mittelstaedt, P. (1997b): The emergence of statistical laws in quantum mechanics. In: M. Ferrero and A. van der Merwe (Eds.): *New Developments on Fundamental Problems in Quantum Physics*, Kluwer, Dordrecht, pp. 265–274

Mittelstaedt, P. (1998a): *The Interpretation of Quantum Mechanics and the Measurement Process*, Cambridge University Press, Cambridge

Mittelstaedt, P. (1998b): The constitution of objects in Kant's philosophy and in modern physics. In: E. Castellani (Ed.): *Interpreting Bodies. Classical and Quantum Objects in Modern Physics*, Princeton University Press, Princeton, pp. 168–180

Mittelstaedt, P. (1999): Individualistic versus statistical interpretation of quantum mechanics. In: M.L. Dalla Chiara et al. (Eds.): *Language, Quantum, Music*, Kluwer, Dordrecht, pp. 231–239

Mittelstaedt, P. (2001): What if quantum mechanics is universally valid? In: E. Agazzi and J. Faye (Eds.): *The Problem of the Unity of Science* World Scientific, Singapore, pp. 177–188

Mittelstaedt, P. (2004): Interpretationsprobleme der Quantenmechanik, Phil. Nat. **42/2**, 227–256

Mittelstaedt, P. (2006): The intuitiveness and truth of modern physics. In: Carson and Huber (2006) pp. 251–266

Mittelstaedt, P., Prieur, A., and Schieder, R. (1987): Unsharp particle wave duality in a photon split-beam experiment, Found. Phys. **17**, 891–903

Mittelstaedt, P., and Weingartner, P. (2005): *Laws of Nature*, Springer, Berlin

Møller, Ch. (1931): Über den Stoß zweier Teilchen unter Berücksichtigung der Retardation der Kräfte, Z. Physik **70**, 786–795

Morgan, M., and Morrison, M. (Eds.) (1999): *Models as Mediators. Perspectives on Natural and Social Science*, Cambridge University Press, Cambridge

Mott, N.F. (1929): The wave mechanics of α-rays, Proc. Roy. Soc. A **126**, 79–84. Repr. in: Wheeler and Zurek (1983)

Mott, N.F., and Massey, H.S. (1965): *The Theory of Atomic Collisions*, Clarendon Press, Oxford

Murdoch, D. (1981): *Niels Bohr's Philosophy of Physics*, Cambridge University Press, Cambridge

Nachtmann, O. (1990): *Elementary Particle Physics. Concepts and Phenomena*, Springer, Berlin

Nagel, E. (1931): Measurement, Erkenntnis **2**, 313–333

Nagel, E. (1961): *The Structure of Science. Problems in the Logic of Scientific Explanation*, Routledge & Kegan Paul, London

Narens, L. (1985): *Abstract Measurement Theory*, MIT Press, Cambridge (Mass.)

Natorp, P. (1910): *Die logischen Grundlagen der exakten Wissenschaften*, 2nd edn., B.G. Teubner, Leipzig, Berlin

Nersessian, N.J. (1987): *The Process of Science*, Reidel, Dordrecht

Newton, I. (1687): *Philosophiae naturalis principia mathematica*, Londinum. Quoted from the new Engl. transl.: *The Principia. Mathematical Principles of Natural Philosophy*, translated by I. Bernard Cohen and Anne Whitman with the assistance of Julia Budenz. University of California Press, Berkeley, London (1999)

Newton, I. (1704): *Opticks. Or, a Treatise of the Reflexions, Refractions, Inflexions and Colours of Light. Also two Treatises of the Species and Magnitude of Curvilinear Figures*, London (1st edn.)

Newton, I. (1729): *Principia. Mathematical Principles of Natural Philosophie and his System of the World*. Translated into English by A. Motte. To which are added the Laws of the Moon's motion, according to gravity, by J. Machin. The preface of Mr. R. Cotes to the second edition, etc., London. Repr.: The translations revised, and supplied with an historical and explanatory appendix by F. Cajori, University of California Press, Los Angeles (1934)

Newton, I. (1730): *Opticks*, London (4th edn.). Reprint with a Preface by I.B. Cohen, Dover, New York (1979)

Pais, A. (1986): *Inward Bound*, Clarendon Press, Oxford

Panofsky, W.K.H. (1968): Rapporteur talk. In: J. Prentki and J. Steinberger (Eds.): Proc. of the Fourteenth Int. Conf. on High-Energy Physics, Vienna 1988. Genf: CERN, 23–39

Papineau, D. (Ed.) (1996): *The Philosophy of Science*, Oxford University Press, Oxford

Particle Data Group (2004): *Particle Physics Booklet*, extracted from: S. Eidelman et al.: Review of Particle Physics, Physics Letters B 592

Peres, A. (1993): *Quantum Theory: Concepts and Methods*, Kluwer, Dordrecht

Peres, A., and Zurek, W.H. (1982): Is quantum theory universally valid? Am. J. Phys. **50**, 807–810

Perkins, D.H. (1987): *Introduction to High Energy Physics*, 3rd edn., Addison-Wesley, Menlo Park (Calif.)

Perkins, D.H. (1989): Cosmic-ray work with emulsions in the 1940s and 1950s. In: Brown et al. (1989) pp. 89–108

Perkins, D.H. (2000): *Introduction to High Energy Physics*, 4th edn., Addison-Wesley, Menlo Park (Calif.)

Perkins, D.H. (2003): *Particle Astrophysics*, Oxford University Press

Pfleegor, R.L., and Mandel, L. (1967): Interference of independent photon beams, Physical Review **159**, 1084–1088

Pickering, A. (1984): *Constructing Quarks*, Edinburgh University Press, Edinburgh

Pinsker, Z.G. (1953): *Electron Diffraction*, Butterworth, London

Planck, M. (1901): Über das Gesetz der Energieverteilung im Normalspectrum, Ann. Phys. **4**, 553–563

Planck, M. (1908): Die Einheit des physikalischen Weltbildes. In: Planck (1965) pp. 28–51

Planck, M. (1965): *Vorträge und Erinnerungen*, 9th edn., Wissenschaftliche Buchgesellschaft, Darmstadt

Poser, H. (1981): Gottfried Wilhelm Leibniz. In: O. Höffe (Ed.): *Klassiker der Philosophie I*, Beck, München

Povh, B., Rith, K., Scholz, C., and Zetsche, F. (1999): *Particles and Nuclei. An Introduction to the Physical Concepts*, Springer, Berlin, London

Powell, C.F., Fowler, P.H., and Perkins, D.H. (1959): *The Study of Elementary Particles by the Photographic Method. An Account of the Principal Techniques and Discoveries Illustrated by an Atlas of Photomicrographs*, Pergamon Press, London

Pringe, H. (2006): Critic of the quantum power of judgement, Dissertation (University of Dortmund)

Psillos, St. (1999): *Scientific Realism. How Science Tracks Truth*, Routledge, London

Putnam, H. (1975): What is mathematical truth? In: *Mathematics, Matters and Methods. Philosophical Papers*, Vol. I, Cambridge University Press, Cambridge, pp. 60–78

Putnam, H. (1980): Models and reality, Journal of Symbolic Logic **45.3**, 464–482. Repr. in: *Realism and Reason. Philosophical Papers*, Vol. III, Cambridge University Press, Cambridge (1983) pp. 1–25

Putnam, H. (1990): *Realism with a Human Face*, ed. by James Conant, Harvard University Press, Cambridge, Mass.

Quine, W.V.O. (1951): Two dogmas of empiricism. In: Quine (1953) pp. 20–46

Quine, W.V.O. (1953): *From a Logical Point of View*, Harvard University Press, Harvard

Quine, W.V.O. (1969): *Ontological Relativity and Other Essays*, Columbia University Press, New York and London

Raether, H. (1957): Elektroneninterferenzen. In: *Handbuch der Physik*, Bd. 32, 443

Ramsay F.P. (1929): Theories. In: R.B. Braithwaite (Ed.): *The Foundations of Mathematics*, Kegan Paul, London (1931) pp. 212–236

Redhead, M. (1983): Quantum field theory for philosophers. In: P.D. Asquith and T. Nickles (Eds.), PSA 2, pp. 57–99

Redhead, M. (1987): *Incompleteness, Nonlocality and Realism: A Prolegomenon to the Philosophy of Quantum Mechanics*, Oxford University Press, Oxford

Redhead, M. (1988): A philosopher looks at quantum field theory. In: Brown and Harré (1988) pp. 9–23

Reichenbach, H. (1957): *The Philosophy of Space and Time* (Philosophie der Raum-Zeit-Lehre), translated by M. Reichenbach and J. Freund, Dover Publications, New York

Rescher N. (1955): Axioms for the part relation, Phil. Stud. **6**, 8–11

Riordan, M. (1987): *The Hunting of the Quark*, Simon & Schuster, New York

Robertson, H.P. (1929): The uncertainty principle, Phys. Rev. **34**, 163–164

Rosenbluth, M.N. (1950): High energy elastic scattering of electrons on protons, Phys. Rev. **79**, 615–619

Rossi, B. (1952): *High-Energy Particles*, Prentice Hall, New York

Rossi, B. (1983): The decay of "mesotrons" (1939–1943). Experimental particle physics in the age of innocence. In: Brown and Hoddeson (1983) pp. 183–205

Rutherford, E. (1911): The scattering of α and β particles by matter and the structure of the atom, Philo. Mag. **21**, 669–688

Rutherford, E., Chadwick, J., and Ellis, C.D. (1930): *Radiations from Radioactive Substances*, Cambridge University Press, Cambridge. Repr.: Cambridge University Press, Cambridge (1951)

Sachs, R.G., and Wali, K.C. (1989): Comments on electromagnetic form factors of the nucleon. In: Brown et al. (1989) pp. 144–146

Scheibe, E. (1973): *The Logical Analysis of Quantum Mechanics*, Pergamon Press, Oxford

Scheibe, E. (1997a): Mißverstandene Naturwissenschaft. In: *Wissenschaft und Aufklärung*, Montagsvorträge der Martin-Luther-Universität Halle-Wittenberg, ed. by R. Enskat, Leske und Budrich, Opladen, pp. 9–29

Scheibe, E. (1997b): *Die Reduktion physikalischer Theorien*. Teil I: *Grundlagen und elementare Theorie*, Springer, Berlin

Scheibe, E. (1999): *Die Reduktion physikalischer Theorien*. Teil II: *Inkommensurabilität und Grenzfallreduktion*, Springer, Berlin

Scheibe, E. (2001): *Between Rationalism and Empiricism. Selected Papers in Philosophy of Physics*, ed. by B. Falkenburg, Springer, New York

Schilpp, P.A. (Ed.) (1949): *Albert Einstein: Philosopher – Scientist*, Library of Living Philosophers, Evanston, Illinois

Schmidt, H.-J. (1993): A definition of mass in Newton–Lagrange mechanics, Phil. Nat. **30**, 189–207

Schrödinger, E. (1927): Über den Comptoneffekt, Annalen der Physik **82**, 257–264

Schweber, S.S. (1994): *QED and the Men Who Made It. Dyson, Feynman, Schwinger, and Tomonaga*, Princeton University Press, Princeton (N.J.), Chichester (1994)

Scully, M.O., and Drühl, K. (1982): Quantum eraser: A proposed photon correlation experiment concerning observation and "delayed choice" in quantum mechanics, Phys. Rev. A **25**, 2208–2213

Scully, M.O., and Walther, H. (1989): Quantum optical test of observation and complementarity in quantum mechanics, Phys. Rev. A **39**, 5229–5236

Scully, M.O., Englert, B.-G., and Schwinger, J. (1989): Spin coherence and Humpty-Dumpty. III. The effects of observation, Phys. Rev. A **40**, 1775–1784

Scully, M.O., Englert, B.-G., and Walther, H. (1991): Quantum optical tests of complementarity, Nature **351**, 111–116

Seibt, J. (2002): Quanta, tropes, and processes: On ontologies for QFT beyond the myth of substance. In: Kuhlmann et al. (2002) pp. 53–93

Shamos, M.H. (Ed.) (1959): *Great Experiments in Physics*, Henry Holt & Co, New York

Shapere, D. (1982): The concept of observation in science and philosophy, Phil. of Sci. **49**, 485–525

Shillady, C., and Busch, P. (2006): Complementarity in Mach–Zehnder interferometry. Publication in preparation

Simons, P. (1987): *Parts. A Study in Ontology*, Clarendon Press, Oxford

Smart, J.J.C. (1963): *Philosophy and Scientific Realism*, Routledge & Kegan Paul, London

Sneed, J.D. (1971): *The Logical Structure of Mathematical Physics*, Reidel, Dordrecht

Sneed, J.D. (1983): Structuralism and scientific realism, Erkenntnis **19**, 345–370

Stegmüller, W. (1970): *Probleme und Resultate der Wissenschaftstheorie und Analytischen Philosophie*, Vol. 2: *Theorie und Erfahrung*, Springer, Berlin, Heidelberg, New York

Stöckler, M. (1984): *Philosophische Probleme der relativistischen Quantenmechanik*, Duncker & Humblot, Berlin

Stoney, G.J. (1891): On the cause of double lines and of equidistant satellites in the spectra of gases, Scientific Transactions of the Royal Dublin Society **4**, 563–608

Storey, E.P., Tan, S.M., Collet, M.J., and Walls, D.F. (1994): Path detection and the uncertainty principle, Nature **367**, 626–628

Storey, E.P., Tan, S.M., Collet, M.J., and Walls, D.F. (1995): Complementarity and uncertainty, Nature **375**, 368

Streater, R.F. (1988): Why should anyone want to axiomatize quantum field theory? In: Brown and Harré (1988) pp. 137–148

Streater, R.F., and Wightman, A.S. (1964): *PCT, Spin and Statistics, and All That*, Benjamin, New York

Suppes, P. (1962): Models of data. In: E. Nagel, P. Suppes, and A. Tarski (Eds.): *Logic, Methodology and Philosophy of Science*, Proceedings of the 1960 International Congress, Stanford University Press, Stanford 252–261

Suppes, P. (1969): *Studies in the Methodology and Foundations of Science*, Reidel, Dordrecht

Suppes, P. (1980): Article "Messung" in: J. Speck (Ed.): *Handbuch wissenschaftstheoretischer Begriffe*, Vol. 2., Vandenhoeck & Ruprecht, Göttingen

Swoyer, C. (1987): The metaphysics of measurement. In: Forge (1987) pp. 235–290

Tarski, A. (1956): *Logic, Semantics, Metamathematics. Papers from 1923 to 1938*, translated by J.H. Woodger, Clarendon Press, Oxford

Teller, P. (1995): *An Interpretive Introduction to Quantum Field Theory*, Princeton University Press, Princeton (N.J.)

Tetens, H. (1987): *Experimentelle Erfahrung*, Meiner, Hamburg

Tetens, H. (1994): *Geist, Gehirn, Maschine: Philosophische Versuche über ihren Zusammenhang*, Reclam Verlag, Stuttgart

380   References

Thomson, J.J. (1897): Cathode rays, Phil. Mag. **44**, 293–316
Thomson, J.J. (1899): On the masses of the ions in gases at low pressures, Phil.
    Mag. **48**, 547–567
Townsend, J.S. (1897): On electricity in gases and the formation of clouds in
    charged gases, Proceedings of the Cambridge Philosophical Society **9**, 244–258
Trigg, G.L. (1971): *Crucial Experiments in Modern Physics*, Crane Russah & Co,
    New York
van Fraassen, B.C. (1980): *The Scientific Image*, Clarendon Press, Oxford
van Fraassen, B.C. (1987): The semantic approach to scientific theories. In: Ners-
    essian (1987) pp. 105–124
van Fraassen, B.C. (1991): *Quantum Mechanics. An Empiricist View*, Clarendon
    Press, Oxford
von Neumann, J. (1932): *Mathematische Grundlagen der Quantenmechanik*, Sprin-
    ger, Berlin. Engl. transl. by R. Beyer: *Mathematical Foundation of Quantum
    Mechanics*, Princeton University Press, Princeton (1955)
von Weizsäcker, C.F. (1971): *Die Einheit der Natur*, Hanser, München, Wien
von Weizsäcker, C.F. (1985): *Der Aufbau der Physik*, Hanser, München, Wien
von Wright, G.H. (1971): *Explanation and Understanding*, Cornell University Press,
    Ithaca
Walborn, St.P., Terra Cunha, M.O., Pádua, S., and Monken, C.H. (2003): Quan-
    tum erasure, American Scientist **91**, 336–343
Walls, D.F., and Milburn, G.J. (1994): *Quantum Optics*, Springer, Berlin
Weingard, R. (1988): Virtual particles and the interpretation of quantum field the-
    ory. In: Brown and Harré (1988) pp. 43–59
Wheaton, B.R. (1983): *The Tiger and the Shark*, Cambridge University Press,
    Cambridge
Wheeler, J.A. (1979–1981): Law without law. In: Wheeler and Zurek (1983) pp. 182–
    213
Wheeler, J.A., and Feynman, R.P. (1945): Interaction with the absorber as the
    mechanism of radiation, Rev. Mod. Phys. **17**, 157–181
Wheeler, J.A., and Zurek, W.H. (1983): *Quantum Theory and Measurement*, Prince-
    ton University Press, Princeton (N.J.), Guildford
Wigner, E.P. (1939): On unitary representations of the inhomogeneous Lorentz
    group, Annals of Mathematics **40**, 149–204
Wigner, E.P. (1964): Symmetry and conservation laws, Proceedings of the National
    Academy of Sciences USA **51**, 956–965. Repr. in: *Philosophical Reflections and
    Syntheses*, Springer, Heidelberg (1997) pp. 297–333
Wigner, E.P. (1967): *Symmetries and Reflections*, Indiana University Press, Bloom-
    ington. Repr.: Woodbridge, Connecticut (1979)
Wind, E. (1934): *Das Experiment und die Metaphysik. Zur Auflösung der kosmo-
    logischen Antinomien*, Mohr, Tübingen. Repr. ed. by B. Buschendorf and in-
    troduced by B. Falkenburg, Suhrkamp Verlag, Frankfurt am Main (2001). Engl.
    transl.: *Experiment and Metaphysics. Towards a Resolution of the Cosmological
    Antinomies*, transl. by C. Edwards and introduced by M. Rampley, European
    Humanities Research Centre, University of Oxford, Oxford (2001)
Wiseman, H., and Harrison, F. (1995): Uncertainty over complementarity? Nature
    377, 584
Wooters, W.K., and Zurek, W.H. (1979): Complementarity in the double-slit ex-
    periment: Quantum nonseparability and a quantitative statement of Bohr's
    principle, Phys. Rev. D **19**, 473–484

Worrall, J. (1989): Structural realism: The best of both worlds? Dialectica **43**, 99–124. Repr. in: Papineau (1996) pp. 139–165

Wu, Ch.-Sh. (1957): Experimental test of parity conservation in beta decay, Phys. Rev. **105**, 1413–1415

Zajonc, A.G. (1983): Proposed quantum-beats, quantum-eraser experiment. In: Mandel, L., and Wolf, E. (Eds.): *Coherence and quantum optics V: Proceedings of the Fifth Rochester Conference on Coherence and Quantum Optics*, held at the University of Rochester, June 13–15, 1983, Plenum Press, New York (1984) pp. 323–329

Zou, X.Y., Wang, L.J., and Mandel, L. (1991): Induced coherence and indistinguishability in optical interference, Phys. Rev. Lett. **67**, 318–321

Name Index

Abachi, S.   255
Abe, F.   255
Afshar, S.S.   314–315
Allison, S.K.   154
Anderson, C.D.   80, 112–115, 117, 173, 178
Anderson, P.W.   239–241, 244
Andrade, E.N.   128
Archimedes   42
Aristotle   11, 17, 19, 39, 41, 50, 71, 75
Armstrong, D.M.   347
Aspect, A.   26, 192, 261, 282, 290
Aston, F.W.   99
Auyang, S.Y.   209, 262

Bacon, F.   52, 57
Balzer, W.   8, 17
Bell, J.S.   26, 28, 282
Beller, M.   175, 267
Berkeley, G.   1, 53
Bernays, P.   168
Bethe, H.   111, 174, 178–185, 193, 204
Bjorken, J.D.   142–145, 147, 224, 252
Blackett, P.M.S.   115, 116
Bloch, F.   111, 116, 174, 183–185
Bloom, E.D.   252
Bogen, J.   60
Bogoliubov, N.N.   236
Bohm, D.   28, 161, 216, 270, 319, 330
Bohr, N.   X, XI, 25, 30, 36, 46, 75, 82, 88–90, 110, 136, 148, 162, 167, 172, 174, 178, 188–191, 193, 198, 203, 205, 206, 215, 217, 219, 220, 223, 265, 267, 269, 272–277, 279, 289, 296, 298, 301, 302, 304, 318, 325, 332, 335, 336, 338, 339
Boltzmann, L.   2, 5, 46, 211, 247

Born, M.   XI, 26, 102, 111, 174–176, 178, 180, 183, 193, 194, 203, 212, 215, 265, 267, 269–272, 276, 278, 279, 282, 332, 339
Boscović, R.   247
Bothe, W.   88, 90, 102, 121, 267
Boyle, R.   247
Breidenbach, M.   252
Breit, G.   137, 143, 151, 156
Bridgman, P.W.   45, 168, 347, 349
Brown, H.I.   113
Brown, H.R.   132, 209, 262
Brown, J.R.   17, 56
Busch, P.   161, 165, 216, 288, 297, 303
Butts, R.E.   17, 56

Cahn, R.N.   252
Callender, C.   196
Campbell, N.R.   343, 349
Cao, T.Y.   209, 262
Carnap, R.   5, 8, 11, 12, 16, 47, 108, 191, 343
Carrier, M.   14, 64
Cartwright, N.   10, 15, 24, 64, 65, 100, 161, 186, 207
Casimir, H.B.G.   226
Cassidy, D.   117
Cassirer, E.   26, 28, 37, 213
Chadwick, J.   94, 97, 104
Chevalley, C.   36, 217
Clarke, S.   6, 35, 247, 359
Clifton, R.   218, 221, 262, 281
Cohen, H.   17, 53
Compton, A.H.   90, 102, 154
Copernicus, N.   2, 235
Cowan, G.   170, 186
Cramer, J.G.   28, 34
Crookes, W.   94, 128